Discovery and Development of Antidiabetic Agents From Natural Products

Natural Product Drug Discovery

Discovery and Development of Antidiabetic Agents From Natural Products

Natural Product Drug Discovery

Goutam Brahmachari
Department of Chemistry
Visva-Bharati (a Central University)
Santiniketan, West Bengal, India

ELSEVIER

AMSTERDAM • BOSTON • HEIDELBERG • LONDON
NEW YORK • OXFORD • PARIS • SAN DIEGO
SAN FRANCISCO • SINGAPORE • SYDNEY • TOKYO

Elsevier
Radarweg 29, PO Box 211, 1000 AE Amsterdam, Netherlands
The Boulevard, Langford Lane, Kidlington, Oxford OX5 1GB, United Kingdom
50 Hampshire Street, 5th Floor, Cambridge, MA 02139, United States

Library of Congress Cataloging-in-Publication Data
A catalog record for this book is available from the Library of Congress

British Library Cataloguing-in-Publication Data
A catalogue record for this book is available from the British Library

ISBN: 978-0-12-809450-1

Publisher: John Fedor
Acquisition Editor: Anneka Hess
Editorial Project Manager: Anneka Hess
Production Project Manager: Paul Prasad Chandramohan
Designer: Maria Inês Cruz

Typeset by TNQ Books and Journals

Dedicated To

Natural Product Chemists of the Past, Present, and Future

Contents

4. Glucose Transporter 4 Translocation Activators From Nature

K. Dev, E. Ramakrishna and R. Maurya

5. Carbohydrate-Based Antidiabetic Agents From Nature

S. Mishra, A.S. Singh, N. Mishra, H. Pandey and V.K. Tiwari

List of Contributors

G. Abbas University of Nizwa, Nizwa, Sultanate of Oman

A.S. Al-Harrasi University of Nizwa, Nizwa, Sultanate of Oman

G. Brahmachari Visva-Bharati University, Santiniketan, West Bengal, India

D. Chaturvedi Amity University Uttar Pradesh, Lucknow, Uttar Pradesh, India

S.J. Chen Graduate School of Kyushu University, Fukuoka, Japan

K. Dev CSIR-Central Drug Research Institute, Lucknow, India; Academy of Scientific and Innovative Research, New Delhi, India

P.K. Dwivedi Amity University Uttar Pradesh, Lucknow, Uttar Pradesh, India

Joseph M. Hayes School of Physical Sciences & Computing, University of Central Lancashire, Preston, United Kingdom

H. Hussain University of Nizwa, Nizwa, Sultanate of Oman

T. Matsui Graduate School of Kyushu University, Fukuoka, Japan

R. Maurya CSIR-Central Drug Research Institute, Lucknow, India; Academy of Scientific and Innovative Research, New Delhi, India

J.S. Meshram Rashtrasant Tukadoji Maharaj Nagpur University, Nagpur, India

N. Mishra Banaras Hindu University, Varanasi, Uttar Pradesh, India

S. Mishra Banaras Hindu University, Varanasi, Uttar Pradesh, India

I. Mohanram Rashtrasant Tukadoji Maharaj Nagpur University, Nagpur, India

H. Pandey Sam Higginbottom Institute of Agriculture, Technology & Sciences, Allahabad, India

E. Ramakrishna CSIR-Central Drug Research Institute, Lucknow, India

K. Rashid Bose Institute, Calcutta, West Bengal, India

H.M. Sampath Kumar Vaccine Immunology Lab, Natural Products Chemistry Division CSIR-Indian Institute of Chemical Technology, Hyderabad, India

P.C. Sil Bose Institute, Calcutta, West Bengal, India

A.S. Singh Banaras Hindu University, Varanasi, Uttar Pradesh, India

V. Sravanthi Vaccine Immunology Lab, Natural Products Chemistry Division CSIR-Indian Institute of Chemical Technology, Hyderabad, India

V.K. Tiwari Banaras Hindu University, Varanasi, Uttar Pradesh, India

Editor Biography

Professor (Dr) Goutam Brahmachari currently holds the position of full professor of chemistry at the Department of Chemistry, Visva-Bharati (a Central University), Santiniketan, India. He was born at Barala in the district of Murshidabad (West Bengal, India) in 1969. He received BSc (Honors) in Chemistry and MSc with specialization in Organic Chemistry from Visva-Bharati (a Central University), India in 1990 and 1992, respectively. Thereafter, he received PhD (Organic Chemistry) in 1997 from the same University. In 1998, he joined his alma mater as assistant professor. He became associate professor in 2008 and promoted to full professor in 2011. At present, he is responsible for teaching courses in organic chemistry, natural products chemistry, and physical methods in organic chemistry. Several students received their PhD degree under the supervision of Prof. Brahmachari during this period, and couples of research fellows are presently working with him both in the fields of natural products and synthetic organic chemistry. Prof. Brahmachari's research is supported by several funding organizations including SERB-DST (New Delhi), CSIR (New Delhi), DBT (New Delhi), and UGC (New Delhi). He is a *Who's Who in the World*-2015 and 2016 Listee and also a recipient of Academic Brilliance Award-2015 (Excellence in Research).

Prof. Brahmachari's research interests include (1) isolation, structural determination, and/or detailed NMR study of new natural products from medicinal plants; (2) synthetic organic chemistry with special emphasis on green chemistry; (3) semisynthetic studies with natural products; and (4) evaluation of biological activities and pharmacological potential of natural and synthetic compounds. With more than 17 years of teaching experience, he has also produced so far nearly 160 publications including original research papers, review articles, and invited book chapters in edited books in the field of natural products and organic synthesis from internationally reputed presses. Prof. Brahmachari has authored/edited a number of text and reference books that include *Organic Name Reactions: A Unified Approach* (Narosa Publishing House, New Delhi; Co-published by Alpha Science International, Oxford, 2006), *Chemistry of Natural Products: Recent Trends & Developments* (Research Signpost, 2006), *Organic Chemistry Through Solved Problems* (Narosa Publishing House, New Delhi; Co-published by Alpha Science International, Oxford, 2007), *Natural Products: Chemistry, Biochemistry and Pharmacology* (Narosa Publishing House, New Delhi; Co-published by Alpha Science International, Oxford, 2009), *Handbook of Pharmaceutical Natural Products—2 volume-set* (Wiley-VCH Verlag GmbH & Co. KGaA, Weinheim, Germany, 2010), *Bioactive Natural Products: Opportunities & Challenges in Medicinal Chemistry* (World Scientific Publishing Co. Pte. Ltd., Singapore, 2011), *Chemistry and Pharmacology of Naturally Occurring Bioactive Compounds* (CRC Press, Taylor & Francis group, USA, 2013), *Natural Bioactive Molecules: Impacts & Prospects* (Narosa Publishing House, New Delhi; Copublished by Alpha Science International, Oxford, 2014), *Green Synthetic Approaches for Biologically Relevant Heterocycles* (Elsevier Inc., USA, 2014), *Bioactive Natural Products—Chemistry & Biology* (Wiley-VCH Verlag GmbH & Co. KGaA, Weinheim, Germany, 2015), *Room Temperature Organic Synthesis* (Elsevier Inc., USA, 2015), *Biotechnology of Microbial Enzymes: Production, Biocatalysis and Industrial Applications* (Academic Press, London, 2016), and few are forthcoming.

Prof. Brahmachari serves as a member of the Indian Association for the Cultivation of Science (IACS) and Indian Science Congress Association (ISCA), Kolkata, and as an Editor-in-Chief, *Signpost Open Access Journal of Organic and Biomolecular Chemistry*. He also serves as an editorial advisory board member for several international journals. He is regularly consulted as a referee by leading international journals including Elsevier, Royal Society of Chemistry, American Chemical Society, Wiley, Taylor & Francis, Springer, Bentham Science, Indian Chemical Society, Indian Journal of Chemistry (Sec. B), Korean Chemical Society, Pakistan Chemical Society, Brazilian Chemical Society, Bulgarian Academy of Sciences and so on, and also by various financial commissions.

Goutam Brahmachari enjoys Songs of Rabindranath Tagore and finds interests in Literature as well!

Foreword

NATURAL PRODUCT DRUG DISCOVERY—THE ROAD AHEAD

Natural products have a long and distinguished record as drugs and drug leads. This is illustrated both by older drugs such as morphine and quinine and also by newer drugs such as the anticancer agents Taxol, Oncovin, Yondelis, and Eribulin; by most of the currently used antibiotics; and by the antimalarial drug artemisinin. New drugs continue to be developed from natural products; thus the active principle of the actinic keratosis treatment Picato is ingenol mebutate, and withaferin A has recently been shown to reverse resistance to leptin in obese mice, suggesting that it could serve as a weight loss treatment or an antidiabetic drug (*Nat. Med.* **2016**, DOI: 10:1038/nm.4145). These and other triumphs of natural products as drugs are thoroughly documented in the comprehensive reviews published in the *Journal of Natural Products* from time to time by David Newman and Gordon Cragg; the most recent version (Newman, D.J.; Cragg, G.M. *J. Nat. Prod.* **2016**, *79*, 629) indicates that 49% of all anticancer drugs and 33% of all small molecule drugs are natural products or compounds directly derived from natural products.

Given the significance of natural products as sources of drugs and drug leads, the publication of the present book series on ***Natural Product Drug Discovery***, edited by Professor Goutam Brahmachari and published by Elsevier, is a timely and welcome addition to the literature. The inaugural volume ***Discovery and Development of Antidiabetic Agents From Natural Products*** is a useful book that covers subjects ranging from specific compounds such as andrographolide to herbal remedies from Indian plants.

This book series is a useful addition to the literature on natural products as drugs and drug leads and provides a good starting point for scientists interested in finding new therapeutic agents for diabetes. I wish the series success, and I look forward to seeing additional volumes published in the future.

David G.I. Kingston
Department of Chemistry
Virginia Tech
Blacksburg, Virginia 24061, USA

This inaugural volume ***Discovery and Development of Antidiabetic Agents From Natural Products*** in the book series on ***Natural Product Drug Discovery***, edited by Professor Goutam Brahmachari, is a timely, highly significant, and useful book for readers engaged in chemical, biological, pharmacological, and medicinal studies, as well as for scientists deeply involved in diabetes research. Diabetes mellitus is a major health problem, chronically affecting millions of people around the world, and the prevalence of this disease deserves intelligent literature written in a compact manner. Such a book has been missing for quite a while and the present book solves this problem. It has been well organized and edited by Professor Brahmachari and will go a long way toward providing solutions to this problem. He has expertly gathered a group of excellent scientists to write chapters dealing with potential solutions to the problem of diabetes. These solutions involve herbal antidiabetic drugs, natural products, and their semisynthetic analogs. The natural products include antioxidants, activators of glucose transport translocation, enzyme inhibitors, herbal antidiabetics, and functional food components. Also covered in this book are immune modulators, sesquiterpene lactones, their analogs, and carbohydrate-based agents. Many scientists, around the world, both academic and industrial, are very interested in the problem of diabetes, as well as possible solutions. The quality and timeliness of this book in a market of competitive research will stimulate present and future generations of scientists who are interested in improving the quality of lives through work on natural products and derivatives stemming from them.

The editor has a long history of publishing books that have received tremendous scientific attention. His dedication in the selection of the authors, subject matter, and organization of the chapters proves his experience and scholarly activity in this particular area. Each chapter is highly informative and precisely described. The present book will increase the possibility that scientists from pharmaceutical companies, biotechnology organizations, and academia will come up with potential solutions for this prevalent disease.

I strongly recommend this state-of-the-art book on discovery and development of antidiabetic agents from natural products to students, researchers, and professionals interested in developing new and efficient antidiabetic agents to improve the quality of our lives. At the same time, I wish success for this book series and hope to see more of such volumes in the future.

Arnold L. Demain
Research Institute for Scientists Emeriti (RISE)
Drew University, Madison, New Jersey, United States

Preface

The book series titled ***Natural Product Drug Discovery*** is an endeavor to access the ongoing developments and recent cutting-edge research advances in the field of medicinally responsible natural products in regard to their identification, isolation, and overall chemistry and pharmacology, as well as to underline how natural product research continues to make significant contributions in the domain of discovery and development of new medicinal entities, through the publication of series of *stand-alone open-ended* volumes within reasonable time intervals.

Natural products usually refer to chemical substances produced by a living organism or found in nature possessing distinctive biological and pharmacological effects, which have played a crucial role in modern drug development and still constitute a prolific source of novel lead compounds, or pharmacophores, for ongoing drug discovery programs. Natural products are classified based upon their origins, biological functions, and structures, encompassing a wide variety of chemical compound classes that include alkaloids, antibiotics, terpenoids, flavonoids, xanthonoids, phenolics, carbohydrates, lipids, proteins and amino acids, and nucleic acids. This huge diversity in chemical structures of natural products is an outcome of biosynthetic processes that have been modulated over the millennia through genetic effects. With the advent of modern techniques, particularly the rapid improvements in spectroscopic as well as accompanying advances in high-throughput screening techniques, it has become possible to have an enormous repository of bioactive natural compounds, thus opening up exciting new opportunities in the field of new drug development to the pharmaceutical industry. Medicinal chemistry of such bioactive compounds encompasses a vast area that includes their isolation and characterization from natural sources, structure modification for optimization of their activity and other physical properties, and also total and semisynthesis for a thorough scrutiny of structure–activity relationships.

Despite substantial progresses in the fields of chemistry and biology, natural products chemistry still remains a central division of chemical and biological research with no boundaries. Natural product chemistry may nowadays be looked upon as a hybrid discipline! Due to the availability of huge numbers of bioactive natural products, synthetic organic chemistry, computational chemistry, medicinal chemistry, biochemistry, and analytical chemistry as well as molecular biology, pharmacognosy, biotechnology, and clinical science have become the major scientific areas in the past few decades. It has been extensively demonstrated that the search for natural products, or products obtained from them through synthesis, is directed toward identifying new molecules for diseases and investigating their mechanism of actions and their specific targets of interactions (for example, DNA, RNA, protein, and enzymes). This fact is corroborated with the very recent literature estimating 49% of all anticancer drugs and 33% of all small molecule drugs are natural products or compounds directly derived from natural products.

The inaugural volume titled ***Discovery and Development of Antidiabetic Agents From Natural Products*** of this book series is dedicated to the field of diabetes, one of the major dreadful diseases worldwide, which can lead to further complicated morbid. Though certain useful treatments are currently available against this disease, still remedies out of this complicated disease manifestations are poor and not up to the mark. As a result, there is a continuing thirst of developing novel and effective molecules to treat the disorder satisfactorily. Hence, it is the high time to turn our attention to the natural resources in search of appropriate drug entities to address the complications of diabetes and to get rid of them in an effective manner. This present volume, which comprises a variety of 10 chapters written by active researchers and leading experts working in this remarkable field, brings together an overview of current discoveries and trends in this direction. Chapter 1 by Brahmachari offers a comprehensive discussion on the isolation, purification, and spectroscopic characterization of andrographolide, the principal bioactive chemical constituent of *Andrographis paniculata*, from its natural source, total synthesis, semisyntheses, antidiabetic studies with the biomolecule and its semi-synthetic derivatives, and also its pharmacokinetics and metabolism including safety aspects. Hayes has presented an excellent and up-to-date overview on the computer-aided discovery of glycogen phosphorylase (GP) inhibitors in Chapter 2, focusing mainly on the role of different in silico methods based on varying natural products and their analogues in the discovery and understanding of the mechanism of action of these compounds. Chapter 3 by Rashid and Sil deals with a thorough and vivid discussion on the protective role of various naturally occurring active principles such as curcumin, mangiferin, and arjunolic acid on the oxidative stress-mediated diabetic pathophysiology; natural antioxidants could find an immense application in treating this deadly disease. Maurya and his group have overviewed a comprehensive literature on the glucose transporter 4 (GLUT4) translocation activators from natural sources in Chapter 4; a large number of natural products of different classes, such as alkaloids, flavonoids, isoflavonoids, steroids, terpenoids, phenols, poylphenols, saponins, and tannins, isolated from living organisms have been nicely documented herein along with their glucose uptake activity either via the activation of GLUT4 translocation or protein expression. Carbohydrate-based glycosidase inhibitors and sodium glucose cotransporter 2 (SGLT2) inhibitors are regarded as the major active classes of compounds for the management of diabetes—Tiwari and coauthors have focused on this interesting area in Chapter 5 delineating the recent progress in designing, synthesizing, or isolating carbohydrate-based antidiabetic agents from nature with an emphasis on their mechanism of action, structure, and synthetic approach to some important leading antidiabetic agents of this class.

Chapter 6 by Chaturvedi and Dwivedi deals with a comprehensive account on the recent developments in the field of naturally occurring sesquiterpene lactones and their semisynthetic analogues as potential antidiabetic agents reported so far. Prediction of prediabetes is an important issue in the management of diabetes and related complications—Chen and Matsui offers an overview of prediabetes and the current research outcomes involved with functionally natural compounds in decreasing the risk of developing diabetes at prediabetic stage, in Chapter 7. The authors have also illustrated in detail about analytical prediction of prediabetes by monitoring levels of advanced glycation end products (AGEs), typical diabetes-related metabolites, and preventive strategy of prediabetes by natural products through the inhibition of AGEs accumulation. Type 1 diabetes (T1D) is one of the most prevailing autoimmune disorders in the world. Chapter 8 by Sravanthi and Sampath Kumar focuses mainly on phytochemicals used for treatment and prophylaxis of T1D, and many of these molecules act as immunomodulators either by avoiding β-cell destruction or by promoting their growth and development as pointed out by the authors in their discussion. Abbas and coauthors have presented a comprehensive account on naturally occurring α-glucosidase enzyme inhibitors reported during the period 2000–2015, in Chapter 9. Chapter 10 by Mohanram and Meshram offers an exhaustive list of potentially important antidiabetic plant species from different regions of India used traditionally to cure diabetes mellitus; in addition, major phytochemicals of the respective plant species responsible for the pharmacological profile have also been documented in this chapter.

This timely volume encourages interdisciplinary work among chemists, biologists, pharmacologists, botanists, and agronomists with an interest in bioactive natural products, very particularly responsible for their antidiabetic potential. The volume offers a handful of information to the researchers deeply engaged in the domain of drug discovery and development. It is also an outstanding source of information with regard to the industrial application of natural products for medicinal purposes. The broad interdisciplinary approach dealt with in this book would surely make the work much more interesting for scientists deeply engaged in the research and/or use of bioactive natural products. It will serve not only as a valuable resource for researchers in their own fields to predict promising leads for developing pharmaceuticals to treat diabetes and related complications but also to motivate young scientists to the dynamic field of bioactive natural products research.

Representation of facts and their discussions in each chapter are exhaustive, authoritative, and deeply informative; hence, the book would serve as a key reference for recent developments in the frontier research on bioactive natural products to have antidiabetic leads and would also be of much utility to scientists working in this area. I would like to express my sincere thanks once again to all the contributors for the excellent reviews they have produced, and it is their participation that makes my effort to organize such a book possible. Their masterly accounts will surely provide the readers with a strong awareness of current cutting-edge research approaches being followed in the promising field of drug discovery and development of antidiabetic natural products.

I would like to express my sincere thanks and deep sense of gratitude to Professor David G.I. Kingston, Department of Chemistry, Virginia Tech, Virginia (USA) and Professor Arnold L. Demain, Drew University in Madison, New Jersey (USA) for their keen interests in the manuscript and for writing foreword to the book.

I would also like to express my deep sense of appreciation to all of the editorial and publishing staff members associated with Elsevier Inc. for their keen interest in publishing the works as well as their all-round help so as to ensure that the highest standards of publication have been maintained in bringing out this inaugural volume.

Goutam Brahmachari
Chemistry Department
Visva-Bharati (a Central University)
Santiniketan, India
August 2016

Chapter 1

Andrographolide: A Molecule of Antidiabetic Promise

G. Brahmachari
Visva-Bharati University, Santiniketan, West Bengal, India

1. INTRODUCTION

Andrographolide, an *ent*-labdane diterpenoid lactone, is the major principal bioactive chemical constituent of *Andrographis paniculata* (Burm. F.) Nees (family: Acanthaceae); this prime constituent is mainly concentrated in leaves and can easily be isolated from the crude plant extracts as crystalline solid (Rajani et al., 2000, p. 204; Lomlim et al., 2003, p. 24; Kulyal et al., 2010, p. 356). *A. paniculata* grows abundantly in many Asian countries, including China, India, Thailand, and Sri Lanka, and has a long history of therapeutic usages in traditional Chinese and Indian medicine (Sabu et al., 2000, p. 637; Panossian et al., 2002, p. 598; Balachandran and Govindarajan, 2005, p. 19; Rao, 2006, p. 47; Khare, 2007; Patarapanich et al., 2007; Woo et al., 2008, p. 226). The traditional uses and pharmacological aspects of *A. paniculata* have been well documented in the recently published reviews (Mishra et al., 2007, p. 283; Jarukamjorn and Nemoto, 2008, p. 370; Niranjan et al., 2010, p. 125). A number of active principles are reported from this plant, which mainly include diterpene lactones, flavonoids, and polyphenols (Li et al., 2007, p. 455; Rao et al., 2004, p. 2317). However, the major constituent, andrographolide, has been found to be responsible for its key therapeutic properties as reported by Brahmachari, (2012, p. 335). This chapter is aimed to highlight on its antidiabetic potential as a promising lead molecule in modern drug discovery process.

2. THE MOLECULE

Andrographolide (**1**; molecular formula: $C_{20}H_{30}O_5$) (Fig. 1.1), a colorless and crystalline labdane diterpenoid lactone (α-alkylidene γ-butyrolactone) with very much bitter in taste, was first isolated from *A. paniculata* in the year 1911 by Gorter (1911, p. 151). Later on the compound has been isolated by various investigators in different times from the same plant source (Chakravarti and Chakravarti, 1951, p. 96; Cava et al., 1962, p. 397; Lu et al., 1981, p. 182; Rajani et al., 2000,

Discovery and Development of Antidiabetic Agents From Natural Products.
http://dx.doi.org/10.1016/B978-0-12-809450-1.00001-6

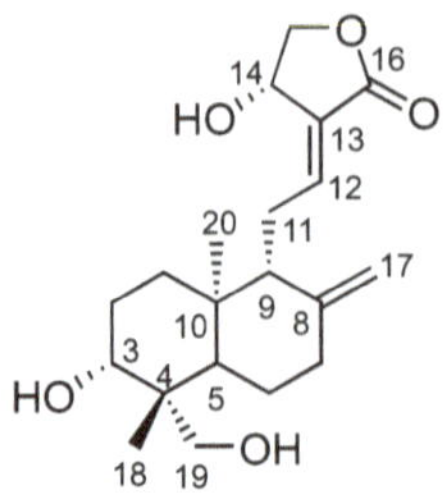

FIGURE 1.1 Andrographolide (**1**).

p. 204; Lomlim et al., 2003, p. 24; Kulyal et al., 2010, p. 356). Andrographolide (**1**) is chemically designated as (4*S*,*E*)-4-hydroxy-3-(2-((1*R*,5*R*,6*R*,8a*S*)-6-hydroxy-5-(hydroxymethyl)-5,8a-dimethyl-2-methylenedecahydronaphthalen-1-yl) ethylidene)dihydrofuran-2(3*H*)-one or 3α,14,15,18-tetrahydroxy-5β,9β*H*,10α-labda-8,12-dien-16-oic acid γ-lactone. The structure of the compound was also analyzed by X-ray crystallographic method, and the molecular stereochemistry, bond distances, bond angles including other parameters were determined by Smith et al. (1982, p. 309) and also by Fujita et al. (1984, p. 2117).

3. EXTRACTION, PURIFICATION, AND CHARACTERIZATION DATA OF ANDROGRAPHOLIDE

3.1 Extraction and Purification of Andrographolide From *Andrographis paniculata* Nees

Andrographolide is mainly concentrated in leaves of *A. paniculata* (Burm. F.) Nees (Acanthaceae) and can easily be isolated from the crude plant extracts as crystalline solid as reported by many groups of researchers. Herein the extraction method of Fujita et al. (1984, p. 2117) for the isolation and purification of andrographolide is documented. Dried leaves (6.1 kg) of *A. paniculata* were extracted with methanol (190 L) for 4 days at room temperature. The extract then was concentrated in vacuo to about 8 L, and active charcoal (330 g) was added with constant stirring. After standing for 1 day, the charcoal was filtered off and the filtrate was concentrated in vacuo. On standing white crystals (Cryst. I) were precipitated out and collected by simple filtration. Now, the recovered charcoal was refluxed with methanol (2 L) for 2 h; on removal of the charcoal by filtration, the resulting filtrate was concentrated in vacuo to about 0.6 L, which on standing yielded white crystals (Cryst. II) collected by filtration. Crude crystals (Cryst. I and II) were combined and recrystallized from methanol to afford pure andrographolide (25.2 g; 0.6% yield) as colorless plates.

3.2 Physical and Spectral Data of Andrographolide

Natural andrographolide (Fujita et al., 1984): colorless plates; mp 218–221°C (from methanol), $[\alpha]^{25}_{D}$—96.2° (pyridine, *c* 1.0); UV (MeOH): λ_{max} 223 nm (ε 13200); IR (KBr): 3340-3200, 1725, 1667, 909 cm^{-1}; ^{1}H-NMR (100 MHz, dimethylsulfoxide (DMSO)-d_6): δ 0.86 (3H, s, C_{10}-CH_3), 1.07 (3H, s, C_4-CH_3), 3.18 (1H, m, 3β-*H*), 3.23 and 3.82 (each 1H, d, J_{AB}=11 Hz, 19-H_2), 3.98 (1H, dd, *J*=2 and 10 Hz, 19-H_1), 4.34 (1H, dd, *J*=6 and 10 Hz, 19-H_1), 4.58 and 4.77 (each 1H, s, 17-H_2), 4.85 (1H, br d, *J*=5.5 Hz, 14-H), and 6.49 (1H, t, *J*=8 Hz, 12-H). Analysis calculated for $C_{20}H_{30}O_5$: C, 68.54; H, 8.65. Found: C, 68.49; H, 8.73.

Synthetic andrographolide (Gao et al., 2014): colorless plates, R_f=0.11 (petroleum ether/EtOAc = 1: 2), mp 225–227°C, $[\alpha]^{25}{}_D$—108° (MeOH, *c* 0.4); IR (KBr): 3388, 3316, 2925, 1723, 1672, 1217, 1032, 979, 907 cm^{-1}; ^{1}H-NMR (300 MHz, DMSO-d_6): δ 6.62 (1H, t, *J*=6.5 Hz), 5.71 (1H, d, *J*=5.7 Hz), 5.04 (1H, d, *J*=4.8 Hz), 4.91 (1H, t, *J*=5.9 Hz), 4.81 (1H, s), 4.63 (1H, s), 4.39 (1H, dd, *J*=6.2, 9.8 Hz), 4.12 (1H, dd, *J*=2.6, 7.4 Hz), 4.04 (1H, dd, *J*=2.1, 9.9 Hz), 3.84 (1H, dd, *J*=2.7, 10.8 Hz), 3.20–3.26 (2H, m), 2.32 (1H, d, *J*=12.3 Hz), 1.85–1.97 (2H, m), 1.62–1.76 (4H, m), 1.28–1.38 (1H, m), 1.15–1.23 (3H, m), 1.08 (3H, s), 0.66 (3H, s); ^{13}C-NMR (75 MHz, DMSO-d_6): δ 169.9, 147.6, 146.3, 129.0, 108.3, 78.6, 74.3, 64.6, 62.7, 55.5, 54.4, 42.3, 38.6, 37.5, 36.5, 27.9, 24.0, 24.0, 23.1, 14.8; ^{1}H-NMR (300 MHz, C_5D_5N): δ 7.18 (1H, t, *J*=6.6 Hz), 5.38 (1H, br s), 4.96 (3H, br s), 4.89 (1H, s), 4.86 (1H, s), 4.62 (1H, dd, *J*=5.7, 9.9 Hz), 4.52 (1H, dd, *J*=2.4, 9.9 Hz), 4.45 (1H, d, *J*=10.5 Hz), 3.62–3.83 (2H, m), 2.73 (2H, t, *J*=6.6 Hz), 2.34 (1H, d, *J*=13.5 Hz), 1.78–2.03 (4H, m), 1.69 (1H, dd, *J*=3.3, 12.9 Hz), 1.52 (3H, s), 1.17–1.47 (4H, m), 0.69 (3H, s); ^{13}C-NMR (75 MHz, C_5D_5N): δ 170.7, 148.0, 147.0, 130.2, 108.8, 79.9, 75.4, 66.0, 64.2, 56.4, 55.3, 43.3, 39.2, 38.2, 37.3, 29.0, 25.0, 24.4, 23.7, 15.2; HRMS (ESI): *m/z* calculated for $C_{20}H_{30}O_5$: Na 373.1985; Found 373.1981 for $[M+Na]^+$.

4. TOTAL SYNTHESIS OF ANDROGRAPHOLIDE

The first total synthesis of (−)-andrographolide (**1**) was achieved by Li and coworkers (Gao et al., 2014, p. 9436) via the biomimetic cyclization (cation–olefin annulation) of an epoxy homoiodo allylsilane precursor **14** prepared starting from geraniol (**2**) (Scheme 1.1). 3,7-Dimethyl-8-((tetrahydro-2*H*-pyran-2-yl)oxy)octa-2,6-dien-1-ol (**8**) was first prepared from geraniol (**2**), in an overall yield of 26% over eight steps, following standard procedure. The allylic alcohol **11**, obtained from **8**, was then subjected to Sharpless epoxidation to generate the epoxide **12** (colorless oil), which underwent etherification with *p*-methoxybenzyl bromide to provide epoxide **13** (86% ee) in 79% yield over the two steps. The investigators developed the optimized three-step route to synthesize the homoiodo allylsilane **16** from cyclopropyl ketone **13**, involving (1) chemoselective 1,2-addition of (phenyldimethylsilyl)methylcerium chloride to **13** in tetrahydrofuran (THF) at 0°C, (2) exposure of the resulting cyclopropyl carbinol **14** to freshly prepared MgI_2 etherate to cleave both the cyclopropyl and epoxide rings to have bis-iodo allylsilane intermediate **15**, and (3) treatment of the crude intermediate **15** with potassium carbonate in methanol to furnish the

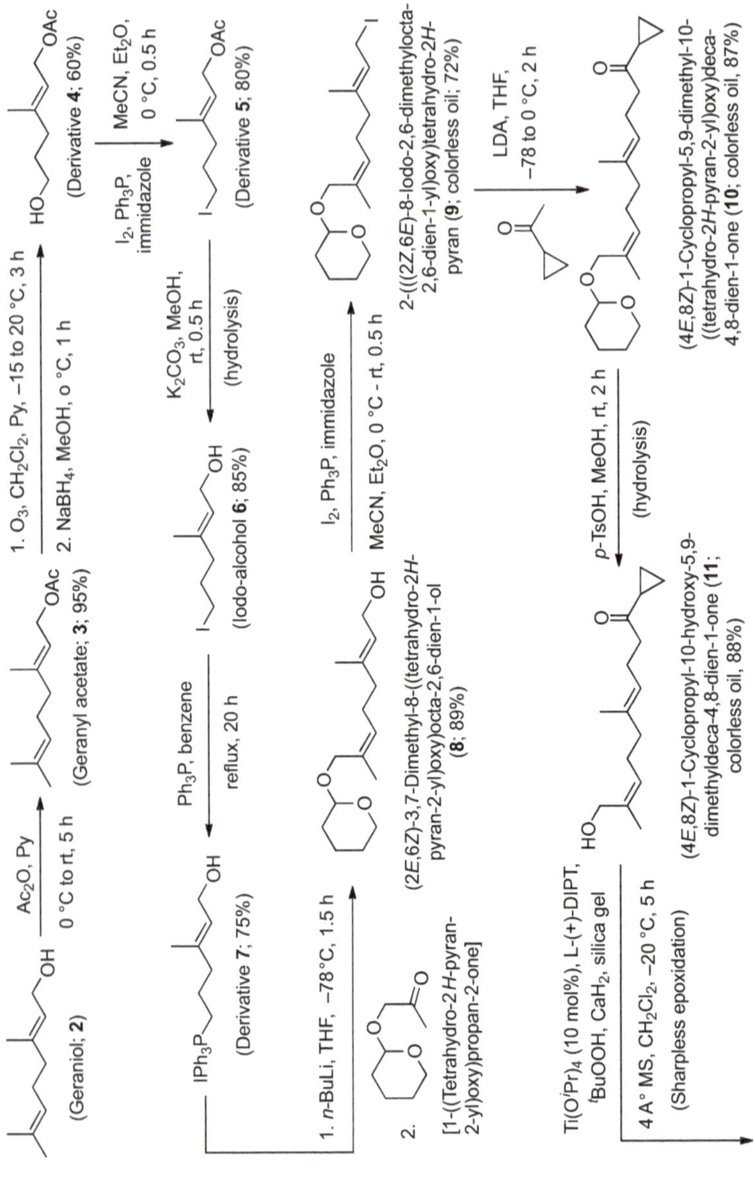

SCHEME 1.1 Li's total synthesis of (−)-andrographolide (**1**) (Gao et al., 2014, p. 9436).

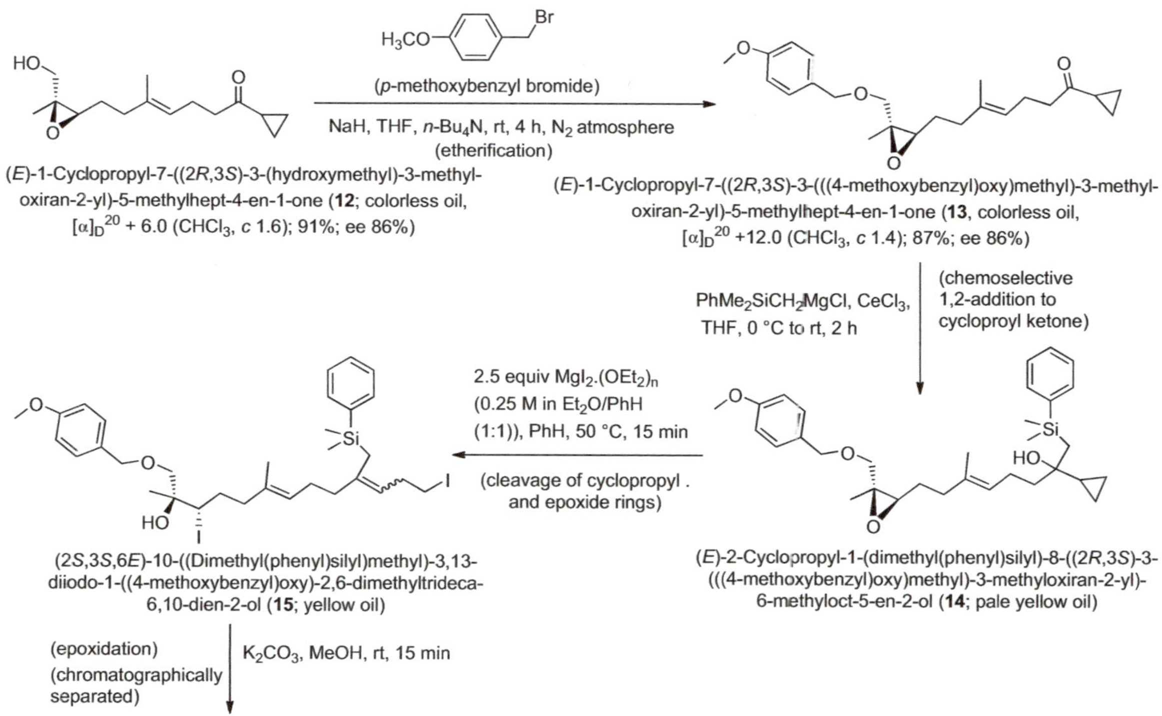

SCHEME 1.1 Cont'd.

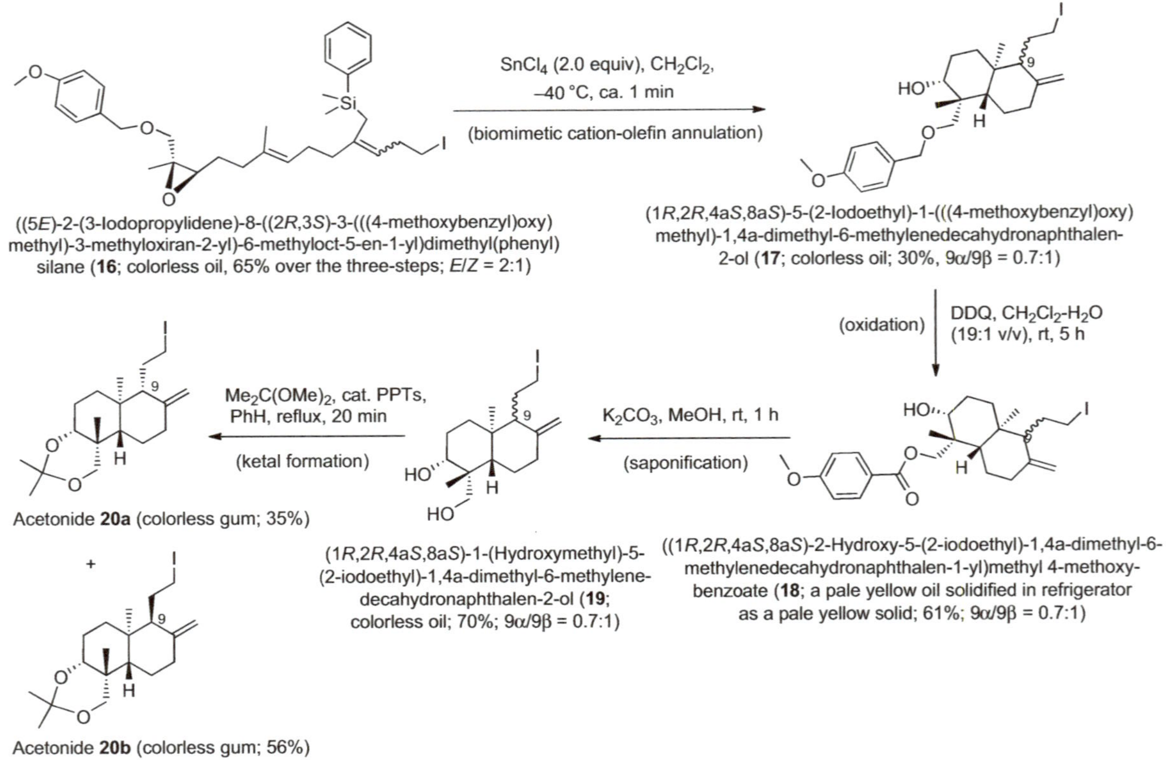

SCHEME 1.1 Cont'd.

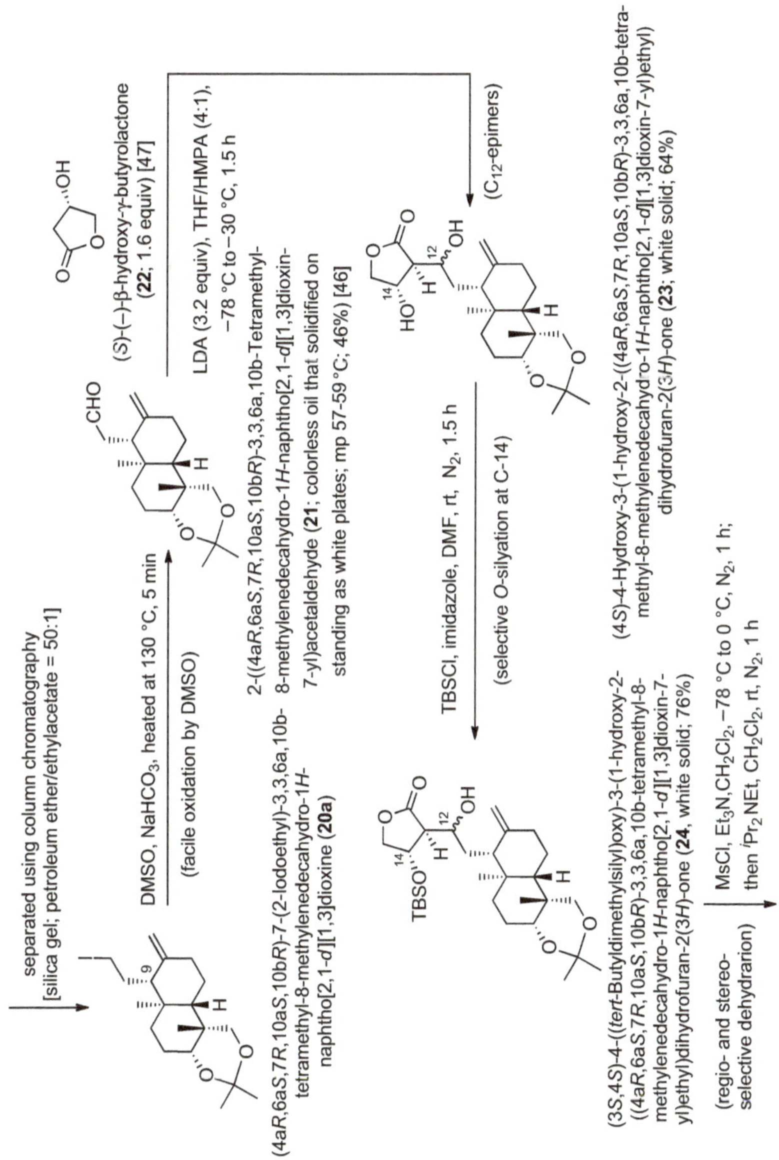

SCHEME 1.1 Cont'd.

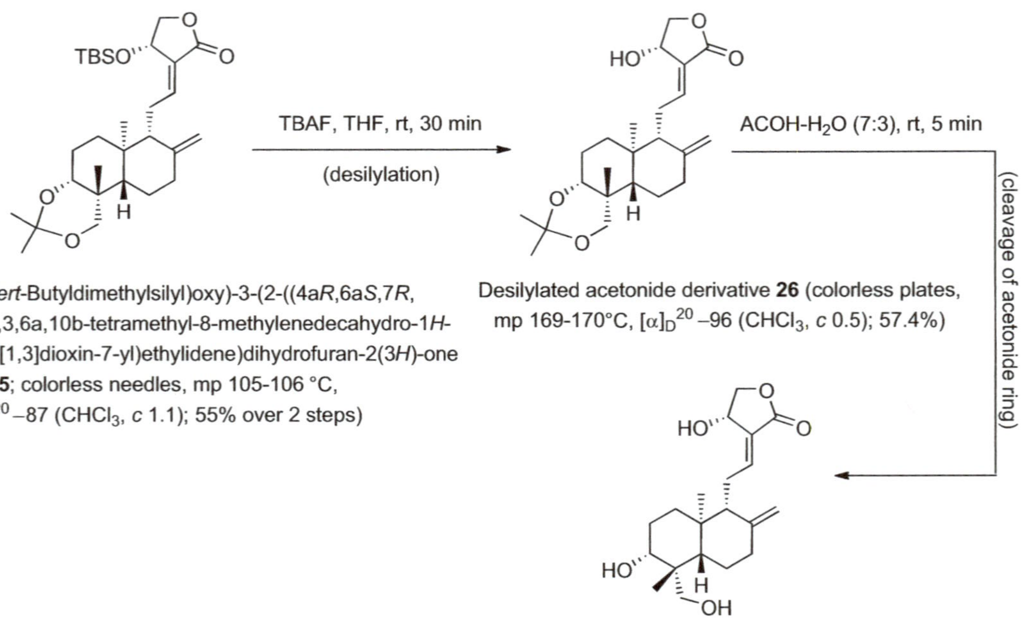

(*S*,*E*)-4-((*tert*-Butyldimethylsilyl)oxy)-3-(2-((4a*R*,6a*S*,7*R*,10a*S*,10b*R*)-3,3,6a,10b-tetramethyl-8-methylenedecahydro-1*H*-naphtho[2,1-*d*][1,3]dioxin-7-yl)ethylidene)dihydrofuran-2(3*H*)-one (**25**; colorless needles, mp 105-106 °C, $[\alpha]_D^{20}$ −87 ($CHCl_3$, *c* 1.1); 55% over 2 steps)

Desilylated acetonide derivative **26** (colorless plates, mp 169-170°C, $[\alpha]_D^{20}$ −96 ($CHCl_3$, *c* 0.5); 57.4%)

(−)-Andrographolide **1** (colorless plates, mp. 225-227°C, $[\alpha]_D^{20}$ −108 (MeOH, *c* 0.4)

SCHEME 1.1 Cont'd.

epoxy homoiodo allylsilane **16** in 65% yield from **13** as a mixture of geometric isomers (*E*/*Z* 2:1).

In the next step, this epoxy homoiodo allylsilane precursor **16** was subjected to undergo the biomimetic cation–olefin annulation on treating with freshly distilled $SnCl_4$ in dichloromethane and subsequent aqueous work-up, to form a bicyclic alcohol **17** as a mixture of C_9 epimers ($\alpha/\beta = 0.7:1$) as colorless oil in 30% yield. This crude product was then converted into the readily separable acetonides **20a** and **20b** by a sequence of reactions that involve the following: (1) 2,3-dichloro-5,6-dicyanobenzoquinone (DDQ) oxidation in aqueous dichloromethane at room temperature conditions to give the ester **18** in 61% yield, (2) saponification of the resulting ester derivative **18** to form the iodo-diol, 1-(hydroxymethyl)-5-(2-iodoethyl)-1,4a-dimethyl-6-methylene-decahydronaphthalen-2-ol (**19**; colorless oil; 70%; 9α/9β = 0.7:1), and (3) reacetonization of the corresponding diol **19**, followed by column chromatographic separation of the acetonide **20a**. Compound **20a** underwent facile oxidation with DMSO on heating at 130°C to afford 2-((4a*R*,6a*S*,7R,10aS,10bR)-3,3,6a,10b-tetramethyl-8-methylenedecahydro-1*H*-naphtho[2,1-*d*][1,3]dioxin-7-yl)acetaldehyde (**21**; colorless oil that solidified on standing as white plates; mp 57–59°C; 46%), which served as the key intermediate for the attachment of the B-ring lactone side chain. This aldehydic derivative **21** on aldol condensation with the corresponding lithium enolate of (*S*)-(−)-β-hydroxybutyrolactone (**22**) gave dihydroxy lactone **23** (as a mixture of C_{12} epimers), which was selectively *O*-silylated at C-14 and dehydrated regio- and stereoselectively via the corresponding mesylate intermediate to give the *E*-configurated lactone **25** in 55% yield over the two steps as colorless needles. Finally, the investigators were successful in isolating (−)-andrographolide (**1**) on standard desilylation and acetonide cleavage of **25**, as colorless crystals (mp 225–227°C) from methanol, which was found to be identical spectroscopically with the natural andrographolide (Fujita et al., 1984, p. 2117).

5. ANTIDIABETIC POTENTIAL OF NATURAL ANDROGRAPHOLIDE AND ITS SEMISYNTHETIC DERIVATIVES

5.1 Antidiabetic Studies With Natural Andrographolide and Deoxyandrographolide

Yu et al. (2003, p. 1075) thoroughly studied the antidiabetic activity of natural andrographolide on streptozotocin (STZ)-induced diabetic rat model. Oral administration of the drug was found to reduce plasma glucose concentrations of STZ diabetic rats in a dose-dependent manner. The investigators also observed that similar treatment with andrographolide decreases plasma glucose in normal rats, although the maximal effect was more pronounced in STZ diabetic rats. Andrographolide at an effective dose of 1.5 mg/kg body weight significantly attenuated the increase of plasma glucose induced by an intravenous glucose challenge test in normal rats. The investigators further observed that the drug enhances uptake of glucose in a concentration-dependent manner in isolated soleus muscle of STZ diabetic rats. Moreover, the *m*RNA and protein levels of subtype 4 form of the glucose transporter protein subtype 4 (GLUT4) in soleus muscle were increased after repeated intravenous administration of andrographolide in STZ diabetic rats for 3 days. Hence, it may be concluded that andrographolide can increase glucose utilization to lower plasma glucose in diabetic rats lacking insulin (Yu et al., 2003, p. 1075). To have an idea about the mode of action of this drug molecule, the investigators carried out an in-depth study on the mechanism(s) for glucose-lowering action of andrographolide in STZ diabetic rats (Yu et al., 2008, p. 529); it was observed that the drug lowers plasma glucose concentrations in a dose-dependent manner and increases plasma β-endorphin-like immunoreactivity (BER) dose dependently in diabetic rats. They observed that both of these responses to andrographolide are abolished in animals pretreated with prazosin or *N*-(2-(2-cyclopropylmethoxy)ethyl)-5-choro-α-dimethyl-1*H*-indole-3-thylamine (RS17053) at doses sufficient to block α_1-adrenoceptors (ARs). Also, andrographolide enhanced BER release from isolated rat adrenal medulla in a concentration-related manner that could be abolished by α_1-ARs antagonists. Bilateral adrenalectomy in STZ diabetic rats eliminated the activities of andrographolide, including the plasma glucose lowering effect and the plasma BER-elevating effect. However, andrographolide failed to lower plasma glucose in the presence of opioid μ-receptor antagonists and in the opioid μ-receptor knockout diabetic mice. From their detailed study, they suggested that andrographolide may activate α_1-ARs to enhance the secretion of β-endorphin, which can stimulate the opioid μ-receptors to reduce hepatic gluconeogenesis and to enhance the glucose uptake in soleus muscle, resulting in a decrease of plasma glucose in STZ diabetic rats. However, the roles of other endogenous opioid peptides or the mixture of several opioid peptides in the activation of opioid μ-receptors associated with the plasma glucose lowering action of andrographolide should be considered and requires more investigation.

Diabetes mellitus is generally associated with several other complications such as diabetic retinopathy (DR) and diabetic nephropathy, which greatly reduce the quality of life and the survival of diabetic patients. DR arises out of the chronic vascular complication with the development of diabetes mellitus (Fantes et al., 2010, p. 213; Willard and Herman, 2012, p. 209). Vision loss from DR has been a major and leading cause of blindness in adult. It is reported that nearly all persons with type 1 diabetes and about 60% of persons with type 2 diabetes will develop DR when living with DM for the first two decades (Chibber et al., 2007, p. 3; Abu El-Asrar, 2013, p. 273). Recently, Yu et al. (2015, p. 824) demonstrated that andrographolide possesses considerable beneficial effect on STZ-induced DR in mice. Andrographolide at a dose of 10 mg/kg body weight was found to ameliorate DR via attenuating retinal

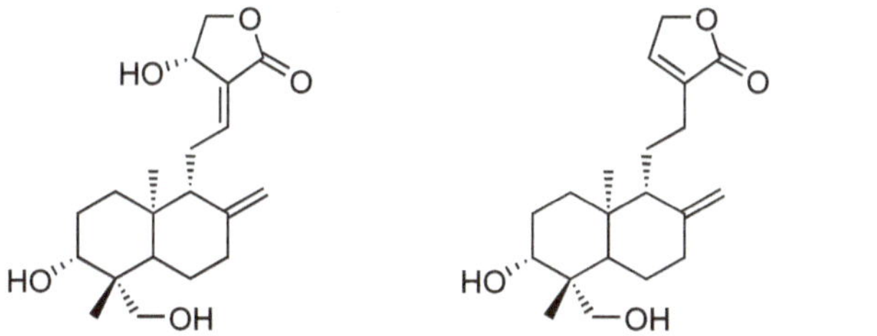

Andrographolide **1** Deoxyandrographolide (DeoAn) **1A**

FIGURE 1.2 Chemical structures of andrographolide **1** and deoxyandrographolide (DeoAn) **1A**.

angiogenesis and inflammation during the development of DR, and VEGF (vascular endothelial growth factor), NF-κB, and Egr-1 (earlygrowthresponse-1) signals all play important roles in such process. This study provides strong evidence for the potential application of this drug molecule for the treatment of DR in clinic. Arha et al. (2015, p. 207) isolated deoxyandrographolide (DeoAn; **1A**) (Fig. 1.2) from *A. paniculata* Nees and studied in vivo antihyperglycemic potential of the molecule by investigating on its glucose utilization in skeletal muscle using STZ-induced diabetic rats and genetically diabetic db/db mice. It is well known that skeletal muscle is the principal site for postprandial glucose utilization and augmenting the rate of glucose utilization in this tissue may help to control hyperglycemia associated with diabetes mellitus. The investigators showed that DeoAn (**1A**) possesses postprandial blood glucose lowering effect in STZ-induced diabetic rats along with preventing the rises in the fasting blood glucose, serum insulin, triglycerides, and LDL-cholesterol levels of db/db mice. The drug molecule was found to stimulate glucose uptake in L6 myotubes dose dependently by enhancing the translocation of GLUT4 to cell surface, without affecting the total cellular GLUT4 and GLUT1 content, and importantly, these effects of DeoAn were additive to insulin. Further analysis revealed that DeoAn activates PI-3-K- and AMPK-dependent signaling pathways, which account for the augmented glucose transport in L6 myotubes. These findings suggest the therapeutic efficacy of the DeoAn (**1A**) for type 2 diabetes mellitus and can be potential phytochemical for its management.

5.2 Antidiabetic Studies With Semisynthetic Derivatives of Andrographolide

An andrographolide–lipoic acid conjugate (AL-1, **30**) was synthesized (Scheme 1.2) by covalently linking andrographolide (**1**) with α-lipoic acid (**28**) (Jiang et al., 2009, p. 2936) and the investigators demonstrated that this conjugate molecule possesses significant therapeutic properties against bacterial infections (Jiang et al., 2009, p. 2936) and influenza virus (Yuan et al., 2016, p. 769). It has been mentioned that andrographolide, the primary active component of *A. paniculata*, decreased the plasma glucose concentrations of STZ diabetic rats

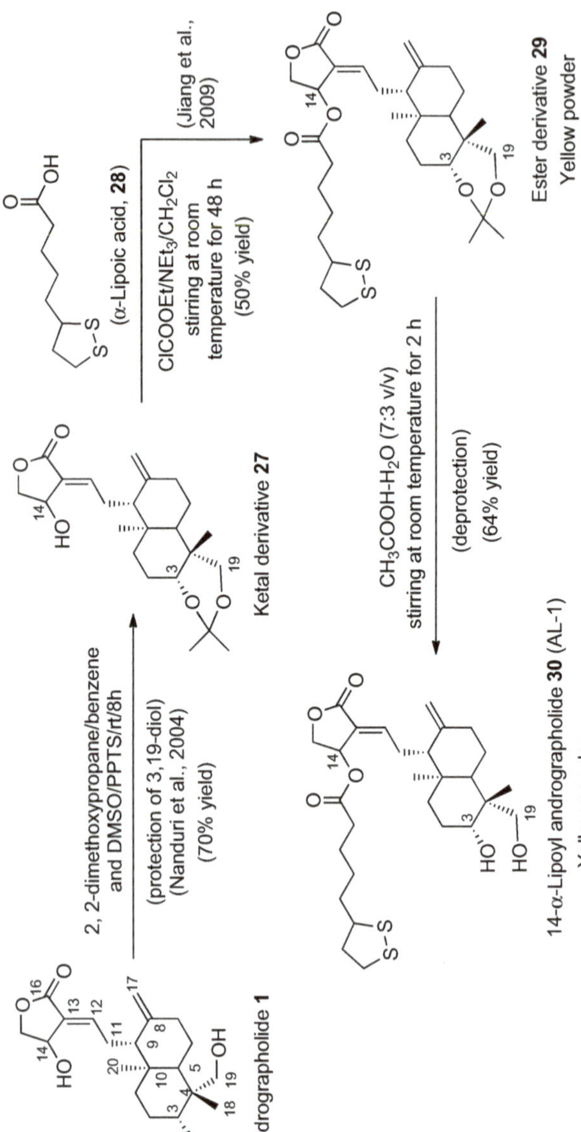

SCHEME 1.2 Synthesis of andrographolide–lipoic acid conjugate (AL-1, **30**) (Jiang et al., 2009, p. 2936).

dose dependently (Yu et al., 2003, p. 69), and also its analogue decreased the blood glucose level in db/db diabetic mice (Koranyi et al., 1990, p. 962). Again, α-lipoic acid (**28**), a dithiol compound derived from octanoic acid, is one of the most potent antioxidants. Evidence from previous studies showed that α-lipoic acid treatment improved glucose uptake and metabolism in animal models of diabetes and in type 2 diabetic patients (Ruhnau et al., 1999, p. 1040; Smith et al., 2004, p. 1135). In this regard, α-lipoic acid has been used to manage diabetes-associated neuropathy in Germany for more than 20 years. Andrographolide–lipoic acid conjugate (AL-1, **30**) was found to possess significant antidiabetic potential in alloxan-induced diabetic mice (model type-I diabetes); the andrographolide analogue lowered blood glucose, increased insulin, and prevented loss of β-cells and their dysfunction, which stimulated GLUT4 membrane translocation in soleus muscles. Pretreatment of RIN-m cells in vitro with the test compound prevented H_2O_2-induced cellular damage, quenched glucose and glibenclamide-stimulated reactive oxidative species production, and inhibited cytokine-stimulated NF-κB activation. Hence, the conjugate derivative was found to possess both hypoglycemic and β-cell protective effects, which translated into antioxidant and NF-κB inhibitory activity, thereby, establishing the andrographolide derivative (AL-1) as a potential new antidiabetic agent (Zhang et al., 2009, p. 62).

It had also been investigated that the antioxidative and/or antiinflammatory activity of AL-1 contributes to its cytoprotective effects (Yan et al., 2013, p. e63656; Zhu et al., 2013, p. 169). Recently, the investigators from the same laboratory reported the effect of AL-1 on insulin resistance in a high-fat diet/STZ-induced animal diabetic model and the mechanisms responsible for its action (Li et al., 2015, p. 3151). AL-1 (**30**), at doses of 40 and 80 mg/kg, had a significant hypoglycemic effect; it significantly reduced the level of cholesterol and increased HDL as well. AL-1 thus facilitated the responsiveness to insulin, decreased insulin resistance and blood glucose level in experimental rats with high-fat diet/STZ-induced diabetes. In addition, they also investigated the effect of AL-1 on the NF-κB signaling pathway in rat islet-derived insulinoma cells (RIN-m cells) with a focus to explore the link between reactive oxygen species (ROS)-associated inflammation and insulin resistance. The experimental outcomes established that the drug improves the morphology of pancreatic islets and their function and suppresses high glucose-induced phosphorylation of p65 and IκBα in RIN-m cells. It is thus revealed AL-1 (**30**) protected islets from high glucose-induced oxidative damage by downregulating the NF-κB signaling pathway. The overall results underline that maintaining a stable environment for β-cells might be a new insight for the prevention and treatment of diabetes (Li et al., 2015, p. 3151). In their follow-up study in 2016, the same group of investigators (Yan et al., 2016, p. 499) has focused on the underlying mechanism of the protective effect of AL-1 on high glucose-induced pancreatic β-cell injury. From their experimental observations, it has been revealed that high glucose-induced overproduction of ROS and nitric oxide (NO) is responsible for pancreatic β-cell damage, and the tested drug (AL-1) protected these β-cells damage under high glucose conditions via activating antioxidant enzymes and NF-E2-related factor 2 (Nrf2) signaling pathway. All these findings warrant for further in-depth investigations with AL-1 that might be developed into medicinal preparations for treatment and/or prevention of diabetes.

Xu et al. (2007a, p. 4247) reported on the preparation of a good number of other semisynthetic derivatives of andrographolide and examined their α-glucosidase inhibitory activity in vitro. The series of 3,19-dinicotinate of 15-alkylidene derivatives of andrographolide (**34a–g**; Scheme 1.3) was appeared to be a novel family of α-glucosidase inhibitors. All the compounds [*viz.* 14-deoxy-11,12-didehydro-3,19-dinicotinate-15-(2,8-dimethanyl-2,7-octanedieneylidene)-andrographolide (**34a**), 15-benzylidene14-deoxy-11,12-didehydro-3,19 dinicotinateandrographolide (**34b**), 15-(*p*-chlorinebenzylidene)-14-deoxy-11,12-didehydro-3,19-dinicotinateandrographolide (**34c**), 15-(*m*-bromoacylbenzylidene)-14-deoxy-11,12-didehydro-3,19-dinicotinateandrographolide (**34d**), 14-deoxy-11,12-didehydro-3,19-dinicotinate-15-(*p*-(*N,N*-dimethylamino)benzylidene)-andrographolide (**34e**), 14-deoxy-11,12-didehydro-3,19-dinicotinate-15-(2-furanmethanylidene)-andrographolide (**34f**) and 14-deoxy-11,12-didehydro-3,19-dinicotinate-15-*p*-methoxylbenzylidene andrographolide (**34g**)] were obtained in good to excellent yields and characterized based on detailed spectral studies including FTIR, NMR, and mass and were evaluated for their in vitro α-glucosidase inhibitory potential using acarbose as positive control (56.5% inhibition at 1 mM concentration). The inhibitory activity against α-glucosidase enzyme for the test compounds **34a–g** was determined as 68.9%, 100%, 100%, 100%, 74.8%, 95.9%, and 100% with respective IC_{50} values of 28, 16, 6, 14, 25, 36, and 11 μM. Hence, **34c**, **34d**, and **34g** appeared to be the most potent lead molecules among the series and demand further in-depth studies with these useful lead molecules.

The same group of investigators (Xu et al., 2007b, p. 73) further developed another such kind of series of modified semisynthetic derivatives of andrographolide (Fig. 1.3), among those compounds **35a–c** showed promising α-glucosidase inhibitory activity with IC_{50} values of 8.3, 14.1, and 15.7 μM, respectively. Among the inhibitors, **35a** (8.3 μM) should prove useful in developing new drugs such as diabetes, antitumor, and antiviral medications.

It is pertinent to be mentioned herein that Xu et al. (2010, p. 880) reported a detailed report on the QSAR studies of various andrographolide derivatives as α-glucosidase inhibitors, where the 2D QSAR and 3D QSAR models were successfully established to quantitatively describe the relationship between structures and activities of andrographolide derivatives as α-glucosidase inhibitors. The developed QSAR model could be used to predict the inhibitory activities of unknown andrographolide derivatives on α-glucosidase, and application of this model would greatly facilitate the discovery of better α-glucosidase inhibitors in future endeavors.

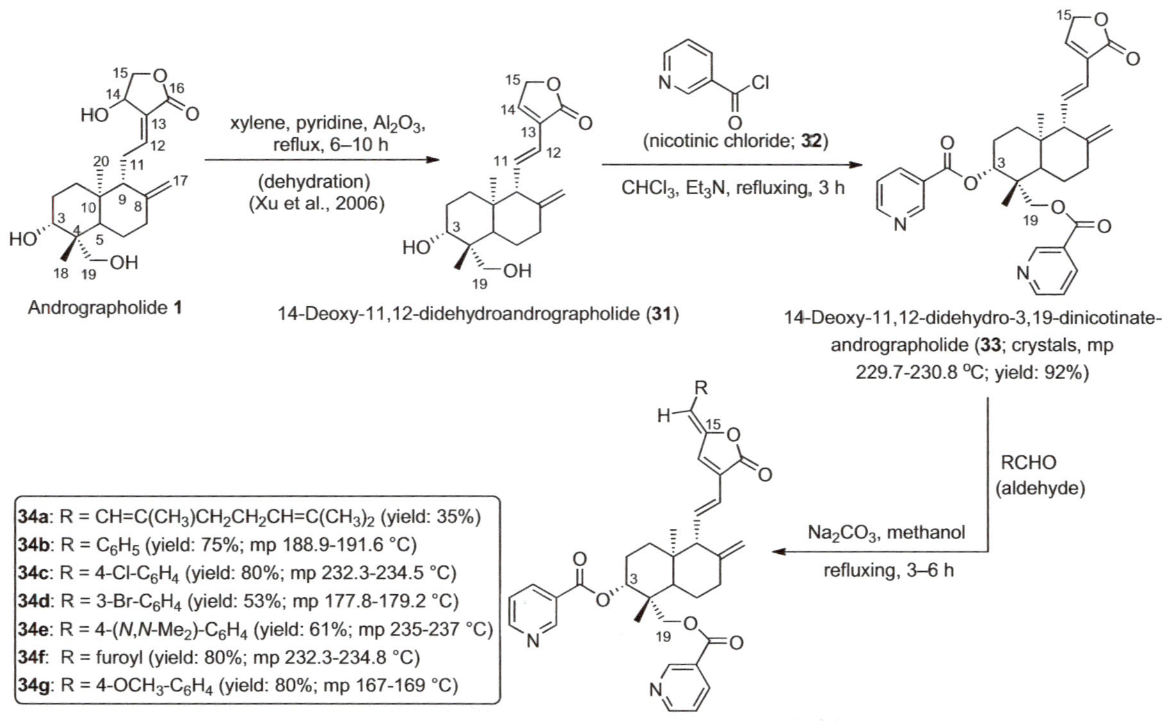

SCHEME 1.3 Synthesis of 3,19-dinicotinate of 15-alkylidene derivatives of andrographolide (**34a–g**) (Xu et al.2007a, p. 4247).

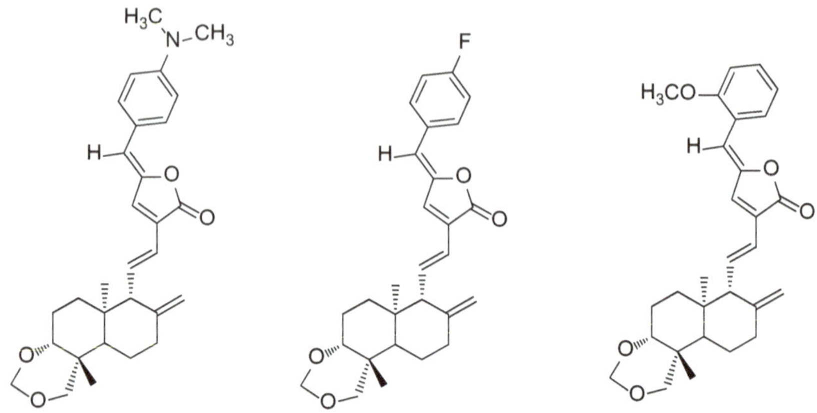

35a (IC_{50} 8.3 μM) **35b** (IC_{50} 14.1 μM) **35a** (IC_{50} 15.7 μM)

FIGURE 1.3 α-Glucosidase inhibitors prepared from andrographolide (Xu et al., 2007a,b, p. 73).

6. STUDIES ON PHARMACOKINETICS AND METABOLISM

Andrographolide has been found to be quickly absorbed and extensively metabolized both in rats and humans as revealed from pharmacokinetic studies (Zhu and Kang, 1981, p. 266; Panossian et al., 2000, p. 351). Panossian et al. (2000, p. 351) determined the amount of andrographolide in the blood plasma of rats and human volunteers following oral administration of *A. paniculata* extract and *Kan Jang* tablets (having fixed combination of *A. paniculata*); andrographolide was found to be quickly and almost completely absorbed into the blood following the oral administration of the extract at a dose of 20 mg/kg body weight in rats. Its bioavailability was, however, observed to be decreased fourfold when a 10 times higher dose was used. Since a large part (~55%) of andrographolide is usually bound to plasma proteins and only a limited amount can enter into the cells, the present investigators described the pharmacokinetics of the drug with the help of a one-compartment model. Renal excretion is not the main route for eliminating the drug molecules, and thus it is most likely to be intensely and dose dependently metabolized. Following the oral administration of four *Kan Jang* tablets (a single therapeutic dose, equal to 20 mg of andrographolide) to humans, maximum plasma level was quantified as approximately 393 ng/mL (~1.12 μM) after 1.5–2 h. Half-life and mean residence times were determined as 6.6 and 10.0 h, respectively. These investigators explained the pharmacokinetics of andrographolide in humans well by an open two-compartment model. The calculated steady-state plasma concentration of the compound for multiple doses of *Kan Jang* (after the normal therapeutic dose regimen, 3 × 4 tablets/day, about 1 mg andrographolide/kg body weight/day) was approximately 660 ng/mL (~1.9 μM), enough to reveal any anti-PAF (platelet activating factor) effect, particularly after drug uptake when the concentration of andrographolide in blood is about 1342 ng/mL (~3.8 μM, while for anti-PAF effect EC_{50}: 5 μM) (Panossian et al., 2000, p. 351).

Pharmacokinetics of andrographolide was also evaluated by Wangboonskul et al. (2006, p. 209) in healthy Thai male volunteers. From their detailed study, it was evident that andrographolide is rapidly absorbed and extensively metabolized, and the saturated metabolism process of the drug occurs at high dose. The individual variation of absorption among subjects was observed, which may lead to the variation of therapeutic efficacy. The same group of investigators studied on the possible drug interaction of andrographolide with commonly used drugs such as acetaminophen, amoxicillin, aspirin, chlorpheniramine, and norfloxacin using artificial lipophilic membrane and everted rat intestine (Daodee et al., 2007, p. 2078). Transport of andrographolide in both artificial lipophilic membrane and everted rat intestine was found not to be affected by different pH of the medium (stimulated gastric and intestine fluids), different concentrations of andrographolide as well as the coadministered drugs examined. The present investigators also examined the participation of ATP-dependent efflux transporters such as P-glycoprotein and MRP2 in andrographolide transport in rat intestine in vivo; andrographolide was found to be related to P-glycoprotein and caused P-glycoprotein-mediated drug interaction—hence, it is necessary to pay attention in clinical pharmacotherapy with andrographolide from the viewpoint of P-glycoprotein-mediated drug interaction also (Wangboonskul et al., 2006, p. 209; Daodee et al., 2007, p. 2078).

Suo et al. (2007, p. 730) also studied the pharmacokinetics of andrographolide [10 mg/kg, intravenous (i.v.)] in rats and observed that the blood concentration of andrographolide was approximately 11 μg/mL (~30 μM). Recently, Chen et al. (2014, p. 1) investigated the pharmacokinetics and bioavailability of andrographolide in rats and studied whether andrographolide enhances antioxidant defense in different tissues and protects against carbon tetrachloride-induced oxidative damage. After a single 50 mg/kg administration, the maximum plasma concentration of andrographolide was observed as 1 μM that reached at 30 min showing its bioavailability of 1.19%. After treatment for 5 days, the investigators also found that andrographolide activates Nrf2 and the expression of numerous Nrf2-dependent antioxidant genes and attenuates CCl_4-induced liver damage. Similar changes in antioxidant defense are observed in the heart and kidneys. It is recognized that increased oxidative stress in accumulated fat is an early indicator of metabolic syndrome. Therefore, modulation of the redox state in adipose tissue is a potentially therapeutic target for obesity-associated metabolic syndrome. Based on the results of the present study, andrographolide has excellent antioxidant activity and it might be used in the management of oxidative stress-mediated diseases (Chen et al., 2014, p. 1).

A good number of studies on metabolic pathways of andrographolide were reported so far. In 2003, He et al. (2003a, p. 6603) isolated and identified four sulfonate metabolites of andrographolide, such as 14-deoxy-12(*R*)-sulfo andrographolide (**36**), 14-deoxy-12(*S*)-sulfo andrographolide (**37**),

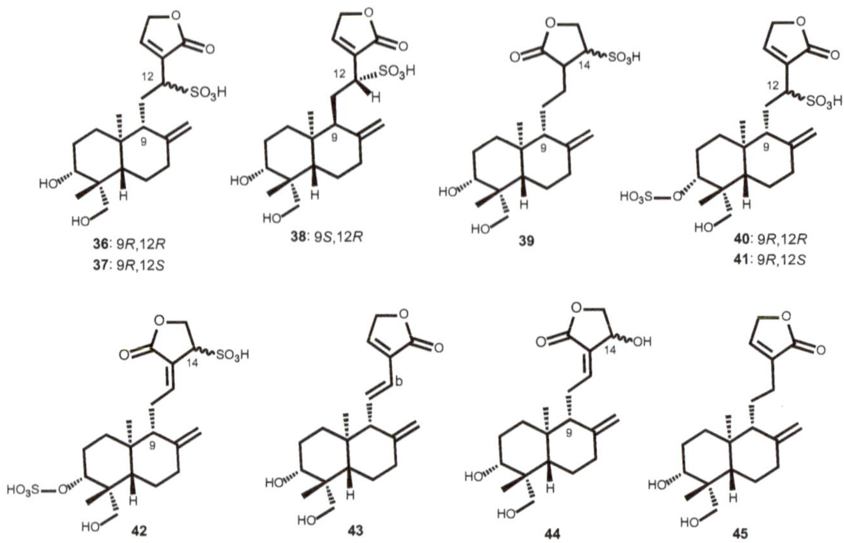

FIGURE 1.4 Metabolites of andrographolide in rats (He et al., 2003a,c, p. 6603, 586).

14-deoxy-12(*R*)-sulfo-9(*S*)-andrographolide (**38**), and 14-sulfo isoandrographolide (**39**), from urine and feces in rats. These four metabolites were formed through a rare metabolic reaction and were all new compounds. In another reporting, the same investigators mentioned 14-deoxy-12(*R*)-sulfo andrographolide (**36**) as the main metabolite of andrographolide in rats (Fig. 1.4); it was proposed that the β-carbon of α,β-unsaturated carbonyl is attacked by sulfonic acid to form the sulfonate compound (**36**) in a rare metabolic reaction, which might be the main metabolic pathway of andrographolide in rats (Scheme 1.4) (He et al., 2003b p. 983). The polarity of the sulfonate metabolite increased greatly and could be easily eliminated from body (He et al., 2003b p. 983). 14-Deoxy-12(*R*)-sulfo andrographolide (**36**) was found to be identical to an antiinflammatory drug (Lianbizhi) used in the clinic as an injection in China (Meng, 1981, p. 571). He et al., (2003c, p. 586) isolated and identified six more (**40–45**) metabolites from rat urine, feces, and the contents of the small intestine; the structures of these six metabolites were determined to be 14-deoxy-12(*R*)-sulfo andrographolide 3-sulfate (**40**), 14-deoxy-12(*S*)-sulfo andrographolide 3-sulfate (**41**), 14-sulfo isoandrographolide 3-sulfate (**42**), 14-deoxy-11,12-didehydroandrographolide (**43**), isoandrographolide (**44**), and 14-deoxy andrographolide (**45**), respectively, based on chemical evidence and spectroscopic analysis.

Later on Cui et al. (2004, p. 772) reported isolation and identification of four new metabolites (**46–49**) of andrographolide from human urine (collected between 0 and 72h) of eight healthy volunteers aged 21–28 years and weighed 50–80 kg (all males) each of whom was given orally three tablets three times per day (50 mg of andrographolide per tablet) for 2 days and the

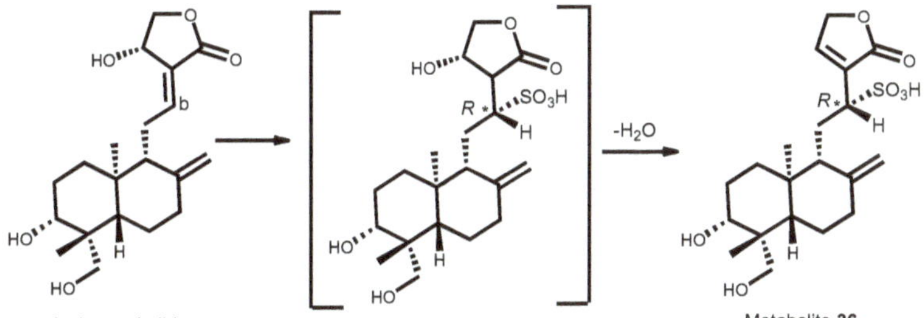

SCHEME 1.4 Proposed metabolic path of andrographolide in rats (He et al., 2003b, p. 983).

urine was collected between 0 and 72h. All of the metabolites (**46–49**) were characterized as sulfate and one of them also as a cysteine *S*-conjugate (**49**). The structures were determined to be andrographolide-3-*O*-sulfate (**46**), isoandrographolide-3-*O*-sulfate (**47**), 14-deoxyandrographolide-3-*O*-sulfate (**48**), 14-deoxy-12-(cysteine-*S*-yl)-andrographolide-3-*O*-sulfate (**49**), respectively, based on chemical evidence and spectroscopic analyses. To gain additional insight into the metabolism of andrographolide in human, the same group of investigators (Cui et al., 2005a,b, p. 369, 555) further examined the biotransformation of andrographolide in urine after oral administration in humans and reported to identify seven more metabolites (**50–56**), which were found to be the glucuronide conjugates of andrographolide; they were characterized as andrographolide-19-*O*-β-D-glucuronide (**50**), isoandrographolide-19-*O*-β-D-glucuronide (**51**), 14-deoxy-12-hydroxy-andrographolide-19-*O*-β-D-glucuronide (**52**), andrographolide-19-*O*-[6′-methyl-β-D-glucuronide] (**53**), 14-deoxy-12(13)-en-andrographolide-19-*O*-β-D-glucuronide (**54**), 14-deoxyandrographolide-19-*O*-β-D-glucuronide (**55**), and 3-oxoandrographolide-19-*O*-β-D-glucuronide (**56**), respectively (Fig. 1.5). These results are potentially important to understand its in vivo metabolic fate and disposition of andrographolide in humans. Structures of these andrographolide metabolites in human urine and possible metabolic pathways for their production are shown in Scheme 1.5. The same group of investigators (Cui et al., 2008, p. 261) further reported the existence of four more andrographolide metabolites in human urine, which are the urea adducts (**57–60**). These urea adducts were characterized as 14-deoxy-12-carbamidoandrographolide metabolite (**57**), 14-deoxy-12-carbamidoandrographolide-19-*O*-sulfate (**58**), 14-deoxy-12(*R/S*)-carbamidoandrographolide-19-*O*-β-D-glucuronide (**59**), and 14-deoxy-12(*S/R*)-carbamidoandrographolide-19-*O*-β-D-glucuronide (**60**), respectively; the mechanistic pathway for such urea adducts of andrographolide was presumed that β-carbon (C-12 of andrographolide) of α,β-unsaturated carbonyl was attacked by one nitrogen atom of urea, which has lone pair of electrons, followed by the double bond migration from 12(13) to 13(14) and elimination of the hydroxyl group at C-14 to form these urea adducts (Cui et al., 2008, p. 261). However, further investigation is needed to explore whether some enzymes in human body were involved in the reaction and where these adducts were formed.

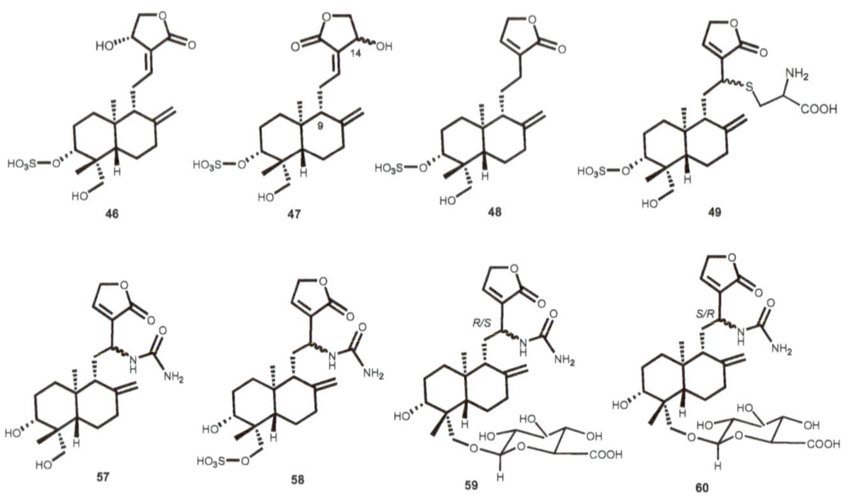

FIGURE 1.5 Metabolites of andrographolide in human (Brahmachari, 2012, p. 335; Cui et al., 2004, p. 772; 2005a,b, p. 369, 555; 2008, p. 261).

7. SAFETY ASPECTS OF ANDROGRAPHOLIDE

Numerous studies on the toxicity of *A. paniculata* have been performed so far indicating that it is extremely nontoxic, even at high doses (Jayakumar et al., 2013). Sakila et al. (2009, p. 10) conducted an antifertility study and found no toxicity, even at a high dose of *A. paniculata* that was administered to rats. Singha et al. (2007, p. 13) showed that pretreatment of *A. paniculata* and andrographolide at 500 mg/kg body weight and 125 mg/kg body weight, respectively, could minimize the toxicity when compared with the ethanol-treated group, as evidenced by different enzymatic assays in the liver and kidney tissues; the results were comparable with those of administering silymarin. It was also demonstrated that andrographolide at concentrations of 22 μg/kg and 55 μg/kg markedly lowered the mortality rate in mice challenged with adenosine diphosphate (ADP; 700 mg/kg) from 90% to 60%, respectively, indicating that andrographolide effectively prevents thromboembolism (Lu et al., 2012, p. 214; Jayakumar et al., 2013). Handa and Sharma (1990, p. 276) evaluated a LD_{50} of andrographolide as 11.46 g/kg determined in male mice through the intraperitoneal route.

In a study of HIV-positive patients, a dose of 1500–2000 mg of andrographolide was administered daily for 6 weeks. The study was discontinued early despite some improvements in $CD4^+$ counts (Calabrese et al., 2000, p. 333), and the side effects were common. Intravenous administration of andrographolide (10 mg/kg) to rabbits showed no abnormal cardiovascular responses; results from liver enzyme tests indicated that the heart, liver, kidney, and spleen of these rabbits were found to be normal (Guo et al., 1988, p. 422). Mice receiving

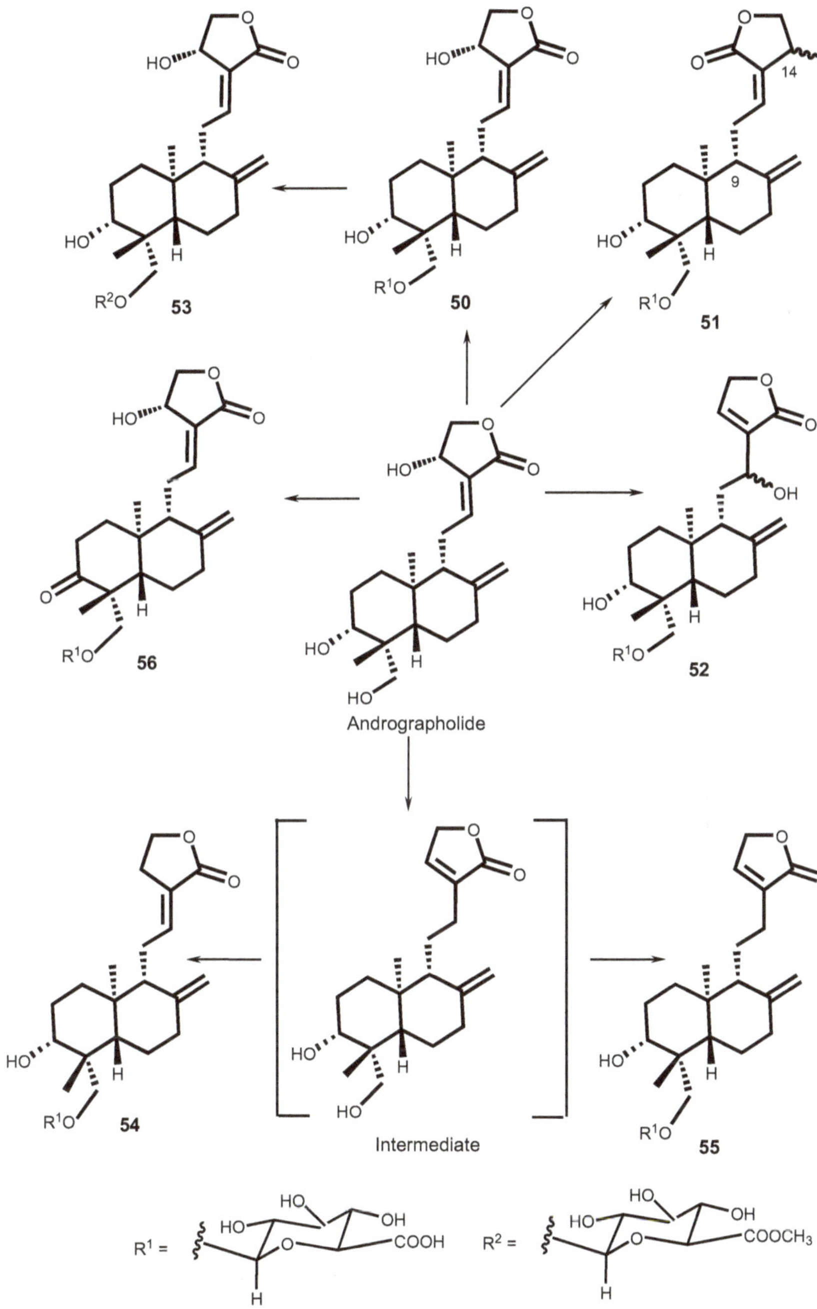

SCHEME 1.5 Structures of glucuronide conjugates of andrographolide in human urine and possible metabolic pathways for their production (Brahmachari, 2012, p. 335; Cui et al., 2008, p. 261).

an oral plant extract (10 g/kg) once a day for 7 days proved that no mortality was observed. In another test for toxicity, rats or rabbits receiving 1 g/kg of andrographolide orally showed no changes in body weight, blood count, or the functions of the liver, kidney, or other vital organs (Akbarsha and Murugaian, 2000, p. 432). Moreover, administering andrographolide causes no cytotoxic effects on platelets at concentrations between 35 and 150 mM (Lu et al., 2011, p. 1263). Therefore, andrographolide is recommended to be clinically tested as a pharmaceutical agent.

8. CONCLUDING REMARKS

This chapter offers a detailed overview of promising antidiabetic activity shown by andrographolide, the key bioactive chemical constituent of *A. paniculata*; pharmacokinetics and metabolism including safety aspects of this drug molecule are also discussed herein. In addition, extraction, purification, and spectroscopic characterization of this biomolecule from the source plant are presented in detail. The discussion also encompasses total synthetic route for andrographolide and also outlines varying semisyntheses leading to the preparation of different series of semisynthetic derivatives of the molecule. The potential role of andrographolide and its semisynthetic derivatives in controlling diabetic complications has created a stir among the scientific community at large to undertake extensive research for exploring the possibility of their prospective use as "lead molecules" in the antidiabetic drug discovery process. This should be mentioned herein that this natural diterpene was chosen as a template for the generation of a combinatorial library due to a favorable arrangement of functional groups and its abundant availability, and utilizing andrographolide itself in parallel solution-phase synthesis leads to a 360-membered library (Mang et al., 2006, p. 268). It is expected that interest in this molecule grows even more, and further research on this molecule is strongly recommended to assess its potential to become a drug molecule in near future.

LIST OF ABBREVIATIONS

ADP Adenosine diphosphate
BER β-Endorphin-like immunoreactivity
DDQ 2,3-Dichloro-5,6-dicyanobenzoquinone
DMF *N,N*-dimethylformamide
DMSO Dimethylsulfoxide
DR Diabetic retinopathy
EBV Epstein–Barr virus
Egr-1 Earlygrowthresponse-1
GLUT4 Glucose transporter subtype 4
HIV Human immunodeficiency virus
HMPA Hexamethylphosphoramide
i.p. Intraperitoneal

i.v. Intravenous
LDA Lithium diisopropylamide
$NaBH_4$ Sodium borohydride
NO Nitric oxide
PAF Platelet-activating factor
Ph_3P Triphenylphosphine
PPTs Pyridinium *p*-toluenesulfonate
ROS Reactive oxygen species
rt Room temperature
STZ Streptozotocin
TBAF Tetra-*n*-butylammonium fluoride
^{t}BHP *Tert*-butylhydroperoxide
TBSCl *Tert*-butyldimethylsilyl chloride
THF Tetrahydrofuran
TIMP-1 Tissue inhibitors of metalloproteinase-1
TNF-α Tumor necrosis factor-α
VEGF Vascular endothelial growth factor

ACKNOWLEDGMENTS

The author deeply acknowledges the help as he received from the literature studies during preparation of this manuscript. Financial support from CSIR, New Delhi (No. 02(0260)/16/EMR-II) is also gratefully acknowledged.

REFERENCES

Abu Ei-Asrar, A.M., 2013. Evolving strategies in the management of diabetic retinopathy. Middle East African Journal of Ophthalmology 20, 273–282.

Akbarsha, M.A., Murugaian, P., 2000. Aspects of the male reproductive toxicity/male antifertility property of andrographolide in albino rats: effect on the testis and the cauda epididymidal spermatozoa. Phototherapy Research 14, 432–435.

Arha, D., Pandeti, S., Mishra, A., Srivastava, S.P., Srivastava, A.K., Narender, T., Tamrakar, A.K., 2015. Deoxyandrographolide promotes glucose uptake through glucose transporter-4 translocation to plasmamembrane in L6myotubes and exerts antihyperglycemic effect in vivo. European Journal of Pharmacology 768, 207–216.

Balachandran, P., Govindarajan, R., 2005. Cancer – an ayurvedic perspective. Pharmacology Research 51, 19–30.

Brahmachari, G., 2012. Andrographolide: a plant-derived natural molecule of pharmaceutical promise. In: Brahmachari, G. (Ed.), Bioactive Natural Products: Opportunities & Challenges in Medicinal Chemistry. World Scientific Publishing Co., Singapore, pp. 335–367.

Calabrese, C., Berman, S.H., Babish, J.G., Ma, X., Shinto, L., Dorr, M., Wells, K., Wenner, C.A., Standish, L.J., 2000. A phase I trial of andrographolide in HIV positive patients and normal volunteers. Phytotheraphy Research 14, 333–338.

Cava, M.P., Chan, W.R., Haynes, L.J., Johnson, L.F., Weinstein, B., 1962. The structure of andrographolide. Tetrahedron 18, 397–403.

Chakravarti, R.N., Chakravarti, D., 1951. Andrographolide, the active constituent of *Andrographis paniculata* Nees; a preliminary communication. Indian Medical Gazette 86, 96–97.

Chen, H.-W., Huang, C.-S., Li, C.-C., Lin, A.-H., Huang, Y.-J., Wang, T.-S., Yao, H.-T., Lii, C.-K., 2014. Bioavailability of andrographolide and protection against carbon tetrachloride-induced oxidative damage in rats. Toxicology and Applied Pharmacology 280, 1–9.

Chibber, R., Ben-Mahmud, B.M., Chibber, S., Kohner, E.M., 2007. Leukocytes in diabetic retinopathy. Current Diabetes Review 3, 3–14.

Cui, L., Qiu, F., Wang, N.L., Yao, X.S., 2004. Four new andrographolide metabolites in human urine. Chemical & Pharmaceutical Bulletin 52, 772–775.

Cui, L., Qiu, F., Wang, N.L., Yao, X.S., 2005a. A new glucuronidated metabolite of andrographolide in human. Chinese Chemistry Letters 16, 369–371.

Cui, L., Qiu, F., Yao, X.S., 2005b. Isolation and identification of seven glucuronide conjugates of andrographolide in human urine. Drug Metabolism Disposition 33, 555–562.

Cui, L., Chan, W., Qiu, F., Cai, Z., Yao, X.S., 2008. Identification of four urea adducts of andrographolide in humans. Drug Metabolism Letters 2, 261–268.

Daodee, S., Wangboonskul, J., Jarukamjorn, K., Sripanidkulchai, B., Murakami, T., 2007. Membrane transport of andrographolide in artificial membrane and rat small intestine. Pakistan Journal of Biological Sciences 10, 2078–2085.

Fantes, R.J., Durairaj, V.D., Oliver, S.C., 2010. Diabetic retinopathy: an update on treatment. American Journal of Medicine 123, 213–216.

Fujita, T., Fujitani, R., Takeda, Y., Takaishi, Y., Yamada, T., Kido, M., Miura, I., 1984. On the diterpenoids of *Andrographis paniculata*: X-ray crystallographic analysis of andrographolide and structure determination of new minor diterpenoids. Chemical & Pharmaceutical Bulletin 32, 2117–2125.

Gao, H.-T., Wang, B.-L., Li, W.-D.Z., 2014. Synthetic applications of homoiodo allylsilane II. Total syntheses of (−)-andrographolide and (+)-rostratone. Tetrahedron 70, 9436–9448.

Gorter, M.K., 1911. The bitter constituent of *Andrographis paniculata* Nees. Recueil des Travaux Chimiques 30, 151–160.

Guo, S.Y., Li, D.Z., Li, W.S., Fu, A.H., Zhang, L.H., 1988. Study of the toxicity of andrographolide in rabbits. The Journal of Beijing Medical University 5, 422–428.

Handa, S.S., Sharma, A., 1990. Hepatoprotective activity of andrographolide from *Andrographis paniculata* against carbontetrachloride. Indian Journal of Medical Research 92B, 276–283.

He, X., Li, J., Gao, H., Qiu, F., Hu, K., Cui, X., Yao, X., 2003a. Four new andrographolide metabolites in rats. Tetrahedron 59, 6603–6607.

He, X., Li, J., Gao, H., Qiu, F., Hu, K., Cui, X., Yao, X., 2003b. Identification of a rare sulfonic acid metabolite of andrographolide in rats. Drug Metabolism Disposition 31, 983–985.

He, X., Li, J., Gao, H., Qiu, F., Hu, K., Cui, X., Yao, X., 2003c. Six new andrographolide metabolites in rats. Chemical & Pharmaceutical Bulletin 51, 586–589.

Jarukamjorn, K., Nemoto, N., 2008. Pharmacological aspects of *Andrographis paniculata* on health and its major diterpenoid constituent andrographolide. Journal of Health Science 54, 370–381.

Jayakumar, T., Hsieh, C.-Y., Lee, J.-J., Sheu, J.-R., 2013. Experimental and clinical pharmacology of *Andrographis paniculata* and its major bioactive phytoconstituent andrographolide. Evidence-Based Complementary and Alternative Medicine 1–16. Article ID 846740 http://dx.doi.org/10.1155/2013/846740.

Jiang, X., Yu, P., Jiang, J., Zhang, Z., Wang, Z., Yang, Z., Tian, Z., Wright, S.C., Larrick, J.W., Wang, Y., 2009. Synthesis and evaluation of antibacterial activities of andrographolide analogues. European Journal of Medicinal Chemistry 44, 2936–2943.

Khare, C.P., 2007. Indian Medicinal Plants: An Illustrated Dictionary. Springer-Verlag, Berlin/Heidelberg.

Koranyi, L., James, D., Mueckler, M., Permutt, M.A., 1990. Glucose transporter levels in spontaneously obese (db/db) insulin-resistant mice. Journal of Clinical Investigation 85, 962–967.

Kulyal, P., Tiwari, U.K., Shukla, A., Gaur, A.K., 2010. Chemical constituents isolated from *Andrographis paniculata*. Indian Journal of Chemistry 49B, 356–359.

Li, W., Xu, X., Zhang, H., Ma, C., Fong, H., Van Breemen, R., Fitzloff, J., 2007. Secondary metaboloites from *Andrographis paniculata*. Chemical & Pharmaceutical Bulletin 55, 455–458.

Li, Y., Yan, H., Zhang, Z., Zhang, G., Sun, Y., Yu, P., Wang, Y., Xu, L., 2015. Andrographolide derivative AL-1 improves insulin resistance through down-regulation of NF-κB signalling pathway. British Journal of Pharmacology 172, 3151–3158.

Lomlim, L., Jirayupong, N., Plubrukarn, A., 2003. Heat accelerated degradation of solid state andrographolide. Chemical & Pharmaceutical Bulletin 51, 24–26.

Lu, X.J., Zhang, S.L., Wang, Z.S., 1981. Analysis of andrographolide compounds. I. Ion pair high performance liquid chromatographic analysis of andrographolide derivatives. Acta Pharmacologica Sinica 16, 182–189.

Lu, W.J., Lee, J.J., Chou, D.S., Jayakumar, T., Fong, T.H., Hsiao, G., Sheu, J.R., 2011. A novel role of andrographolide, an NF-kappa B inhibitor, on inhibition of platelet activation: the pivotal mechanisms of endothelial nitric oxide synthase/cyclic GMP.Journal of Molecular Medicine 89, 1263–1271.

Lu, W.J., Lin, K.H., Hsu, M.J., Chou, D.S., Hsiao, G., Sheu, J.R., 2012. Suppression of NF-κB signaling by andrographolide with a novel mechanism in human platelets: regulatory roles of the p38 MAPK-hydroxyl radical-ERK2 cascade. Biochemical Pharmacology 84, 914–924.

Mang, C., Jakupovic, S., Schunk, S., Ambrosi, H.-D., Schwarz, O., Jakupovic, J., 2006. Natural products in combinatorial chemistry: an andrographolide-based library. Journal of Combinatorial Chemistry 8, 268–274.

Meng, Z.M., 1981. Studies on the structure of the adduct of andrographolide with sodium hydrogen sulfite. Acta Pharmacologica Sinica 16, 571–575.

Mishra, S.K., Sangwan, N.S., Sangwan, R.S., 2007. *Andrographis paniculata* (Kalmegh): a review. Pharmacognosy Reviews 1, 283–298.

Niranjan, A., Tiwari, S.K., Lehri, A., 2010. Biological activities of Kalmegh (*Andrographis paniculata* Nees) and its active principles – a review. Indian Journal of Natural Products and Resources 1, 125–135.

Panossian, A., Hovhannisyan, A., Mamikonyan, G., Abrahamian, H., Hambardzumyan, E., Gabrielian, E., Goukasova, G., Wikman, G., Wagner, H., 2000. Pharmacokinetic and oral bioavailability of andrographolide from *Andrographis paniculata* fixed combination Kan Jang in rats and human. Phytomedicine 7, 351–364.

Panossian, A., Davtyan, T., Gukassyan, N., Gukasova, G., Mamikonyan, G., Gabrielian, E., Wikman, G., 2002. Effect of andrographolide and Kan Jang–fixed combination of extract SHA-10 and extract SHE-3 on proliferation of human lymphocytes, production of cytokines and immune activation markers in the whole blood cells culture. Phytomedicine 9, 598–605.

Patarapanich, C., Laungcholatan, S., Mahaverawat, N., Chaichantipayuth, C., Pummangura, S., 2007. HPLC determination of active diterpene lactones from *Andrographis paniculata* Nees planted in various seasons and regions in Thailand. Thai Journal of Pharmaceutical Sciences 31, 91–99.

Rajani, M., Shrivastava, N., Ravishankara, M.N., 2000. A rapid method for isolation of andrographolide from *Andrographis paniculata* Nees (kalmegh). Pharmaceutical Biology 38, 204–209.

Rao, Y.K., Vimalamma, G., Rao, C.V., Tzeng, Y., 2004. Flavonoids and andrographolides from *Andrographis paniculata*. Phytochemistry 65, 2317–2321.

Rao, N.K., 2006. Anti-hyperglycemic and renal protective activities of *Andrographis paniculata* roots chloroform extract. Iranian Journal of Pharmacology & Therapeutics 5, 47–50.

Ruhnau, K.J., Meissner, H.P., Finn, J.R., Reljanovic, M., Lobisch, M., Schütte, K., Nehrdich, D., Tritschler, H.J., Mehnert, H., Ziegler, D., 1999. Effects of 3-week oral treatment with the antioxidant thioctic acid (alpha-lipoic acid) in symptomatic diabetic polyneuropathy. Diabetes Medication 16, 1040–1043.

Sabu, K.K., Padmesh, P., Seeni, S., 2000. Intraspecific variation in active principle content and isozymes of *Andrographis paniculata* Nees (Kalmegh): a traditional hepatoprotective medicinal herb. Journal of Medicinal and Aromatic Plant Sciences 23, 637–647.

Sakila, S., Begum, N., Kawsar, S., Begum, Z.A., Zoha, M.S., 2009. Relationship of anti-fertility effects of *Andrographis paniculata* and hormonal assay in female rats. Bangladesh Journal of Medical Science 8, 10–14.

Singha, P.K., Roy, S., Dey, S., 2007. Protective activity of andrographolide and arabinogalactan proteins from *Andrographis paniculata* Nees. against ethanol-induced toxicity in mice. Journal of Ethnopharmacology 111, 13–21.

Smith III, A.B., Toder, B.H., Carroll, P.J., Donohue, J., 1982. Andrographolide: an X-ray crystallographic analysis. Journal of Crystallographic and Spectroscopic Research 12, 309–319.

Smith, A.R., Shenvi, S.V., Widlansky, M., Suh, J.H., Hagen, T.M., 2004. Lipoic acid as a potential therapy for chronic diseases associated with oxidative stress. Current Medicinal Chemistry 11, 1135–1146.

Suo, X.B., Zhang, H., Wang, Y.Q., 2007. HPLC determination of andrographolide in rat whole blood: study on the pharmacokinetics of andrographolide incorporated in liposomes and tablets. Biomedical Chromatography 21, 730–734.

Wangboonskul, J., Daodee, S., Jarukamjorn, K., Sripanidkulchai, B., 2006. Pharmacokinetic study of *Andrographis paniculata* tablets in healthy Thai male volunteers. Thai Pharmaceutical and Health Science Journal 1, 209–218.

Willard, A., Herman, I.M., 2012. Vascular complications and diabetes: current therapies and future challenges. Journal of Ophthalmology 209–538.

Woo, A.Y.H., Waye, M.M.Y., Tsui, S.K.W., Yeung, S.T.W., Cheng, C.H.K., 2008. Andrographolide up-regulates cellular reduced glutathione level and protects cardiomyocytes against hypoxia/reoxygenation injury. Journal of Pharmacology and Experimental Therapeutics 325, 226–325.

Xu, H.-W., Dai, G.-F., Liu, G.-Z., Wang, J.-F., Liu, H.-M., 2007a. Synthesis of andrographolide derivatives: a new family of α-glucosidase inhibitors. Bioorganic & Medicinal Chemistry 15, 4247–4255.

Xu, H.-W., Liu, G.-Z., Dai, G.-F., Wu, C.-L., Liu, H.-M., 2007b. Modification of 15-akylidene andrographolide derivatives as alpha-glucosidase inhibitor. Drug Discovery Therapy 1, 73–77.

Xu, J., Huang, S., Luo, H., Li, G., Bao, J., Cai, S., Wang, Y., 2010. QSAR studies on andrographolide derivatives as α-glucosidase inhibitors. International Journal of Molecular Sciences 11, 880–895.

Yan, G.R., Zhou, H.H., Wang, Y., Zhong, Y., Tan, Z.L., He, Q.Y., 2013. Protective effects of andrographolide analogue AL-1 on ROS-induced RIN-mbeta cell death by inducing ROS generation. PLoS One 8, e63656. http://dx.doi.org/10.1371/journal.pone.0063656.

Yan, H., Li, Y., Yang, Y., Zhang, Z., Zhang, G., Sun, Y., Yu, P., Wang, Y., Xu, L., 2016. Protective effects of andrographolide derivative AL-1 on high glucose-induced oxidative stress in RIN-m cells. Current Pharmaceutical Design 22, 499–505.

Yu, B.-C., Hung, C.-R., Chen, W.-C., Cheng, J.-T., 2003. Antihyperglycemic effect of andrographolide in streptozotocin induced diabetic rats. Planta Medica 69, 1075–1079.

Yu, B.-C., Chang, C.-K., Su, C.-F., Cheng, J.-T., 2008. Mediation of β-endorphin in andrographolide induced plasma glucose lowering action in type I diabetes like animals. Naunyn-Schmiedeberg's Archives of Pharmacology 377, 529–540.

Yu, Z., Lu, B., Sheng, Y., Zhou, L., Ji, L., Wang, Z., 2015. Andrographolide ameliorates diabetic retinopathy by inhibiting retinal angiogenesis and inflammation. Biochimica et Biophysica Acta 1850, 824–831.

Yuan, L., Zhang, C., Sun, H., Liu, Q., Huang, J., Sheng, L., Lin, B., Wang, J., Chen, L., 2016. The semi-synthesis of novel andrographolide analogues and anti-influenza virus activity evaluation of their derivatives. Bioorganic & Medicinal Chemistry Letters 26, 769–773.

Zhang, Z., Jiang, J., Yu, P., Zeng, X., Larrick, J.W., Wang, Y., 2009. Hypoglycemic and beta cell protective effects of andrographolide analogue for diabetes treatment. Journal of Translational Medicine 7, 62–73.

Zhu, S.P., Kang, B.A., 1981. Distribution and excretion of [^{35}S]$NaHSO_3$-andrographolide by autoradiography. Zhongguo Yao Li Xue Bao 2, 266–269.

Zhu, Y.Y., Yu, G., Zhang, Y., Xu, Z., Wang, Y.Q., Yan, G.R., He, Q.Y., 2013. A novel andrographolide derivative AL-1 exerts its cytotoxicity on K562 cells through a ROS-dependent mechanism. Proteomics 13, 169–178.

Chapter 2

Computer-Aided Discovery of Glycogen Phosphorylase Inhibitors Exploiting Natural Products

Joseph M. Hayes
School of Physical Sciences & Computing, University of Central Lancashire, Preston, United Kingdom

1. INTRODUCTION

The importance of computation in drug design efforts is now well established. Continuous improvements in software and algorithms together with increases in computational power have made computer-aided drug design (CADD) an important component of modern day drug discovery projects. For a given drug target, these methods can speed up the identification of new drug candidates with considerable savings in terms of cost and time. Equally, there is an immediate need for more efficient routes to drug candidates given the dramatic rise in the prevalence of a number of health conditions worldwide, many of which require new and more effective treatments. One such condition is type 2 diabetes (T2D). Latest figures from the World Health Organization (WHO) show that approximately 422 million people worldwide have diabetes (90% of cases are T2D). It is predicted to become the seventh leading cause of death by the year 2030 so that urgent action is required. For some T2D patients, diet and exercise may be sufficient to control blood glucose levels. However, when lifestyle changes are no longer effective then antidiabetic drugs are prescribed. The major oral hypoglycemic type drugs include biguanides, sulfonylureas, α-glucosidase inhibitors, sodium–glucose cotransporter-2 (SGLT2) inhibitors, thiazolidinediones, and DPP-4 inhibitors. Unfortunately, there are several adverse side effects associated with current medications, which includes a risk of hypoglycemia. New therapeutic strategies leading to alternative T2D treatments are therefore welcomed (Jones, 2012). An important target currently being investigated is liver glycogen phosphorylase (GP) (Henke, 2012).

Discovery and Development of Antidiabetic Agents From Natural Products.
http://dx.doi.org/10.1016/B978-0-12-809450-1.00002-8

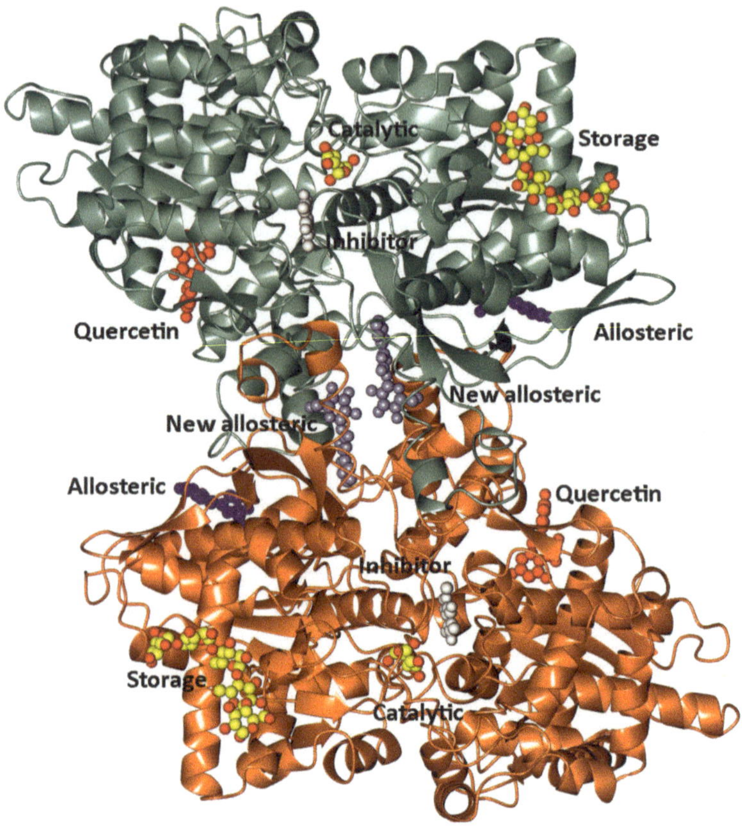

FIGURE 2.1 The dimer structure of GP highlighting the key binding sites.

Outside of diet, the liver is the primary source of glucose production either through the gluconeogenesis or glycogenolysis pathways. Gluconeogenesis involves the formation of glucose from noncarbohydrate substrates such as lactate, pyruvate, glycerol, and glucogenic amino acids. Glycogenolysis is the pathway involving the breakdown of glycogen, which eventually leads to glucose. GP (EC 2.4.1.1) catalyzes the first and rate-determining step of glycogenolysis, the cleavage of glucose units from glycogen in the form of glucose-1-phosphate (GIP):

$$Glycogen_{(n)} + P_i \Leftrightarrow Glycogen_{(n-1)} + Glucose - 1 - P_i$$

The enzyme exists as a dimer (Fig. 2.1) and its activity is modulated allosterically and by phosphorylation. While the Ser14 phosphorylated form of GP (GPa) is predominantly active (R state), the unphosphorylated GP (GPb) is predominantly inactive (T state). GP is extremely well structurally characterized with >200 structures deposited in the RCSB Protein Data Bank (PDB) (www.rcsb.org). Most of these solved structures are high-resolution

GP–inhibitor complexes. Seven different binding sites have been discovered to date (Fig. 2.1): the catalytic, allosteric, new allosteric, inhibitor, benzimidazole, glycogen storage, and quercetin binding sites. This vast structural knowledge has facilitated many successful structure-based inhibitor design efforts. A large number of inhibitors for the different binding sites have been reported and are reviewed regularly (Hayes et al., 2014; Somsak et al., 2008; Somsak, 2011). Some of these inhibitors have demonstrated blood glucose lowering effects in animal models (c.f. for example: Docsa et al., 2011; Lu et al., 2003; Nagy et al., 2013; Thomson et al., 2009) and some have entered clinical trials (Henke, 2012). Many of the inhibitors discovered are natural products or natural product in origin. This chapter will present the role that different computational methods have played in their discovery with a focus mainly on more recent studies. A brief description of CADD methods will be presented in Section 2 and their applications to GP inhibitor design in Section 3. As each GP binding site has its own characteristic features such as key binding site residues, site flexibility, solvent exposure, and hydrophobicity, which influence the applicability of different in silico methods, Section 3 will be presented by binding site.

2. COMPUTER-AIDED DRUG DESIGN METHODS

The ultimate goal of CADD methods is the accurate prediction of ligand binding affinities to a given drug target. If the structure of the macromolecular target is known, structure-based drug design (SBDD) methods are facilitated. Even in the case of target structure being unknown, homology modeling can sometimes be used to construct a model suitable for calculations. However, this often requires a considerable amount of expertise with the accuracy of these models highly dependent on the % identity and homology between query and template proteins. Ligand-based drug design (LBDD) methods can alternatively be used, where no information regarding the target structure is required and which instead relies on the availability of known ligands. In LBDD, machine learning-based or similarity-based virtual screening is often applied. Another popular approach is pharmacophore modeling where for a set of active compounds, pharmacophore hypotheses (Sanders et al., 2012) are generated as a search "query" for 3-D databases to propose new ligands. The simplicity of the pharmacophore definition enables the complexity of protein–ligand intermolecular interactions to be reduced to a handful of features such as hydrogen bond donors, hydrogen bond acceptors, hydrophobic and aromatic groups. A number of programs such as Catalyst, LigandScout, Phase, Unity, MOE, and Pharmer are widely used in pharmacophore modeling (Sanders et al., 2012). QSAR methods are commonly used to predict the activities of novel chemical structures and are based on the theory that the differences in biological activity for a series of compounds can be quantitatively correlated with differences in their molecular structure. Therein, biological activities are related to specific molecular features (descriptors) by regression techniques and the relative importance of those features contributing to the biological effect estimated. Whereas 1D- and 2D-QSAR correlate activities with physiochemical properties such as log S, log P, etc. and structural motifs, respectively; 3D-QSAR considers the three-dimensional structure; and 4D-QSAR the three-dimensional structure with multiple representations of ligand orientations and conformations. Analogous to structure-based or ligand-based approaches, QSAR studies can be grouped as receptor-dependent (RD) or receptor-independent (RI) QSAR analyses (Andrade et al., 2010). In RI QSAR, the geometry of the target is not available (or is neglected) in the QSAR analysis. In RD QSAR, models are derived from the 3D- structure of multiple ligand–receptor complex conformations in an explicit molecular dynamics (MD) simulation of the induced-fit process where both ligand and receptor are flexible and the resulting binding interaction energies used as descriptors. The synergy between the application of ligand-based and structure-based approaches to drug design has been previously discussed (Clark, 2009). In the case of GP, both approaches have been effectively used, but given the wealth of GP structural data available to be exploited, structure-based inhibitor design is normally used.

In practical applications of CADD, accurate *relative* rather than *absolute* binding affinity predictions are normally sufficient. This allows the ranking of predicted potencies among a set or database of ligands. Suggested candidates for synthesis and/or in vitro experimental validation studies can then be based on these rankings and other criteria such as structural novelty and/or predicted absorption, distribution, metabolism, excretion, and toxicity (ADMET) properties. In SBDD, visual inspection of predicted target–ligand interactions can provide vital clues as to increasing ligand potency. In terms of predicting binding free energies, the most accurate methods are free energy perturbation (FEP) or thermodynamic integration (TI) combined with Monte Carlo or MD simulations of protein–ligand complexes. These "pathway" methods are based on thermodynamic cycles in which the modeled system is converted from one state to the other through a set of finite or infinitesimal changes to the energy function. They are extremely computationally demanding but through improvements in algorithms and computational power in recent years, their applications have increased. Calculations of this type are still generally limited to small sets of structurally similar compounds in a low-throughput screening approach to lead optimization.

On the other hand, high-throughput virtual screening methods can screen 100s of thousands of compounds (or more) in practical times using reasonable computational resources. They are standardly applied to obtain hit compounds for a given target and the availability of a large number of compound databases has considerably enhanced their applications. The ZINC database (http://zinc.docking.org/) is widely used and contains over 35 million purchasable compounds in ready to screen 3-D formats. Natural product catalogs (12 in total) have been incorporated into this database (Irwin et al., 2012). A general survey of natural product sources and databases was previously published (Füllbeck et al., 2006). Today, natural product databases are often readily downloadable for virtual screening purposes and are available from diverse sources such as Thai medicinal plants (Sangma et al.), Cameroonian flora (Ntie-Kang et al., 2013), African medicinal plants (Ntie-Kang, 2014), Chinese traditional medicinal herbs, and from the marine (Lei et al., 2002; Blunt et al., 2013). Docking calculations are commonly used to screen the databases for hit compounds, which involve the scoring and ranking of ligand binding poses (orientations/conformations) in a protein binding site (Meng et al., 2012). Comparison of docking scores for ligands allows the predicted relative potencies to be ranked. Docking programs use different search algorithms and scoring functions (empirical-, genetic-, statistical-, or physics-based). Among the most popular programs are Glide (Friesner et al., 2006), GOLD (Verdonk et al., 2003), DOCK (Allen et al., 2015), AutoDock (Morris et al., 2009), and AutoDock Vina (Trott et al., 2010) but more than 60 docking programs and 30 scoring functions have previously been recorded (Moitessier et al., 2008). Some of these, such as DOCK, AutoDock, and AutoDock Vina are free for academic users and there are numerous other free resources available to assist structure-based virtual screening experiments (Villoutreix et al., 2007). Accuracy of different docking programs may vary with the features of binding site under consideration and it has been demonstrated that deriving ligand atomic partial charges from quantum mechanics (QM) methods has the potential to improve docking results (Cho et al., 2005; Raha et al., 2007). QM/MM (molecular mechanics) docking (also known as QM-polarized ligand docking or QM-PLD) involves reparameterizing ligand partial charges in the field of the receptor using QM/MM for use in docking. This method has emerged as a very promising but more computationally expensive approach to docking calculations. Generally, the choice of 3-D virtual screening method and the calculation settings used, together with a rigorous statistical analysis of results has a significant bearing on the quality of results (Kirchmair et al., 2008). It is crucial that a screening protocol is optimized for successful predictions. The Directory of Useful Decoys: Enhanced (DUD-E) is derived from the ZINC database and is available for benchmarking virtual screening studies (http://dude.docking.org/). DUD-E, which is an enhanced and rebuilt version of the original DUD database, contains a total of 22,886 active compounds against a total of 102 targets, which includes GP. For each active, 50 decoys (presumed inactives) with similar physicochemical properties (MW, charge, calculated log P, etc.) but dissimilar 2-D topology is available (or can be generated separately).

Postdocking calculations are additionally often used due to their potential to improve the accuracy of the original docking predictions. The most common methods are molecular mechanics/generalized Born surface area (MM-GBSA) and molecular mechanics/Poisson–Boltzmann surface area (MM-PBSA), which calculate binding free energies combining MM (force fields) with continuum (implicit) solvation models (Hayes and Archontis, 2012; Kollman et al., 2000). The related QM/MM-PBSA methodology is considerably more computationally expensive but in theory more accurate, given that the more important protein–ligand binding site interactions can be better described using QM, with the rest of the protein modeled using MM. In the case of structurally diverse and novel ligands (such as natural products) where accurate force field parameters may not be available, this can be expected in theory to improve the accuracy of predictions. Halogen bonding through the σ-hole effect also requires a QM description (Kantsadi et al., 2012), although recently some docking algorithms and force fields have incorporated more accurate descriptions. MM-GB(PB)SA and QM/MM-PBSA are known as "endpoint" methods, with the binding free energy calculated as the difference in energy for models of the bound (PL) and unbound (P and L) states of the protein(P)–ligand(L) complexes:

$$\Delta G_{\text{bind}} = G(PL) - G(P) - G(L) \tag{2.1}$$

For MM-GB(PB)SA, the free energy of the three systems ($X = PL$, P or L) is given by:

$$G(X) = E_{MM}(X) + G_{\text{solv}}(X) - TS_{MM}(X) \tag{2.2}$$

where E_{MM} is the total MM energy in the gas phase, G_{solv} is a correction term for solvation of X (solvation free energy), and S is the entropy of X. The terms in Eq. (2.2) are "traditionally" averages (<>) obtained over MD trajectories, but the effectiveness of alternatively using selected or single binding conformations such as docking poses which is computationally less demanding has now also been recognized (Hou et al., 2011b; Kuhn et al., 2005; Rastelli et al., 2010; Hayes et al., 2011). Differences (Δ) in the individual terms of Eq. (2.2) between the bound and unbound states are calculated and contribute to the binding free energies according to Eq. (2.1). In postdocking calculations, often a single state of the complex in the form of P-L docking poses (used directly or preminimized) is used to generate 3-D coordinates for all three states (complex, protein, and ligand). This approximation neglects any explicit structural relaxation of the protein and ligand upon binding but may in some cases lead to better prediction of relative binding free energies (Hayes and Archontis, 2012). The entropy term is also sometimes ignored but can be approximated, for example, by estimating the loss of ligand vibrational, rotational, and translational entropy on binding using the rigid rotor harmonic oscillator approximation (Suárez and Diaz, 2015). In QM/MM-PBSA binding free energy calculations (Manta et al., 2012), the E_{MM} terms in Eq. (2.2) are effectively replaced by the *QM/MM* energy:

$$G(X) = E_{QM/MM}(X) + G_{\text{solv}}(X) - TS_{MM}(X) \tag{2.3}$$

Other endpoint methods for binding affinity predictions are based on linear interaction energy (LIE) or the linear response approximation (Aqvist et al., 1994; Zhou et al., 2001). They are classed as "semiempirical," retaining some of the theoretical aspects of FEP but including parameters that are derived from experimental binding data. Recently, they have also had applications as postdocking methods.

While the focus of this chapter is on prediction of GP inhibitor binding, computation can also be used to predict the pharmacokinetic or "drug-like" profile of GP inhibitors at an early stage in the drug design process. This is of particular importance given that the most significant causes of drug attrition in clinical trials today is toxicity together with a lack of efficacy (Smith, 2011). A number of different "user-friendly" programs such as ALOGPS (Tetko and Tanchuk, 2002; Tetko et al., 2005), FAF-Drugs (Lagorce et al., 2015), and QikProp (Schrödinger, 2016) have been used for ADMET property predictions on GP inhibitors (Polyak et al., 2013; Loughlin et al., 2008; Kantsadi et al., 2012; Karis et al., 2009; Schweiker et al., 2014).

3. APPLICATIONS OF COMPUTATION TO GLYCOGEN PHOSPHORYLASE INHIBITOR DESIGN

3.1 Introduction

Of the different GP binding sites, the catalytic site and allosteric site have been the most explored using computation with relatively fewer studies on the new allosteric and inhibitor sites. Applications will be presented later by binding site and in cases where studies on natural products are limited; examples of calculations on synthetic compounds that may assist in future natural product inhibitor discovery will be presented. Of the other binding sites, the quercetin binding site was only recently discovered in a study of polyphenolic extracts from the vinification by-products of Greek varieties of *Vitis vinifera*. Quercetin was found to be the most bioactive ingredient by X-ray crystallography and its binding at this new GP site (PDB ID: 4MRA) promotes the less active T state conformation (Kantsadi et al., 2014). This new binding site is 15 Å from the active site, 43 Å from the allosteric site, and 32 Å from the inhibitor site (Fig. 2.1). The glycogen storage site is located on the surface of GP, approximately 30 Å from the catalytic site (Fig. 2.1) and to date has received little attention in terms of modulating hepatic glucose production. However, millimolar (mM) inhibition (competitive with glycogen) against rabbit muscle GPb (RMGPb) has been reported for α-, β-, and γ-cyclodextrins (CDs) with K_i of 47.1, 14.1, and 7.4 mM, respectively, with the β (PDB ID: 1P2D) and γ-CD (PDB ID: 1P2G) bound complexes at the storage site solved by X-ray crystallography (Pinotsis et al., 2003). The antidiabetic drug acarbose has a K_i of 26 mM for RMGPa and also binds at this site (Goldsmith, 1987). Its antidiabetic action is attributed to inhibition of enzymes needed to digest carbohydrates: α-glucosidases and pancreatic α-amylase. No modeling studies for the storage site have yet been reported.

3.2 Catalytic Site

The catalytic site of GP includes the essential cofactor pyridoxal phosphate (PLP) and is buried at the center of the GP monomer (Fig. 2.1). It is accessible to bulk solvent via a 15 Å long channel. Inhibitors binding at this site stabilize the closed form of the flexible 280s loops (residues 282–287), blocking access of the glycogen substrate and stabilizing the T state conformation. The physiological inhibitor of GP is α-D-glucose (K_i=1.7 mM), while its anomer β-D-glucose has a K_i of 7.4 mM (Fig. 2.2) (Martin et al., 1991). To date, the majority of GP catalytic site inhibitors have been glucose analogues (Hayes et al., 2014; Somsak, 2011). Glucose analogue inhibitors are designed to exploit the core glucose moiety interactions together with additional interactions through α and/or β anomeric carbon substitutions. β-Substitutions are particularly favorable forming interactions in the so-called β-cavity of the catalytic site (Fig. 2.2), a pocket lined by polar and nonpolar residues (Somsák et al., 2008). The choice of "linker" group, which connects the glucose moiety with alkyl/aryl substituents extending into the β-cavity, is crucial to potency (Hayes et al., 2014). Among the most efficient linkers to date are 1,2,4-triazoles (Kun et al., 2014), imidazole (Bokor et al., 2015), and -NHCONHCO- (Somsák et al., 2008). In terms of alkyl or aryl substituents, aryl groups have proved the most effective with in many cases a 2-naphthyl group preferred (Hayes et al., 2014). Some important catalytic site inhibitors are shown in Fig. 2.3. Compound **1** is the most potent catalytic site inhibitor (K_i=31 nM) discovered to date (Bokor et al., 2015), with currently only 10 nanomolar inhibitors reported. The binding of one of these nanomolar inhibitors (**2**; K_i=630 nM) is shown in Fig. 2.4. Analogues of **2** and other glucose derivatives have demonstrated promising decreased hepatic glucose production/blood glucose lowering effects in vivo (Docsa et al., 2011; Goyard et al., 2016; Nagy et al., 2013), highlighting the potential of glucose analogues for the future treatment of T2D. Iminosugars also present promising scaffolds for the design of potent GP inhibitors. These compounds are carbohydrate mimics where the ring "O" has been replaced by "N" (either natural or synthetic). Indeed, 1,4-dideoxy-1,4-imino-D-arabinitol (DAB, **3**) is one of the most potent GP catalytic site inhibitors known to date with an IC_{50} of ~0.4 μM for several different GP enzymes (Oikonomakos et al., 2006). Despite this, GP inhibition by iminosugar derivatives has not been extensively explored and no modeling studies have been reported so far.

On the contrary, computation has played a key role in the discovery of glucose analogue inhibitors (Hayes and Leonidas, 2010). While QSAR methods such as 3D-QSAR, RI and RD–4D-QSAR have been applied in the past (Pan et al., 2003, 2004; Hopfinger et al., 1999; Venkatarangan et al., 1999) and reviewed previously (Hayes and Leonidas, 2010), most recent work has exploited docking and postdocking methods. Critical to the accurate scoring and ranking of inhibitor potencies is that the correct binding conformation is predicted. Given that the glucose moiety positions are well-defined in the catalytic site, positional constraints can be defined reducing the conformational sampling space and correct binding mode predictions are more probable than, for example, the allosteric site for which diverse ligand binding modes have been reported (vide infra). The flexibility of the 280s loop, however, may in some cases complicate modeling of the binding process. Some glucose analogues induce varying shifts

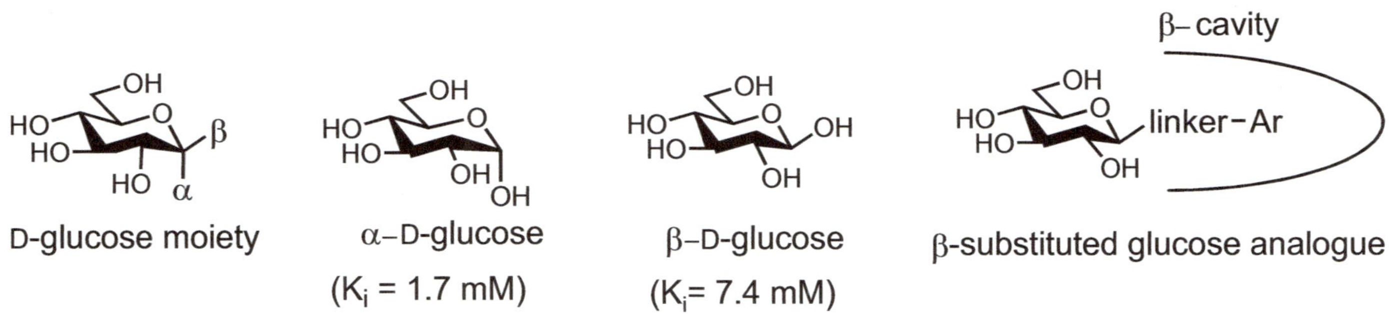

FIGURE 2.2 α- and β-Substitutions at the anomeric carbon of D-glucose have proven very effective in the design of glycogen phosphorylase catalytic site inhibitors. β-Substitutions in particular have the potential to exploit favorable interactions by extending into the GP β-cavity. The choice of linker group, linking the glucose moiety with mainly aryl (Ar) substituents, has proven critical to potency. K_i's values given are for rabbit muscle GPb inhibition.

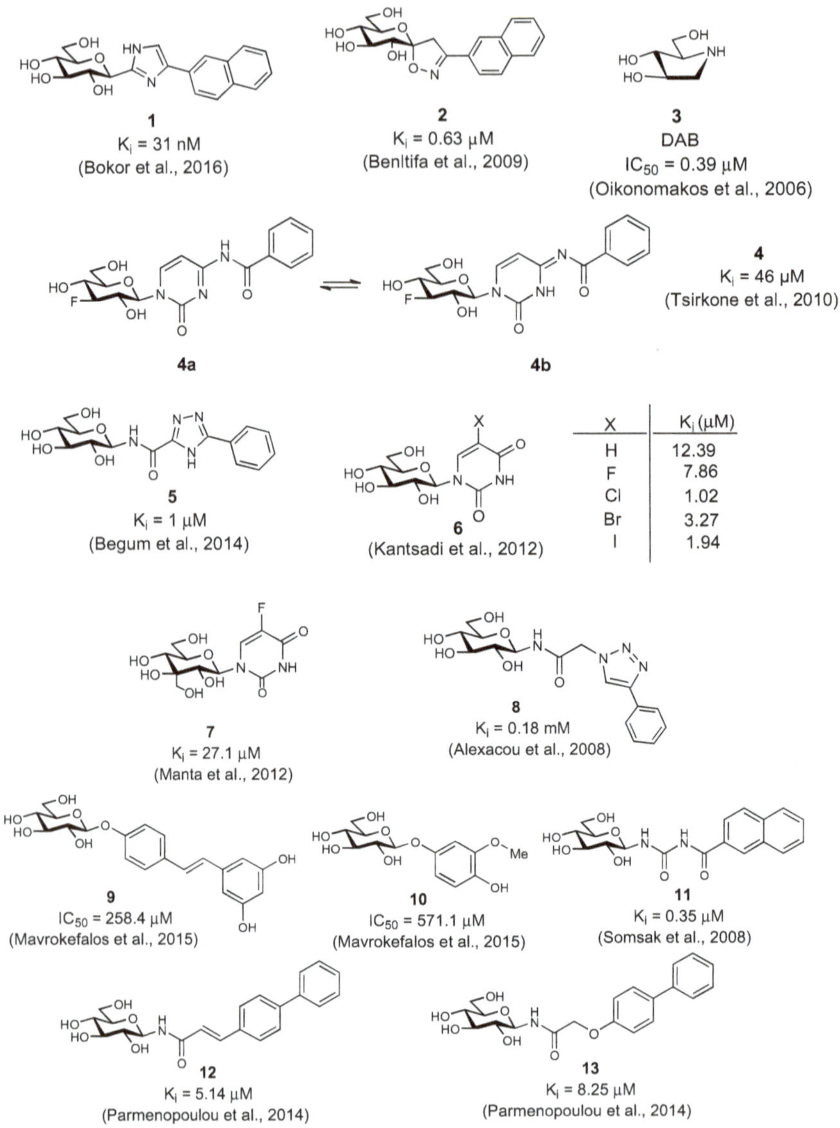

X	K_i (μM)
H	12.39
F	7.86
Cl	1.02
Br	3.27
I	1.94

FIGURE 2.3 Catalytic site inhibitors of glycogen phosphorylase. Inhibition constants are for rabbit muscle GPb. *DAB*, 1,4-dideoxy-1,4-imino-D-arabinitol.

in the 280s loop and different residues have been revealed by X-ray crystallography to adopt different orientations/conformations depending on the bound ligand structure (notably Asn284, Asp283, Ser674, and His377). Additionally, ligand-dependent shifts in the Ala673–Thr676 sequence and the 380s loop have been reported (Hayes and Leonidas, 2010; Pan et al., 2003, 2004). The binding

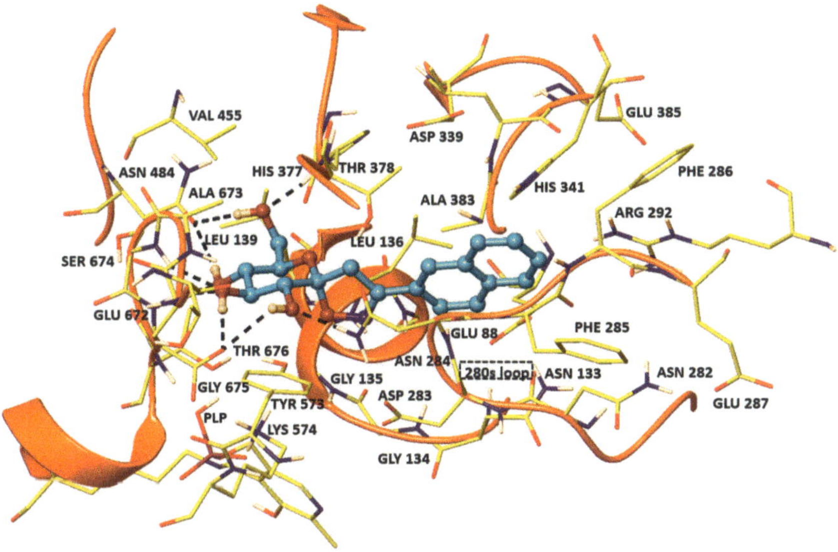

FIGURE 2.4 Ligand **2** bound at the GPb catalytic site (PDB ID: 2QRP).

site histidine residue His341 has the potential to adopt different protonation/tautomeric states and this can directly affect docking accuracy (Kim et al., 2013). Nevertheless, for ligands with the same structural scaffold (congeneric series of compounds), good correlations between experimental and binding affinities of glucose analogues have been obtained in docking calculations (Benltifa et al., 2009; Kantsadi et al., 2015). As mentioned in Section 2, accurate atomic partial charges derived using QM or hybrid QM/MM methods have the potential to improve docking results (Cho et al., 2005; Raha et al., 2007). Benltifa et al. compared the results of QM-PLD using QM electrostatic potential (ESP) fit ligand atomic charges in Glide-XP (extra precision) docking, to results obtained by standard Glide-XP calculations with default OPLS-AA(2001) ligand atomic charges (Benltifa et al., 2009). Using constraints on the glucose moiety for a series of glucopyranosylidene-spiro-isoxazolines (which included **2**), an excellent correlation (R^2) of 0.96 for docking scores against experimental binding free energies was obtained for QM-PLD (with the ligand atomic partial charges calculated in the field of the receptor) compared to 0.72 for standard Glide-XP.

Ligands have also the potential to bind in different ionization/tautomeric states for which QM/MM calculations can be very useful to decipher the correct binding preferences (Cho et al., 2009; Gleeson and Gleeson, 2009; Tsirkone et al., 2010). In a study of β-D-glucopyranosyl pyrimidine derivatives with a fluorine instead a hydroxyl at 3′ position of the glucose moiety, QM/MM calculations exploiting Glide docking poses predicted tautomeric binding preferences (e.g., **4a** vs. **4b**) that were consistent with the solved crystallographic complexes (Tsirkone et al., 2010). More recently, in an example of a successful rational approach to inhibitor design (Begum et al., 2014), Glide docking, QM, and QM/MM calculations motivated the synthesis of *N*-(β-D-glucopyranosyl)-1,2,4-triazolecarboxamide analogues such as **5** where the corresponding oxadiazole linker from previous work (Polyák et al., 2013) was replaced by a 1,2,4-triazole. The QM/MM calculations revealed the most stable solution-phase tautomer in the free state to be also the most favorable for binding to GP. QM molecular ESPs were used to explain the hydrogen bonding preferences of the different tautomers with density functional theory (DFT) at the M06-2X/6-31+G* level of theory used. The best inhibitor from this study, ligand **5** (K_i=1 μM for RMGPb), was one of the top 10 most potent GP catalytic site inhibitors at that particular time.

As mentioned in Section 2, halogen bonding through the σ-hole effect is another area that can benefit from a QM description through QM/MM calculations. C5-halogen-substituted glucopyranosyl nucleoside (1-(β-D-glucopyranosyl)5-X-uracil with X=different halogen atoms) analogues **6** were studied by Glide-XP docking and postdocking QM/MM-PBSA calculations (Kantsadi et al., 2012). While the docking failed to correctly rank the ligand potencies, good agreement with experiment for binding geometries and relative binding free energies was obtained using QM/MM-PBSA. In a separate study, QM/MM-PBSA calculations were used to unravel why compound **7** (K_i=27.1 μM) with a 3′-axial CH_2OH glucopyranose substituent exploiting additional hydrogen bonding with the phosphate group of the PLP cofactor, as revealed by crystallography (PDB ID: 3SYM), was less potent compared to the parent 3′-unsubstituted compound **6** (*X*=*F*) (K_i=7.9 μM) (Manta et al., 2012). The desolvation costs associated with formation of the GPb-**7** complex were predicted by QM/MM-PBSA to outweigh the enthalpy gains from the extra contacts formed. The requirement of an accurate solvation model (PBSA) was highlighted in this particular case, where Glide docking using a crude model of solvation in comparison failed to correctly rank the relative potencies of the two compounds.

Retaining conserved crystallographic waters from the analysis of solved protein–ligand complexes has the potential to improve docking results. One study compared the performance of Glide and GOLD for reproducing the experimental binding features (PDB ID: 2PYI) of *N*-(β-D-glucopyranosyl)-(4-phenyl-1,2,3-triazol-1-yl) acetamide (**8**) at the catalytic site (Alexacou et al., 2008). Both docking programs accurately predicted the binding interactions *with* and *without* the crystallographic binding site water molecules retained. The GOLD docking program uses a unique approach to treatment of retained explicit (crystallographic) binding site water molecules. During docking, their orientations are sampled and the presence of individual waters is automatically toggled on/off depending on the nature of interactions with the respective ligand. While the retention of crystallographic binding site water molecules was not necessary for the accurate docking of compound **8**, MD simulations of the protein–ligand complex in the absence of explicit waters led to the incorrect prediction of binding modes (Alexacou et al., 2008). In contrast, fully explicit solvent molecular dynamics free energy (MDFE) simulations were used to predict the relative binding affinities of three glucopyranosylidene-spiro-hydantoins in good agreement with experiment (Archontis et al., 2005) but as noted in Section 2, there is considerable computational expense associated with these calculations.

A thorough and comprehensive study of the effect of retaining crystallographic water molecules on docking performance was recently reported (Mavrokefalos et al., 2015). Exploring the catalytic site hydration patterns across an ensemble of 125 RMGPb inhibitor complexes, 29 ordered/conserved crystallographic waters were identified using the "ClusterWaters" algorithm of Schrödinger. Optimization of a Glide docking virtual screening protocol exploiting the DUD database (www.dud.docking.org) containing 24 known actives and 1787 decoys for GP revealed the importance of these conserved water molecules for achieving better enrichments (ranking of actives higher than decoys). Docking performance, measured by the enrichment factor values, decreased as the radius of water molecules included surrounding the binding site was reduced. With an optimal protocol in place, virtual screening of 2284 compounds from the National and Kapodistrian University of Athens (NKUA) database consisting of natural products and semisynthetic and synthetic analogues combined with the National Cancer Institute (NCI) diversity set-II was then performed. Of the 18 top-ranked compounds that were selected for in vitro experiments, 16 of these were of natural origin. The best inhibitors, a 4-O-β-D-glucoside of resveratrol (**9**; $IC_{50}=258\,\mu M$) and a phenolic glycoside tachioside (**10**; $IC_{50}=571\,\mu M$), revealed moderate activities.

In another study, retention of four protein–ligand bridging waters common to all *N*-acyl-*N′*-(β-D-glucopyranosyl) urea complexes cocrystallized with GPb also led to improved results in Glide docking and Ligand Biomolecular Association with Energetics (Embrace) calculations with MacroModel v9.9 for a series of 14 *N*-acyl-*N′*-(β-D-glucopyranosyl) urea derivatives (Kantsadi et al., 2015). The most potent ligand in this series is **11** with a K_i of 0.35 µM for RMGPb. A novel Linear Response—Molecular Mechanics Coulomb Surface Area (LR-MM-CBSA) method was developed which related predicted and experimental binding free energies with a correlation coefficient r^2 of 0.89 and leave-one-out cross-validation q^2 value of 0.79 (Kantsadi et al., 2015).

Retention of crystallographic water molecules, however, is not a prerequisite for docking success. Exploiting CombiGlide, virtual screening of 1888 β-D-glucopyranosyl-NH-CO-R analogues was performed with the R groups coming from commercially available acid chlorides (RCOCl) taken from the ZINC docking database (Parmenopoulou et al., 2014). A consensus scoring approach to ranking of ligand potencies was applied using the Glide and GOLD docking programs with different scoring functions. The protein conformation for docking was taken from the GPb–*N*-4-phenylbenzoyl-2-*N′*-β-D-glucopyranosyl urea crystal complex (PDB code: 2QLN). Six candidates, including **12** and **13**, were selected for synthesis. Three of these displayed inhibition in the low µM range and were among the most potent β-D-glucopyranosyl-NH-CO-R analogues discovered. Compounds **12** and **13** also led to GP inactivation in rat hepatocytes at low μM concentrations.

3.3 Allosteric Site

The allosteric site is located at the monomer–monomer interface, partially exposed to solvent, and is approximately 30 Å from the catalytic site (Fig. 2.1). For calculation purposes, therefore, the GP dimer if unavailable should be generated by symmetry operations using monomeric crystal structure coordinates downloaded from the RCSB Protein Data Bank. Inhibitors at this site can act either by direct inhibition of the binding of allosteric activator AMP and/or the indirect inhibition of glycogen substrate binding by stabilization of the T- or T′-(symmetry-related monomer) state conformation. Allosteric activators and inhibitors (such as ATP and Glc-6-P) can alter the equilibrium between the less T state and more active R state (Monod et al., 1965). A remarkable characteristic of this binding site to recognize different classes of ligands using mainly the same residues has been attributed to its conformational plasticity (Tiraidis et al., 2007). Understanding this conformational flexibility is important for the design and optimization of effective inhibitors. Dihydropyridine diacid derivatives (Oikonomakos et al., 1999; Tsitsanou et al., 2000; Zographos et al., 1997), phenyl diacid analogs (Lu et al., 2003), and several dihydropyridine diacid analogs (Ogawa et al., 2003) bind at this site. Other allosteric site inhibitors include phenoxyphthalates (Kristiansen et al., 2004), which inhibit basal and glucagon-induced glucose production in cultured primary hepatocytes, acyl ureas discussed later (Klabunde et al., 2005; Oikonomakos et al., 2005), together with some natural product derivatives (Hayes et al., 2014). Relevant allosteric site inhibitors to this discussion are shown in Fig. 2.5.

Supported by X-ray crystallography (PDB codes: 1WUT and 2ATI) and in a particularly novel study on GP allosteric site inhibitor design, pharmacophore modeling combined with 3D-QSAR methods was successfully used to guide the chemical modification of acyl urea compounds (Klabunde et al., 2005; Li et al., 2007). This process led to the identification of a potent inhibitor **14** of human liver GPa (HLGPa) (IC_{50} = 53 nM) that inhibited glycogen-derived glucose production in rat hepatocytes (IC_{50} = 380 nM) and which demonstrated a significant reduction of glucagon-induced hyperglycemia in Wistar rats (Klabunde et al., 2005). This comprehensive study was consistent with the inhibition of hepatic glycogenolysis by targeting the GP allosteric site and supported further design of GP inhibitors for the potential treatment of T2D.

While docking studies have been used to study binding of ligands at the allosteric site, to our knowledge, there are no reports of docking-based virtual screening of compound databases. Compared to the catalytic site, accuracy of binding pose prediction is complicated by the diversity of binding orientations/conformations that different ligands can adopt at this site, as revealed by

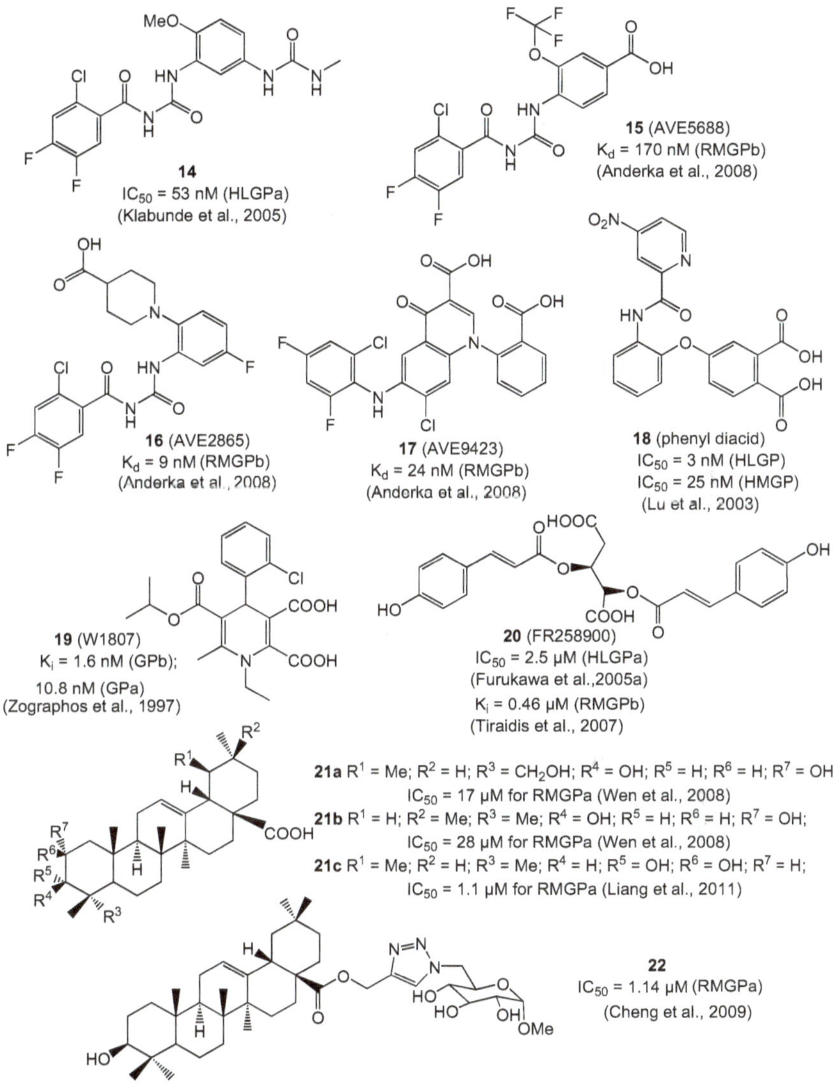

FIGURE 2.5 Allosteric site inhibitors together with their K_i's or IC_{50}'s for glycogen phosphorylase inhibition.

crystallography. This also complicates the thermodynamics of binding. Using isothermal titration calorimetry experiments (Anderka et al., 2008), markedly different thermodynamic contributions (enthalpy–entropy balance) to potency were related to the binding modes adopted by the acyl ureas AVE5688 (**15**; K_d = 170 nM) and AVE2865 (**16**; K_d = 9 nM), and the quinolone AVE9423 (**17**; K_d = 24 nM) ligand. The crystal structure complexes of all three ligands with HLGPa were solved (PDB codes 3CEH, 3CEJ, and 3CEM, respectively), with

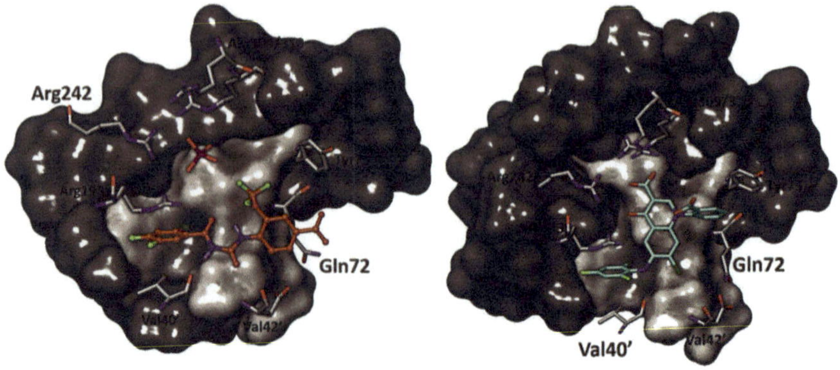

FIGURE 2.6 Binding modes of AVE5688 (**15**; left) and AVE9423 (**17**; right) from crystal structure complexes with human liver GPa, PDB codes 3CEH and 3CEM, respectively.

each inhibitor stabilizing the tense (T) state of the enzyme. Although AVE5688 was a strong enthalpy binder indicating predominantly polar interactions, its binding entropy contribution was poor compared to the more potent AVE2865 and AVE9423 inhibitors. The binding of AVE5688 and AVE9423 is shown in Fig. 2.6, where solvent exclusion in the upper part of the binding cavity and charge–charge interactions with the arginine residues were attributed to the greater potency of AVE9423 (and AVE2865). It was noted that ionic interactions are often entropically favorable without having a significant effect on the enthalpy term (Holdgate, 2001). The change in the computed accessible surface area (ΔASA) of the ligands on binding reflected the relative potencies of AVE5688 (560 Å^2), AVE2865 (670 Å^2), and AVE9423 (670 Å^2). Sanofi-Aventis advanced AVE5688 into human clinical trials.

Despite the complexities of the binding mechanisms at the allosteric site, there are a number of examples of effective CADD efforts. Using the 3-D similarity/superimposition tool SQ (Miller et al., 1999) and exploiting solved crystal structure complexes from the RCSB PDB database (Deng et al., 2005), conformers of lead compound **18** (IC_{50} = 3 nM for HLGP), a phenyl diacid, were first compared to Bayer diacid W1807 (**19**) bound at the AMP allosteric site (PDB ID 3AMV (Oikonomakos et al., 1999)), Pfizer compound CP320626 (**24**; Fig 2.7) bound at the new allosteric site (PDB ID 1C50 (Oikonomakos et al., 2000a)), and caffeine (**28**; Fig. 2.9) bound at the inhibitor site (PDB ID 1GFZ (Oikonomakos et al., 2000b)). Using this method, the allosteric site was predicted as the preferred binding site (Deng et al., 2005). Then, guided by ICM docking (ICM-pro) of compound **18**, hydrophobic/hydrophilic surfaces revealed a vacant hydrophobic region adjacent to the central phenyl ring of **18** to the allosteric site, which on replacement with a naphthyl moiety led to the design of new and more potent analogues.

2,3-Bis(4-hydroxycinnamoyloxy)glutaric acid (FR258900; **20**) is a naturally occurring compound isolated from fungi no. 138354. It inhibits HLGPa

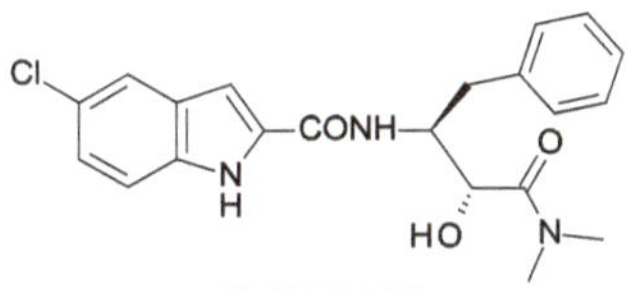

23 (CP91149)
IC_{50} = 110 nM (HLGPa)
(Hoover et al., 1998)

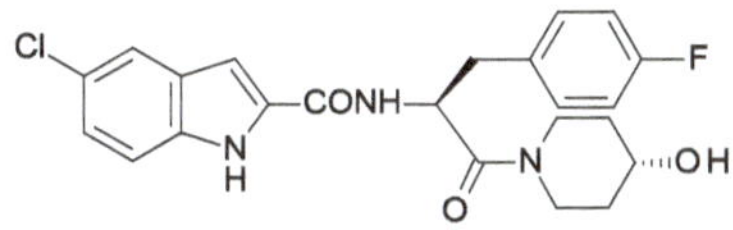

24 (CP320626)
IC_{50} = 200 nM (HLGPa)
IC_{50} = 83 nM (HMGPa)
(Hoover et al., 1998)

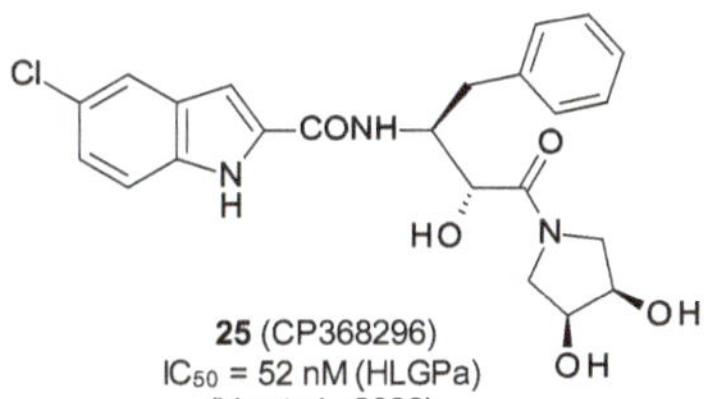

25 (CP368296)
IC_{50} = 52 nM (HLGPa)
(Yu et al., 2006)

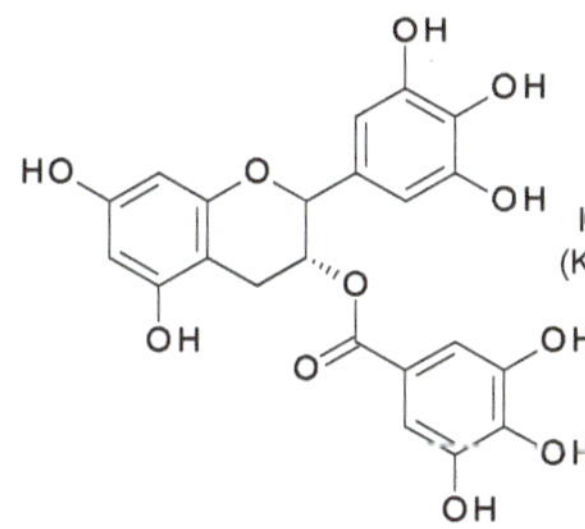

26 (GCG)
IC_{50} = 6.3 μM (HLGPa)
(Kamiyama et al., 2010)

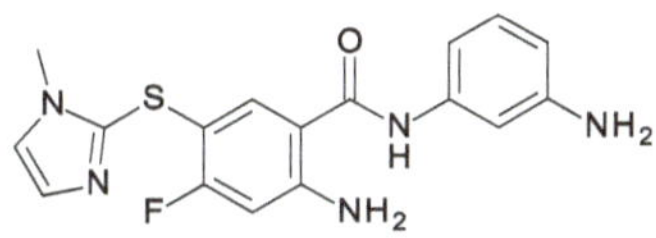

27 (Benzamide derivative)
IC_{50} = 2.68 μM (RMGPa)
(Kamiyama et al., 2010)

FIGURE 2.7 New allosteric inhibitors together with their IC_{50}'s for glycogen phosphorylase inhibition. *HLGPa*, human liver GPa; *RMGPa*, rabbit muscle GPa.

with an IC_{50} of 2.5 μM (Furukawa et al., 2005a) and its crystal structure with RMGPb (PDB code: 2OFF) has been solved (Tiraidis et al., 2007). FR258900 significantly reduced plasma glucose concentrations in diabetic mice models and this was accompanied by increased liver glycogen levels (Furukawa et al., 2005b). Juhász et al. (2014) recently modeled the binding of FR258900 and its analogues using Glide docking and with QSAR models. Different QSAR models were built exploiting descriptors from docking, the MM LIE approximation and QM energies calculated at the M06-2X/LACVP**+ level of theory. Models were derived and validated for tartaric acid diester, monoacylated L-malic acid, 3-hydroxypentanedioic acid, and glutamic acid derivatives. The best model consisted of two descriptors, the Liaison score (modified GlideScore, based on LIE sampling) and a QM descriptor (the inhibitor LUMO energy), which also demonstrated "satisfactory" results for other allosteric inhibitors such as acyl ureas and phthalic acids.

Pentacyclic triterpenes (PTs) represent an important class of natural product GP allosteric binding site inhibitors (Liang et al., 2011; Wen et al., 2008). The crystal structures of asiatic (**21a**; PDB ID: 2QN1) and maslinic acid (**21b**; PDB ID: 2QN2) bound at the RMGPb allosteric site have been solved, facilitating further SBDD efforts (Wen et al., 2008). Both PTs stabilize the inactive T state conformation and inhibit RMGPa with IC_{50}s of 17 μM and 28 μM, respectively

(Wen et al., 2008). PT inhibitors reported to date are generally μM inhibitors with the best of these in the low μM range. The type of PT substituents and their positioning and stereochemistry influence the relative potencies. In this regard, a 3D-QSAR study using CoMFA and CoMSIA for a set of 106 PT derivatives divided into a training set of 80 and a test set of 26 compounds was performed (Liang et al., 2011). Receptor-based alignment of ligands was achieved by docking using Autodock 3.0.3 to the protein structure from the solved GPb–asiatic acid X-ray complex (PDB code: 2QN1). Using CoMFA, models were obtained with statistical indexes $r^2=0.991$ and $q^2=0.623$ for the training set, and $r^2=0.907$ for test set; with CoMSIA, values of $r^2=0.985$ and $q^2=0.508$ for the training set, and $r^2=0.906$ for the test set were obtained. For both CoMFA and CoMSIA methods, after cross-validated partial least square analysis of the training set data, the number of optimal principle components (PC) was 7. Further analysis of models suggested prioritizing electrostatic over steric effects as a means of improving potency. This led to the synthesis of 56 new PT derivatives with 17 of these compounds having IC_{50}s < 10 μM; compound **21c** was the most potent with an IC_{50} of 1.1 μM. In a separate study, a series of glucoconjugates of the PT oleanic acid linked by either triazole or ester linkages were investigated as GP inhibitors (Cheng et al., 2009b). The most potent compound **22** in this series had an IC_{50} of 1.14 μM compared to 14 μM for the parent oleanic acid compound. Glide-XP docking was used to propose binding of **22** exclusively at the allosteric site.

Docking studies have in some cases been used to predict the preferred binding site of new GP inhibitors. Following fragment-based screening of a series of small heterocycles to GP (IC_{50}s against GPa determined), docking calculations with Surflex-Dock were performed to the allosteric, new allosteric, and inhibitor sites to determine the binding preferences of the different scaffolds (Schweiker et al., 2014). In general, the docking suggested nonspecific binding and this was attributed to the low molecular weight of the compounds.

3.4 New Allosteric Site

The new allosteric, also known as the indole binding or dimer interface site, is located at the central cavity formed by the association of the GP dimer subunits, 33 Å from the catalytic site, 15 Å from the allosteric site, and 37 Å from the inhibitor site (Fig. 2.1). The 18 residues (9 from each subunit) that make up the binding site are highly conserved across the isoforms (liver, muscle, and brain) for human and rat GP (Hudson et al., 1993). The new allosteric site inhibitors discussed in this section are shown in Fig. 2.7. Reported inhibitors of this site to date are mostly synthetic; they stabilize the inactive T state of the enzyme synergistic with glucose but also promote the dephosphorylation of HLGPa (Rath et al., 2000). In terms of natural products, gallocatechin gallate (GCG; **26**) has a K_i of 6.3 μM for RMGPb and binds at the site based on multiple inhibition studies using Dixon plots (Kariyama et al., 2010). To our knowledge, no other natural product new allosteric site inhibitors have been reported, however, there are a number of GP inhibitors for which the binding site is unknown (Hayes et al., 2014). Synthetic compounds such as CP91149 (**23**), CP320626 (**24**), and CP368296 (**25**) from Pfizer contain an indole-2-carboxamide moiety, and this class of molecules was the first discovered to bind at the new allosteric site. CP91149 was discovered by high-throughput screening of >300,000 compounds against recombinant human liver GPa (Martin et al., 1998). It has an IC_{50} for HLGPa inhibition of 110 nM in the presence of 7.5 mM glucose and was 5–10 times less potent in the absence of glucose. The related compound CP320626 is also a potent inhibitor of HLGPa (IC_{50} = 205 nM in the presence of glucose) and has demonstrated glucose lowering effects in diabetic *ob/ob* mice without altering plasma insulin levels (Hoover et al., 1998). Its binding features at GPb have been solved (PDB codes: 1C50, 1H54, and 1LWO) and are displayed in Fig. 2.8 (Oikonomakos et al., 2000a). Inhibitor binding occurs not only with small distortions of the structure but also with the displacement

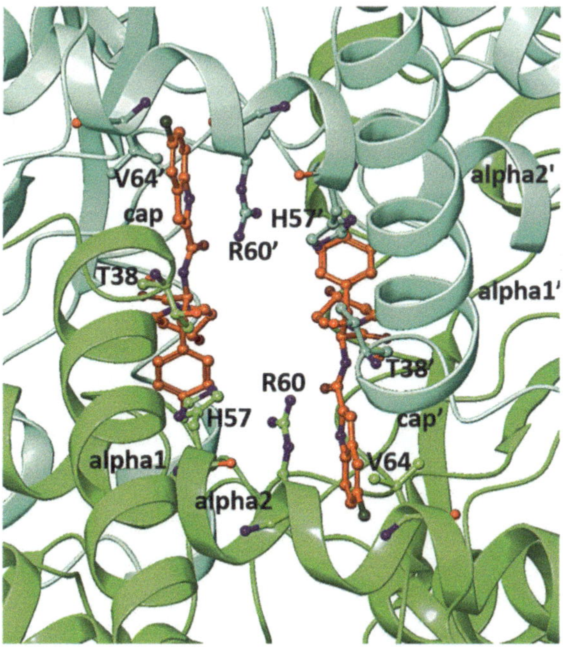

FIGURE 2.8 The structure of the cocrystallized GPb–CP320626 (**24**) complex (Oikonomakos et al., 2000a) showing the ligand bound at the new allosteric site (PDB code 1C50). The site's central cavity is formed at one end by residues from the cap' region and α2 helices (Arg33, His34, Arg60, and Asp61 and their symmetry-related equivalents) and at the other end by the α7 tower helices (residues Asn270, Glu273, Ser276, and their symmetry-related equivalents). The contacts between GPb and CP320626 comprise six hydrogen bonds and extensive van der Waals interactions. Although 33 Å from the catalytic site, the inhibitor exerts its effects by stabilizing the T state at the expense of the R state, thereby shifting the allosteric equilibrium between the two states.

of nine water molecules. The entropy gains from the release of these water molecules, together with the van der Waals, CH/π, halogen/polar, and the specific polar/polar interactions all contribute to the ligands potency. Although solvent exposed, hydrophobic pockets such as those formed by side chains of Leu63, Val64, and Trp67 can make a significant contribution to binding (Onda et al., 2008). One computational study of indole-2-carboxamide inhibitors discussed further estimated that the nonelectrostatic contributions to binding were close to 94% (Liu et al., 2004). Pfizer advanced the CP368296 (ingliforib, **25**) analogue to human clinical trials.

In terms of computational methods applied to binding at this site, the mechanism and thermodynamics of binding can be complicated by two ligand molecules binding at the site (Fig. 2.8). Nevertheless, modeling applications have proved effective in a number of cases. In a study of benzamide derivatives (such as **27**) as inhibitors of GP, analysis of the mapping between pharmacophores created for the different binding sites using LigandScout (Wolber et al., 2005) suggested that they bind at the new allosteric site (Chen et al., 2007). Their binding modes have been explored using Autodock and Glide docking calculations (Chen et al., 2007; Zhang et al., 2009), but no crystallographic evidence of new allosteric site binding has yet been reported. Interestingly, these compounds, as well as inhibiting GP, provide simultaneous activation of glucokinase (GS) and although no cellular or in vivo experiments were reported, it is suggested that such dual-action inhibitors may have potential for controlling hepatic glycogen metabolism in T2D patients (Henke, 2012). Importantly, while activation of GS leads to phosphorylation of glucose in the liver-promoting glycogen synthesis, GS activation in the pancreas promotes β-cell insulin release (Matschinsky, 2002).

3D-QSAR with CoMFA and CoMSIA was successfully applied to study the binding of a set of 25 indole-2-carboxamide inhibitors in two separate studies (Liu et al., 2004; Prathipati et al., 2005). The studies differed in the way the ligand alignments were performed with one approach using Autodock 3.0 docking poses (Liu et al., 2004) and the other exploiting pharmacophoric alignments (Prathipati et al., 2005). Both methods produced statistically significant models with predictive potential. It should also be noted that docking with Autodock 3.0 produced reasonably good correlation between predicted and experimental binding affinities (r^2=0.710). The average electrostatic contribution to ligand binding was just ~6.1%, which is consistent with the significant hydrophobic structural properties of the binding site (Rath et al., 2000).

Accurately accounting for predominantly hydrophobic interactions in itself presents challenges for modeling. For example, the performance of Glide, GOLD, and ICM docking on a set of pharmaceutically relevant protein–ligand complexes with known binding affinities was performed (Perola et al., 2004). In general, with respect to reproducing experimental binding geometries, Glide appeared to perform most consistently with respect to diversity of binding sites and ligand flexibility, while ICM and GOLD performance was more binding site dependent and was significantly poorer when binding was driven predominantly by hydrophobic interactions. A more recent study of eight docking programs (Autodock Vina, Fitted, FlexX, Fred, Glide, GOLD, LibDock, MolDock) using 16 docking and scoring functions compared the rank-order activity of different ligand series for six pharmacologically relevant targets. In general, it was found that hydrophobic targets/binding sites are much more challenging for scoring functions and no program was effective for all six protein–ligand data sets studied. Indeed, the lack of a universal scoring function is well recognized (Plewczynski et al., 2011). Interestingly, in a study comparing the performance of QM/MM docking correlated with different binding sites, it was found that greater improvements over standard docking were obtained for hydrophobic binding sites with substantial hydrophilic interactions (Chung et al., 2009). QM/MM docking calculations have not been reported for the GP new allosteric site to date.

The hydrophobic/hydrophilic nature of a binding site also influences parameter selection in MM-GB(PBSA) binding free energy calculations. In a recent MM-PBSA study, it has been recommended that using a higher solute (internal) dielectric constant $\varepsilon_{in}=4$ for a highly charged binding site, $\varepsilon_{in}=2$ for a moderately charged, and $\varepsilon_{in}=1$ for a hydrophobic binding interface site has the potential to improve ligand ranking results (Hou et al., 2011a). However, the lack of a consistent optimum ε_{in} value should be noted (Aleksandrov et al., 2010; Hayes and Archontis, 2012).

3.5 Inhibitor Site

The inhibitor site, also known as the caffeine binding or purine nucleoside site, is located at the entrance to the catalytic site (Fig. 2.1). Ligands binding at this site are normally intercalated between the aromatic side chains of two residues, Phe285 and Tyr613, with the potential to form sandwich complexes and exploit π-π interaction (Fig. 2.10) and are more potent at high glucose concentrations. The relevant inhibitors discussed here are shown in Fig. 2.9.

In terms of natural product inhibitors, caffeine **28** (first inhibitor found to bind at site) (Kasvinsky et al., 1978; Madsen et al., 1983), indirubins such as indirubin-5-sulfonate **29** ($K_i=13.8\,\mu M$; PDB code: 1UZU) (Kosmopoulou et al., 2004) and some flavonoids are known to bind at this site (Hayes et al., 2014). The flavonoid chrysin (**30**; $K_i=19.01\,\mu M$ for RMGPb) and the structurally related flavopiridol (**31**; $K_i=1.24\,\mu M$) have been studied by in silico calculations, kinetics, and crystallography (Tsitsanou et al., 2013). Crystal structure determination of the GPb complexes with chrysin (PDB code: 3EBO) and flavopiridol (PDB code: 3EBP) revealed close similarities in binding, both intercalating between Phe285 and Tyr613 (common core close to superimposable) and stabilizing the T state conformation through localization of the 280s loop. The binding of chrysin to GPb is shown in Fig. 2.10. While accurate description of π-π stacking cannot be expected using docking or MM methods, QM/MM

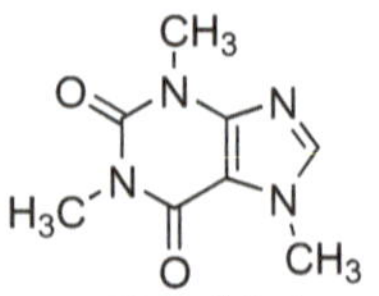

28 (caffeine)
K_i = 0.08 mM (RMGPb)
(Madsen et al., 1983)
K_i = 0.1 mM (RMGPa)
(Kasvinsky et al.,1978)

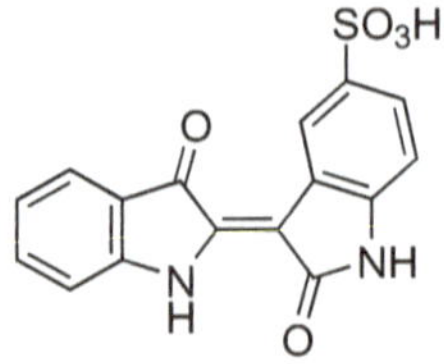

29 (E226)
K_i = 13.8 μM (RMGPb)
(Kosmopoulou et al., 2004)

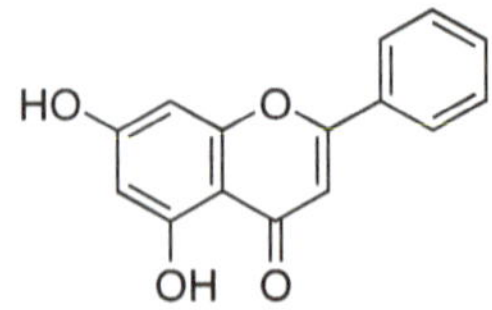

30 (chrysin)
K_i = 19.01 μM (RMGPb)
(Tsitsanou et al., 2013)

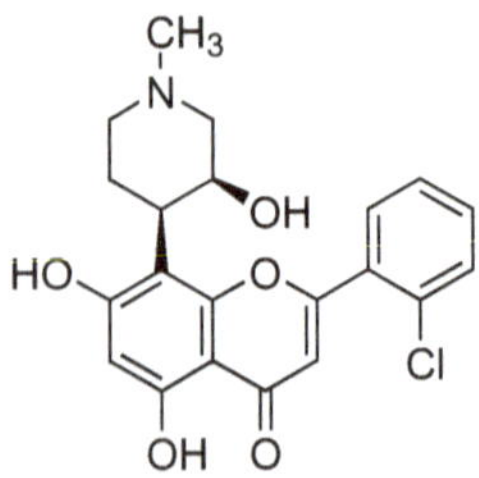

31 (flavopiridol)
K_i = 1.24 μM (RMGPb)
(Tsitsanou et al., 2013)

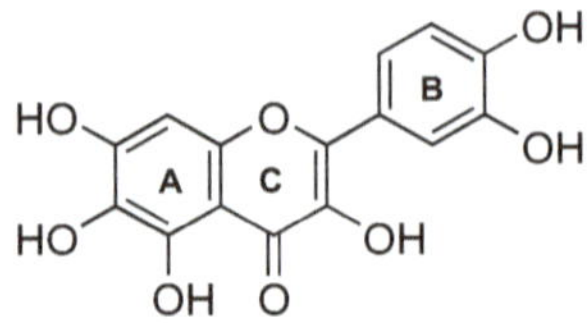

32 (quercetagetin)
IC_{50} = 9.7 μM (RMGPb)
K_i = 3.5 μM (RMGPb)
(Kato et al., 2008)

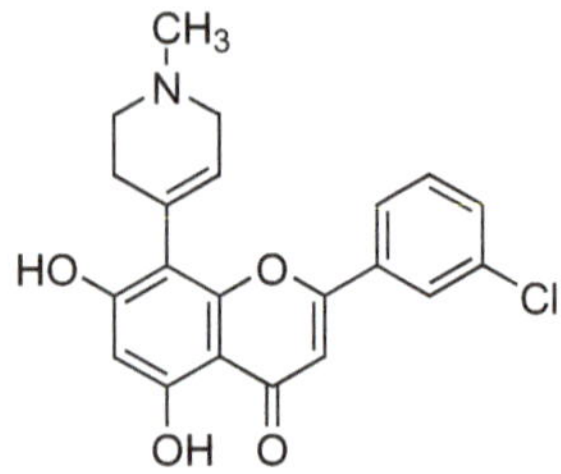

34 (flavopiridol derivative)
K_i = 0.83 μM (RMGPb)
(Hampson et al., 2006)

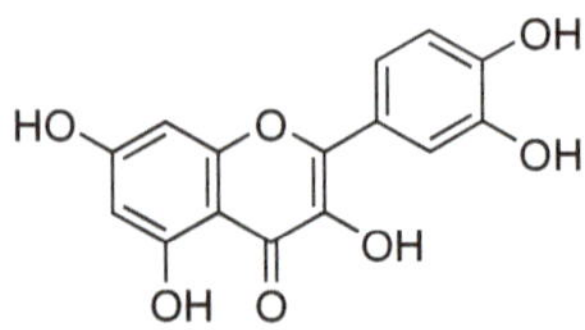

33 (quercetin)
IC_{50} = 20.9 μM (RMGPb)
4.8 μM (RMGPa)
(Jakobs et al., 2006)

FIGURE 2.9 "Inhibitor site" ligands together with their K_i's or IC_{50}'s for glycogen phosphorylase inhibition. *RMGPb*, rabbit muscle GPb.

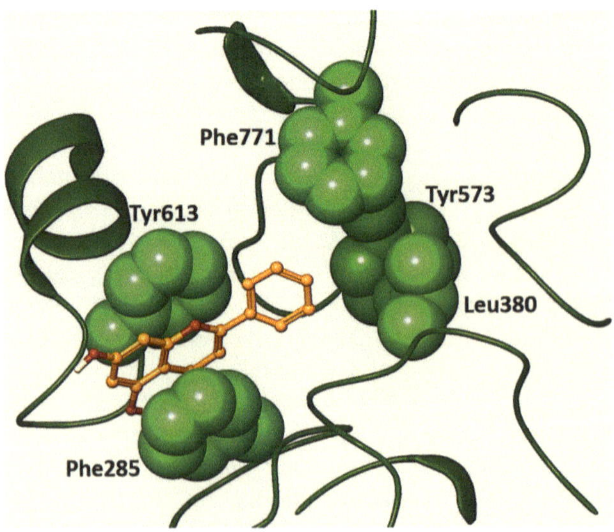

FIGURE 2.10 Binding of chrysin at the "inhibitor site" from PDB code 3EBO (Tsitsanou et al., 2013) highlighting the stacking interactions with Phe285 and Tyr613 and the hydrophobic pocket key residues Leu380, Tyr573, and Phe771.

and QM/MM-PBSA calculations with the binding site described by QM can be applied to obtain a more precise description of binding. Accurate calculation of stacking interactions typically requires computationally expensive QM methods such as CCSD(T) (coupled cluster singles doubles perturbative triples) close to the complete basis set (CBS) limit. However, the development of new DFT in recent years has enhanced their applicability, although some caution is suggested (Riley and Hobza, 2013). To identify the source of flavopiridols' stronger affinity compared to chrysin for GPb (Tsitsanou et al., 2013), Glide docking followed by postdocking QM/MM-PBSA binding free energy calculations using M06-2X/6-31+G* to model the π-π stacking interactions was performed. While docking failed to correctly rank inhibitor-binding conformations, the QM/MM-PBSA method correctly reproduced the experimental results. Breakdown of the thermodynamics of binding attributed the greater potency of flavopiridol to favorable interactions of the 4-hydroxypiperidin-1-yl substituent with GPb, with desolvation effects of the cationic substituent limited by its adopted conformation in the complex. Multiple inhibition and Dixon plots revealed that chrysin binds exclusively at the inhibitor site. On the contrary, QM/MM-PBSA binding free energy calculations for the flavonoid quercetagetin (**32**) suggested poor affinity of this compound for the inhibitor site and led to the proposal that flavonoids with –OH substituents in the *meta*- and *para*-positions of ring B (c.f. Fig. 2.9), and particularly *para*, are likely to be unfavorable due to the hydrophobic pocket formed by residues Phe771, Ile380, and Tyr573 (Fig. 2.10). The subsequent discovery of the quercetin (**33**) binding site (Kantsadi et al., 2014) is consistent with this hypothesis, while quercetagetin is suggested to bind at the allosteric site from multiple inhibition studies using Dixon plots (Kato et al., 2008). Flavonoids have considerable potential to be of benefit to T2D patients, where a number of epidemiological studies have shown link between consuming foods rich in polyphenols and lowering of blood glucose levels (Habtemariam and Varghese, 2014; Williamson, 2013). Flavopiridol derivative **34** is the most potent compound (K_i=0.83 μM) that acts on the inhibitor site discovered to date (Hampson et al., 2006).

4. CONCLUDING REMARKS

While each binding site of GP poses different challenges for molecular modeling, computation has played a key role in the design and discovery of potent GP inhibitors. Given the abundance of GP–inhibitor complexes available in the RCSB Protein Data Bank, structure-based approaches have been mainly used. For the catalytic site, docking and postdocking calculations have recently proved effective, where advancements in algorithms and greater knowledge of the binding mechanisms have facilitated many successful applications. With some inhibitors now being found in the nanomolar range for this site, computation can be used to guide the lead optimization process. Binding cavity water interactions have proven important to accuracy of predictions in some cases.

Although computationally expensive and requiring significant expertise, structural modifications may therefore in some situations be more efficiently guided by explicit water MDFE simulations. Iminosugar derivatives are another class of compounds worthy of further investigation.

Structure-based inhibitor design targeting the allosteric site can be more challenging given the plasticity of the binding site, yet there are many examples of successful applications. PT analogues have proven the most interesting natural product inhibitors binding at this site, with 3D-QSAR analysis exploited to guide the structural modifications of the core PT scaffold (Liang et al., 2011). However, considering the vast structural diversity of natural products, there is an array of unexplored chemical entities ready to be explored for this binding site. Given the conformational flexibility of the site, multiple protein structure (MPS) docking may have a positive impact in docking-based virtual screening experiments. Even docking calculations require skill and know-how to achieve optimal performance, where modification of basic parameters in docking programs together with careful protein and ligand preparation can have a significant effect on docking and virtual screening enrichments (Cross et al., 2009; Kirchmair et al., 2008; Sastry et al., 2013). Some have suggested that the real challenge in protein–ligand binding affinity predictions is polar interactions and their associated desolvation effects (Cheng et al., 2009a), something which we discussed earlier for the catalytic site ligands **6** ($X=F$) and **7** (Manta et al., 2012) but may also have implications for the allosteric site.

The new allosteric site poses different potential challenges due to the significant hydrophobic nature of the site and with two ligands bound at the site sometimes adopting different orientations and even interacting with each other via hydrogen bonds (PDB IDs: 2IEG and 2IEI) (Birch et al., 2007). However, with no report of significance on natural products acting at the site, there is significant opportunities for development of this research area guided by computation. It was proposed that docking and scoring functions should be matched to binding sites and ligand series for optimum results (Xu et al., 2015). As of yet, this has not been investigated for either the allosteric or new allosteric site. Considering the inhibitor site, due to the stacking interactions defining the core-binding features, QM/MM-PBSA provides a valuable approach at present. Once enough ligand binding assay data are available, then application of other methods such as pharmacophore modeling, 3D- and 4D-QSAR will be better facilitated. Given that the other GP binding sites have been relatively unexplored, there is again much potential for computationally motivated exploration of new natural product inhibitors. However, designing optimal virtual screening protocols relies on the availability of benchmarking experimental data which is limited for these sites at present.

In conclusion, many clinical drugs have their origins in natural products (Brahmachari, 2011; Butler, 2005). Natural products provide a huge amount of structural diversity for the development of novel and effective GP inhibitors. The vast majority of these compounds have yet to be explored and notably can be considered to have both nutraceutical (or functional food) and pharmaceutical potential. Among the most potent and promising natural product inhibitors discovered to date are glucose analogues, iminosugars, flavonoids, indirubins, and PTs. Computation has and will continue to play a key role in the discovery and understanding of the mechanism of action of these compounds. Different modeling approaches have proven effective, which is considerably influenced by the characteristics of the GP binding site under consideration.

LIST OF ABBREVIATIONS

CADD Computer-aided drug design
ESP Electrostatic potential
GBSA Generalized Born surface area
HLGP Human liver glycogen phosphorylase
LBDD Ligand-based drug design
LIE Linear interaction energy
MPS Multiple protein structure
PBSA Poisson–Boltzmann Surface Area
QM Quantum mechanics
QM/MM Quantum mechanics/molecular mechanics
QM-PLD Quantum mechanics–polarized ligand docking
QSAR Quantitative structure–activity relationship
RMGP Rabbit muscle glycogen phosphorylase
SBDD Structure-based drug design

ACKNOWLEDGMENTS

Jaida Begum, Ben Chetter, and Demetres D. Leonidas are acknowledged for their assistance in the preparation of some figures.

REFERENCES

Aleksandrov, A., Thompson, D., Simonson, T., 2010. Alchemical free energy simulations for biological complexes: powerful but temperamental. Journal of Molecular Recognition 23 (2), 117–127.

Alexacou, K.M., Hayes, J.M., Tiraidis, C., Zographos, S.E., Leonidas, D.D., Chrysina, E.D., Archontis, G., Oikonomakos, N.G., Paul, J.V., Varghese, B., Loganathan, D., 2008. Crystallographic and computational studies on 4-phenyl-*N*-(β-D-glucopyranosyl)-1*H*-1,2,3-triazole-1-acetamide, an inhibitor of glycogen phosphorylase: comparison with α-D-glucose, *N*-acetyl-β-D-glucopyranosylamine and *N*-benzoyl-*N*′-β-D-glucopyranosyl urea binding. Proteins: Structure, Function and Bioinformatics 71 (3), 301–307.

Allen, W.J., Balius, T.E., Mukherjee, S., Brozell, S.R., Moustakas, D.T., Lang, P.T., Case, D.A., Kuntz, I.D., Rizzo, R.C., 2015. Dock 6: impact of new features and current docking performance. Journal of Computational Chemistry 36 (15), 1132–1156.

Anderka, O., Loenze, P., Klabunde, T., Dreyer, M.K., Defossa, E., Wendt, K.U., Schmoll, D., 2008. Thermodynamic characterization of allosteric glycogen phosphorylase inhibitors. Biochemistry 47 (16), 4683–4691.

Andrade, C.H., Pasqualoto, K.F.M., Ferreira, E.I., Hopfinger, A.J., 2010. 4D-QSAR: perspectives in drug design. Molecules 15 (5), 3281–3294.

Aqvist, J., Medina, C., Samuelsson, J.E., 1994. A new method for predicting binding affinity in computer-aided drug design. Protein Engineering 7 (3), 385–391.

Archontis, G., Watson, K.A., Xie, Q., Andreou, G., Chrysina, E.D., Zographos, S.E., Oikonomakos, N.G., Karplus, M., 2005. Glycogen phosphorylase inhibitors: a free energy perturbation analysis of glucopyranose spirohydantoin analogues. Proteins: Structure Function and Bioinformatics 61 (4), 984–998.

Begum, J., Varga, G., Docsa, T., Gergely, P., Hayes, J.M., Juhasz, L., Somsak, L., 2014. Computationally motivated synthesis and kinetic evaluation of *N*-(β-D-glucopyranosyl)-1,2,4-triazolecarboxamides as glycogen phosphorylase inhibitors. MedChemComm 6 (1), 80–89.

Benltifa, M., Hayes, J.M., Vidal, S., Gueyrard, D., Goekjian, P.G., Praly, J.P., Kizilis, G., Tiraidis, C., Alexacou, K.-M., Chrysina, E.D., Zographos, S.E., Leonidas, D.D., Archontis, G., Oikonomakos, N.G., 2009. Glucose-based spiro-isoxazolines: a new family of potent glycogen phosphorylase inhibitors. Biorganic and Medicinal Chemistry 17, 7368–7380.

Birch, A.M., Kenny, P.W., Oikonomakos, N.G., Otterbein, L., Schofield, P., Whittamore, P.R., Whalley, D.P., 2007. Development of potent, orally active 1-substituted-3,4-dihydro-2-quinolone glycogen phosphorylase inhibitors. Bioorganic and Medicinal Chemistry Letters 17 (2), 394–399.

Blunt, J.W., Copp, B.R., Keyzers, R.A., Munro, M.H.G., Prinsep, M.R., 2013. Marine natural products. Natural Product Reports 30 (2), 237–323.

Bokor, E., Kun, S., Docsa, T., Gergely, P., Somsak, L., 2015. 4(5)-Aryl-2-C-glucopyranosyl-imidazoles as new nanomolar glucose analogue inhibitors of glycogen phosphorylase. ACS Medicinal Chemistry Letters 6 (12), 1215–1219.

Brahmachari, G., 2011. Natural products in drug discovery: impacts and opportunities – an assessment. In: Brahmachari, G. (Ed.), Bioactive Natural Products: Opportunities & Challenges in Medicinal Chemistry. World Scientific Publishing Co., Singapore, pp. 1–199.

Butler, M.S., 2005. Natural products to drugs: natural product derived compounds in clinical trials. Natural Product Reports 22 (2), 162–195.

Chen, L., Li, H., Liu, J., Zhang, L., Liu, H., Jiang, H., 2007. Discovering benzamide derivatives as glycogen phosphorylase inhibitors and their binding site at the enzyme. Bioorganic and Medicinal Chemistry 15 (21), 6763–6774.

Cheng, T., Li, X., Li, Y., Liu, Z., Wang, R., 2009a. Comparative assessment of scoring functions on a diverse test set. Journal of Chemical Information and Modeling 49 (4), 1079–1093.

Cheng, K., Liu, J., Liu, X., Li, H., Sun, H., Xie, J., 2009b. Synthesis of glucoconjugates of oleanic acid as inhibitors of glycogen phosphorylase. Carbohydrate Research 344 (7), 841–850.

Cho, A.E., Guallar, V., Berne, B.J., Friesner, R., 2005. Importance of accurate charges in molecular docking: quantum mechanical/molecular mechanical (QM/MM) approach. Journal of Computational Chemistry 26 (9), 915–931.

Cho, A.E., Chung, J.Y., Kim, M., Park, K., 2009. Quantum mechanical scoring for protein docking. Journal of Chemical Physics 131 (13), 134108.

Chung, J.Y., Hah, J.-M., Cho, A.E., 2009. Correlation between performance of QM/MM docking and simple classification of binding sites. Journal of Chemical Information and Modeling 49 (10), 2382–2387.

Clarke, R.D., 2009. Prospective ligand- and target-based 3D QSAR: state of the art. Current Topics in Medicinal Chemistry 9 (9), 791–810.

Cross, J.B., Thompson, D.C., Rai, B.K., Baber, J.C., Fan, K.Y., Hu, Y., Humblet, C., 2009. Comparison of several molecular docking programs: pose prediction and virtual screening accuracy. Journal of Chemical Information and Modeling 49 (6), 1455–1474.

Deng, Q., Lu, Z., Bohn, J., Ellsworth, K.P., Myers, R.W., Geissler, W.M., Harris, G., Willoughby, C.A., Chapman, K., Mckeever, B., Mosley, R., 2005. Modeling aided design of potent glycogen phosphorylase inhibitors. Journal of Molecular Graphics and Modeling 23 (5), 457–464.

Docsa, T., Czifrak, K., Huse, C., Somsak, L., Gergely, P., 2011. Effect of glucopyranosylidene-spiro-thiohydantoin on glycogen metabolism in liver tissues of streptozotocin-induced and obese diabetic rats. Molecular Medicine Reports 4 (3), 477–481.

Friesner, R.A., Murphy, R.B., Repasky, M.P., Frye, L.L., Greenwood, J.R., Halgren, T.A., Sanschagrin, P.C., Mainz, D.T., 2006. Extra precision glide: docking and scoring incorporating a model of hydrophobic enclosure for protein-ligand complexes. Journal of Medicinal Chemistry 49 (21), 6177–6196.

Füllbeck, M., Michalsky, E., Dunkel, M., Preissner, R., 2006. Natural products: sources and databases. Natural Product Reports 23, 347–356.

Furukawa, S., Tsurumi, Y., Murakami, K., Nakanishi, T., Ohsumi, K., Hashimoto, M., Nishikawa, M., Takase, S., Nakayama, O., Hino, M., 2005a. FR258900, a novel glycogen phosphorylase inhibitor isolated from fungus no. 138354. I. Taxonomy, fermentation, isolation and biological activities. Journal of Antibiotics 58 (8), 497–502.

Furukawa, S., Murakami, K., Nishikawa, M., Nakayama, O., Hino, M., 2005b. FR258900, a novel glycogen phosphorylase inhibitor isolated from fungus no. 138354. II. Anti-hyperglycemic effects in diabetic animal models. Journal of Antibiotics 58 (8), 503–506.

Gleeson, M.P., Gleeson, D., 2009. QM/MM calculations in drug discovery: a useful method for studying binding phenomena. Journal of Chemical Information and Modeling 49 (3), 670–677.

Goldsmith, E.J., Fletterick, R.J., Withers, S.G., 1987. The three dimensional structure of acarbose bound to glycogen phosphorylase. Journal of Biological Chemistry 262 (4), 1449–1455.

Goyard, D., Konya, B., Chajistamatiou, A.S., Chrysina, E.D., Leroy, J., Balzarin, S., Tournier, M., Tousch, D., Petit, P., Duret, C., Maurel, P., Somsak, L., Docsa, T., Gergely, P., Praly, J.-P., Azay-Milhau, J., Vidal, V., 2016. Glucose-derived spiro-isoxazolines are anti-hyperglycemic agents against type 2 diabetes through glycogen phosphorylase inhibition. European Journal of Medicinal Chemistry 108, 444–454.

Habtemariam, S., Varghese, G., 2014. The antidiabetic therapeutic potential of dietary polyphenols. Current Pharmaceutical Biotechnology 15 (4), 391–400.

Hampson, L.J., Arden, C., Agius, L., Ganotidis, M., Kosmopoulou, M.N., Tiraidis, C., Elemes, Y., Sakarellos, C., Leonidas, D.D., Oikonomakos, N.G., 2006. Bioactivity of glycogen phosphorylase inhibitors that bind to the purine nucleoside site. Bioorganic and Medicinal Chemistry 14 (23), 7835–7845.

Hayes, J.M., Archontis, G., 2012. MM-GB(PB)SA calculations of protein-ligand binding free energies. In: Wang, L. (Ed.), Molecular Dynamics – Studies of Synthetic & Biological Macromolecules. InTech, Rijeka, Croatia.

Hayes, J.M., Leonidas, D.D., 2010. Computation as a tool for glycogen phosphorylase inhibitor design. Mini Reviews in Medicinal Chemistry 10 (12), 1156–1174.

Hayes, J.M., Skamnaki, V.T., Archontis, G., Lamprakis, C., Sarrou, J., Bischler, N., Skaltsounis, A.-L., Zographos, S.E., Oikonomakos, N.G., 2011. Kinetics, in silico docking, molecular dynamics, and MM-GBSA binding studies on prototype indirubins, KT5720, and staurosporine as phosphorylase kinase ATP-binding site inhibitors: the role of water molecules examined. Proteins: Structure Function and Bioinformatics 79 (3), 703–719.

Hayes, J.M., Kantsadi, A.L., Leonidas, D.D., 2014. Natural products and their derivatives as inhibitors of glycogen phosphorylase: potential treatment for type 2 diabetes. Phytochemistry Reviews 13 (2), 471–498.

Henke, B.R., 2012. Inhibition of glycogen phosphorylase as a strategy for the treatment of type 2 diabetes. In: Jones, R.M. (Ed.), New Therapeutic Strategies for Type 2 Diabetes. Small Molecule Approaches. RSC Drug Discovery Series No. 27, Cambridge, U.K.

Holdgate, G.A., 2001. Making cool drugs hot: isothermal titration calorimetry as a tool of study binding energetics. BioTechniques 31 (1), 164–184.

Hoover, D.J., Lefkowitz-Snow, S., Burgess-Henry, J.L., Martin, W.H., Armento, S.J., Stock, I.A., Mcpherson, R.K., Genereux, P.E., Gibbs, E.M., Treadway, J.L., 1998. Indole-2-carboxamide inhibitors of human liver glycogen phosphorylase. Journal of Medicinal Chemistry 41 (16), 2934–2938.

Hopfinger, A.J., Reaka, A., Venkatarangan, P., Duca, J.S., Wang, S., 1999. Construction of a virtual high throughput screen by 4D-QSAR analysis: application to a combinatorial library of glucose inhibitors of glycogen phosphorylase b. Journal of Chemical Information and Computer Sciences 39 (6), 1151–1160.

Hou, T., Wang, J., Li, Y., Wang, W., 2011a. Assessing the performance of the MM/PBSA and MM/GBSA methods. I. The accuracy of binding free energy calculations based on molecular dynamics simulations. Journal of Chemical Information and Modeling 51 (1), 69–82.

Hou, T., Wang, J., Li, Y., Wang, W., 2011b. Assessing the performance of the molecular mechanics/Poisson Boltzmann surface area and molecular mechanics/generalized Born surface area methods. II. The accuracy of ranking poses generated from docking. Journal of Computational Chemistry 32 (5), 866–877.

Hudon, J.W., Golding, G.B., Crerar, M.M., 1993. Evolution of allosteric control in glycogen phosphorylase. Journal of Molecular Biology 234 (3), 700–721.

ICM-pro, Molsoft LLC: San Diego, CA.

Irwin, J.J., Sterling, T., Mysinger, M.M., Bolstad, E.S., Coleman, R.G., 2012. Zinc: a free tool to discover chemistry for biology. Journal of Chemical Information and Modeling 52 (7), 1757–1768.

Jones, R.M. (Ed.), 2012. New Therapeutic Strategies for Type 2 Diabetes. Small Molecule Approaches. RSC Drug Discovery Series No. 27, Cambridge, UK.

Juhász, L., Varga, G., Sztankovics, A., Béke, F., Docsa, T., Kiss-Szikszai, A., Gergely, P., Kona, J., Tvaroska, I., Somsák, L., 2014. Structure-activity relationships of glycogen phosphorylase inhibitor FR258900 and its analogues: a combined synthetic, enzyme kinetics, and computational study. ChemPlusChem 79 (11), 1558–1568.

Kamiyama, O., Sanae, F., Ikeda, K., Higashi, Y., Minami, Y., Asano, N., Adachi, I., Kato, A., 2010. In vitro inhibition of α-glucosidases and glycogen phosphorylase by catechin gallates in green tea. Food Chemistry 122 (4), 1061–1066.

Kantsadi, A.L., Hayes, J.M., Manta, S., Skamnaki, V.T., Kiritsis, C., Psarra, A.M., Koutsogiannis, Z., Dimopoulou, A., Theofanous, S., Nikoleousakos, N., Zoumpoulakis, P., Kontou, M., Papadopoulos, G., Zographos, S.E., Komiotis, D., Leonidas, D.D., 2012. The sigma-hole phenomenon of halogen atoms forms the structural basis of the strong inhibitory potency of C5 halogen substituted glucopyranosyl nucleosides towards glycogen phosphorylase b. ChemMedChem 7 (4), 722–732.

Kantsadi, A.L., Apostolou, A., Theofanous, S., Stravodimos, G.A., Kyriakis, E., Gorgogietas, V.A., Chatzileontiadou, D.S.M., Pegiou, K., Skamnaki, V.T., Stagos, D., Kouretas, D., Psarra, A.-M.G., Haroutounian, S.A., Leonidas, D.D., 2014. Biochemical and biological assessment of the inhibitory potency of extracts from vinification byproducts of *Vitis vinifera* extracts against glycogen phosphorylase. Food Chemistry and Toxicology 67, 35–43.

Kantsadi, A.L., Parmenopoulou, V., Bakalov, D., Snelgrove, L., Stravodimos, G.A., Chatzileontiadou, D.S.M., Manta, S., Panagiotopoulou, A., Hayes, J.M., Komiotis, D., Leonidas, D.D., 2015. Glycogen phosphorylase as a target for type 2 diabetes: synthesis, biochemical, structural and computational evaluation of novel *N*-acyl-*N'*-(β-D-glucopyranosyl) urea inhibitors. Current Topics in Medicinal Chemistry 15 (23), 2373–2389.

Karis, N.D., Loughlin, W.A., Jenkins, I.D., Healy, P.C., 2009. Glycogen phosphorylase inhibitory effects of 2-oxo-1,2-dihydropyridin-3-yl amide derivatives. Biorganic and Medicinal Chemistry 17 (13), 4724–4733.

Kasvinsky, P.J., Madsen, N.B., Sygusch, J., Fletterick, R.J., 1978. Regulation of glycogen phosphorylase-a by nucleotide derivatives – kinetic and X-ray crystallographic studies. Journal of Biological Chemistry 253 (9), 3343–3351.

Kato, A., Nasu, N., Takebayashi, K., Adachi, I., Minami, Y., Sanae, F., Asano, N., Watson, A.A., Nash, R.J., 2008. Structure-activity relationships of flavonoids as potential inhibitors of glycogen phosphorylase. Journal of Agricultural and Food Chemistry 56 (12), 4469–4473.

Kim, M.O., Nichols, S.E., Wang, Y., Mccammon, J.A., 2013. Effects of histidine protonation and rotameric states on virtual screening of *M. tuberculois* RmlC. Journal of Computer-Aided Molecular Design 27 (3), 235–246.

Kirchmair, J., Markt, P., Distinto, S., Wolber, G., Langer, T., 2008. Evaluation of the performance of 3D virtual screening protocols: RMSD comparisons, enrichment assessments, and decoy selection—What can we learn from earlier mistakes? Journal of Computer-Aided Molecular Design 22, 213–228.

Klabunde, T., Wendt, K.U., Kadereit, D., Brachvogel, V., Burger, H.J., Herling, A.W., Oikonomakos, N.G., Kosmopoulou, M.N., Schmoll, D., Sarubbi, E., Von Roedern, E., Schonafinger, K., Defossa, E., 2005. Acyl ureas as human liver glycogen phosphorylase inhibitors for the treatment of type 2 diabetes. Journal of Medicinal Chemistry 48 (20), 6178–6193.

Kollman, P.A., Massova, I., Reyes, C., Kuhn, B., Huo, S., Chong, L., Lee, M., Lee, T., Duan, Y., Wang, W., Donini, O., Cieplak, P., Srinivasan, J., Case, D.A., Cheatham III, T.E., 2000. Calculating structures and free energies of complex molecules: combining molecular mechanics and continuum models. Accounts of Chemical Research 33 (12), 889–897.

Kosmopoulou, M.N., Leonidas, D.D., Chrysina, E.D., Bischler, N., Eisenbrand, G., Sakarellos, C.E., Pauptit, R., Oikonomakos, N.G., 2004. Binding of the potential antitumour agent indirubin-5-sulphonate at the inhibitor site of rabbit muscle glycogen phosphorylase b – comparison with ligand binding to pCDK2-cyclin A complex. European Journal of Biochemistry 271 (11), 2280–2290.

Kristiansen, M., Andersen, B., Iversen, L.F., Westergaard, N., 2004. Identification, synthesis, and characterization of new glycogen phosphorylase inhibitors binding to the allosteric AMP site. Journal of Medicinal Chemistry 47 (14), 3537–3545.

Kuhn, B., Gerber, P., Schulz-Gasch, T., Stahl, M., 2005. Validation and use of the MM-PBSA approach for drug discovery. Journal of Medicinal Chemistry 48 (12), 4040–4048.

Kun, S., Bokor, É., Varga, G., Szőcs, B., Páhi, A., Czifrák, K., Tóth, M., Juhász, L., Docsa, T., Gergely, P., Somsák, L., 2014. New synthesis of 3-(β-D-glucopyranosyl)-5-substituted-1,2,4-triazoles, nanomolar inhibitors of glycogen phosphorylase. European Journal of Medicinal Chemistry 76, 567–579.

Lagorce, D., Sperandio, O., Baell, J.B., Miteva, M.A., Villoutreix, B.O., 2015. FAF-Drugs3: a web server for compound property calculation and chemical library design. Nucleic Acids Research 43 (W1), W200–W207.

Lei, J., Zhou, J., 2002. A marine natural product database. Journal of Chemical Information & Computer Sciences 42 (3), 742–748.

Li, J., Liu, H., Yao, X., Liu, M., Hu, Z., Fan, B., 2007. Quantitative structure-activity relationship study of acyl ureas as inhibitors of human liver glycogen phosphorylase using least squares support vector machines. Chemometric and Intelligent Laboratory Systems 87 (2), 139–146.

Liang, Z., Zhang, L., Li, L., Liu, J., Li, H., Chen, L., Cheng, K., Zheng, M., Wen, X., Zhang, P., Hao, J., Gong, Y., Zhang, X., Zhu, X., Chen, J., Liu, H., Jiang, H., Luo, C., Sun, H., 2011. Identification of pentacyclic triterpenes derivatives as potent inhibitors against glycogen phosphorylase based on 3D-QSAR studies. European Journal of Medicinal Chemistry 46 (2), 2011–2021.

Liu, G.X., Zhang, Z.S., Luo, X.M., Shen, J.H., Liu, H., Shen, X., Chen, K.X., Jiang, H.L., 2004. Inhibitory mode of indole-2-carboxamide derivatives against HLGPa: molecular docking and 3D-QSAR analyses. Bioorganic and Medicinal Chemistry 12 (15), 4147–4157.

Loughlin, W.A., Pierens, G.K., Petersson, M.J., Henderson, L.C., Healy, P.C., 2008. Evaluation of novel hyphodermin derivatives as glycogen phosphorylase a inhibitors. Biorganic and Medicinal Chemistry 16 (11), 6172–6178.

Lu, Z., Bohn, J., Bergeron, R., Deng, Q., Ellsworth, K.P., Geissler, W.M., Harris, G., Mccann, P.E., Mckeever, B., Myers, R.W., Saperstein, R., Willoughby, C.A., Yao, J., Chapman, K., 2003. A new class of glycogen phosphorylase inhibitors. Biorganic and Medicinal Chemistry Letters 13 (22), 4125–4128.

Madsen, N.B., Shechosky, S., Fletterick, R.J., 1983. Site-site interactions in glycogen phosphorylase b probed by ligands specific for each site. Biochemistry 22 (19), 4460–4465.

Manta, S., Xipnitou, A., Kiritsis, C., Kantsadi, A.L., Hayes, J.M., Skamnaki, V.T., Lamprakis, C., Kontou, M., Zoumpoulakis, P., Zographos, S.E., Leonidas, D.D., Komiotis, D., 2012. 3′-AXIAL CH2OH substitution on glucopyranose does not increase glycogen phosphorylase inhibitory potency. QM/MM-PBSA calculations suggest why. Chemical Biology and Drug Design 79 (5), 663–673.

Martin, J.L., Veluraja, K., Ross, K., Johnson, L.N., Fleet, G.W.J., Ramsden, N.G., Bruce, I., Orchard, M.G., Oikonomakos, N.G., Papageorgiou, A.C., Leonidas, D.D., Tsitoura, H.S., 1991. Glucose analogue inhibitors of glycogen phosphorylase: the design of potential drugs for diabetes. Biochemistry 30 (42), 10101–10116.

Martin, W.H., Hoover, D.J., Armento, S.J., Stock, I.A., Mcpherson, R.K., Danley, D.E., Stevenson, R.W., Barrett, E.J., Treadway, J.L., 1998. Discovery of a human liver glycogen phosphorylase inhibitor that lowers blood glucose in vitro. Proceeding of the National Academy of Sciences United States of America 95 (4), 1776–1781.

Matschinsky, F.M., 2002. Regulation of pancreatic beta-cell glucokinase: from basics to therapeutics. Diabetes 51 (Suppl. 3), S394–S404.

Mavrokefalos, N., Myrianthopoulos, V., Chajistamatiou, A.S., Chrysina, E.D., Mikros, E., 2015. Discovery of the glycogen phosphorylase-modulating activity of a resveratrol glucoside by using a virtual screening protocol optimized for solvation effects. Planta Medica 81, 507–516.

Meng, X.-Y., Zhang, H.-X., Mezei, M., Cui, M., 2012. Molecular docking: a powerful approach for structure-based drug discovery. Current Computer-Aided Drug Design 7 (2), 146–157.

Miller, M.D., Sheridan, R.P., Kearsley, S.K., 1999. SQ: a program for rapidly producing pharmacophorically relevent molecular superpositions. Journal of Medicinal Chemistry 42 (9), 1505–1514.

Moitessier, N., Englebienne, P., Lee, D., Lawandi, J., Corbeil, C.R., 2008. Towards the development of universal, fast and highly accurate docking/scoring methods: a long way to go. British Journal of Pharmacology 153 (Suppl. 1), S7–S26.

Monod, J., Wyman, J., Changeux, J.P., 1965. On the nature of allosteric transitions: a plausible model. Journal of Molecular Biology 12 (1), 88–118.

Morris, G.M., Huey, R., Lindstrom, W., Sanner, M.F., Belew, R.K., Goodsell, D.S., Olson, A.J., 2009. Autodock4 and AutoDockTools4: automated docking with selective receptor flexiblity. Journal of Computational Chemistry 30 (16), 2785–2791.

Nagy, L., Docsa, T., Szanto, M., Brunyanszki, A., Hegedus, C., Marton, J., Konya, B., Virag, L., Somsak, L., Gergely, P., Bai, P., 2013. Glycogen phosphorylase inhibitor *N*-(3,5-dimethylbenzoyl)-*N*′-(β-D-glucopyranosyl) urea improves glucose tolerance under normoglycemic and diabetic conditions and rearranges hepatic metabolism. PLoS One 8 (7), e69420.

Ntie-Kang, F., Mbah, J.A., Mbaze, L.M., Lifongo, L.L., Scharfe, M., Hanna, J.N., Cho-Ngwa, F., Onguéné, P.A., Owono Owono, L.C., Megnassas, E., Sippl, W., Efange, S.M.N., 2013.

CamMedNP: building the Cameroonian 3D structural natural products database for virtual screening. BMC Complementary and Alternative Medicine 13, 88.

Ntie-Kang, F., Onguéné, P.A., Fotso, G.W., Andrae-Marobela, K., Bezabih, M., Ndom, J.C., Ngadjui, B.T., Ogundaini, A.O., Abegaz, B.M., Meva'a, L.M., 2014. Virtualizing the p-ANAPL Library: a step towards drug discovery from African medicinal plants. PLoS One 9 (3), e90655.

Ogawa, A.K., Willoughby, C.A., Bergeron, R., Ellsworth, K.P., Geissler, W.M., Myers, R.W., Yao, J., Harris, G., Chapman, K.T., 2003. Glucose lowering in a db/db mouse model by dihydropyridine diacid glycogen phosphorylase inhibitors. Bioorganic and Medicinal Chemistry Letters 13 (20), 3405–3408.

Oikonomakos, N.G., Tsitsanou, K.E., Zographos, S.E., Skamnaki, V.T., Goldmann, S., Bischoff, H., 1999. Allosteric inhibition of glycogen phosphorylase a by the potential antidiabetic drug 3-isopropyl 4-(2-chlorophenyl)-1,4-dihydro-1-ethyl-2-methyl-pyridine-3,5,6-tricarboxylate. Protein Science 8 (10), 1930–1945.

Oikonomakos, N.G., Skamnaki, V.T., Tsitsanou, K.E., Gavalas, N.G., Johnson, L.N., 2000a. A new allosteric site in glycogen phosphorylase b as a target for drug interactions. Structure 8 (6), 575–584.

Oikonomakos, N.G., Schnier, J.B., Zographos, S.E., Skamnaki, V.T., Tsitsanou, K.E., Johnson, L.N., 2000b. Flavopiridol inhibits glycogen phosphorylase by binding at the inhibitor site. Journal of Biological Chemistry 275 (44), 34566–34573.

Oikonomakos, N.G., Kosmopoulou, M.N., Chrysina, E.D., Leonidas, D.D., Kostas, I.D., Wendt, K.U., Klabunde, T., Defossa, E., 2005. Crystallographic studies on acyl ureas, a new class of glycogen phosphorylase inhibitors, as potential antidiabetic drugs. Protein Science 14 (7), 1760–1771.

Oikonomakos, N.G., Tiraidis, C., Leonidas, D.D., Zographos, S.E., Kristiansen, M., Jessen, C.U., Norskov-Lauritsen, L., Agius, L., 2006. Iminosugars as potential inhibitors of glycogenolysis: structural insights into the molecular basis of glycogen phosphorylase inhibition. Journal of Medicinal Chemistry 49 (19), 5687–5701.

Onda, K., Suzuki, T., Shiraki, R., Yonetoku, Y., Negoro, K., Momose, K., Katayama, N., Orita, M., Yamaguchi, T., Ohta, M., Tsukamoto, S., 2008. Synthesis of 5-chloro-*N*-aryl-1*H*-indole-2-carboxamide derivatives as inhibitors of human liver glycogen phosphorylase a. Bioorganic and Medicinal Chemistry 16 (10), 5452–5464.

Pan, D.H., Tseng, Y.F., Hopfinger, A.J., 2003. Quantitative structure-based design: formalism and application of receptor-dependent RD-4D-QSAR analysis to a set of glucose analogue inhibitors of glycogen phosphorylase. Journal of Chemical Information and Computer Sciences 43 (5), 1591–1607.

Pan, D.H., Liu, J.Z., Senese, C., Hopfinger, A.J., Tseng, Y., 2004. Characterization of a ligand-receptor binding event using receptor dependent four-dimensional quantitative structure-activity relationship analysis. Journal of Medicinal Chemistry 47 (12), 3075–3088.

Parmenopoulou, P., Kantsadi, A.L., Tsirkone, V.G., Chatzileontiadou, D.S.M., Manta, S., Zographos, S.E., Kollatos, N., Archontis, G., Agius, G., Hayes, J.M., Leonidas, D.D., Komiotis, D., 2014. Structure based inhibitor design targeting glycogen phosphorylase b. Virtual screening, synthesis, biochemical and biological assessment of novel N-(β-D-glucopyranosyl) amides. Biorganic and Medicinal Chemistry 22 (17), 4810–4825.

Perola, E., Walters, W.P., Charifson, P.S., 2004. A detailed comparison of current docking and scoring methods on systems of pharmaceutical relevance. Proteins – Structure, Function & Bioinformatics 56 (2), 235–249.

Pinotsis, N., Leonidas, D.D., Chrysina, E.D., Oikonomakos, N.G., Mavridis, I.M., 2003. The binding of beta- and gamma-cyclodextrins to glycogen phosphorylase b: kinetic and crystallographic studies. Protein Science 12 (9), 1914–1924.

Plewczynski, D., Lazniewski, M., Augustyniak, R., Ginalski, K., 2011. Can we trust docking results? Evaluation of seven commonly used programs on PDBbind database. Journal of Computational Chemistry 32 (4), 742–755.

Polyak, M., Varga, G., Szilagyi, B., Juhasz, L., Docsa, T., Gergely, P., Begum, J., Hayes, J.M., Somsak, L., 2013. Synthesis, enzyme kinetics and computational evaluation of *N*-β-D-glucopyranosyl oxadiazolecarboxamides as glycogen phosphorylase inhibitors. Biorganic and Medicinal Chemistry 21 (18), 5738–5747.

Prathipati, P., Pandey, G., Saxena, A.K., 2005. CoMFA and docking studies on glycogen phosphorylase a inhibitors as antidiabetic agents. Journal of Chemical Information and Modeling 45 (1), 136–145.

Raha, K., Peters, M.B., Wang, B., Yu, N., Wollacott, A.M., Westerhoff, L.M., Merz, K.M., 2007. The role of quantum mechanics in structure-based drug design. Drug Discovery Today 12 (17–18), 725–731.

Rastelli, G., Del Rio, A., Degliesposti, G., Sgobba, M., 2010. Fast and accurate predictions of binding free energies using MM-PBSA and MM-GBSA. Journal of Computational Chemistry 31 (4), 797–810.

Rath, V.L., Ammirati, M., Danley, D.E., Ekstrom, J.L., Gibbs, E.M., Hynes, T.R., Mathiowetz, A.M., Mcpherson, R.K., Olson, T.V., Treadway, J.L., Hoover, D.J., 2000. Human liver glycogen phosphorylase inhibitors bind at a new allosteric site. Chemical Biology 7 (9), 677–682.

Riley, K.E., Hobza, P., 2013. On the importance and origin of aromatic interactions in chemistry and biodisciplines. Accounts of Chemical Research 46 (4), 927–936.

Sanders, M.P.A., Barbosa, A.J.M., Zarzycha, B., Nicolaes, G.A.F., Klomp, J.P.G., De Vlieg, J., Del Rio, A., 2012. Comparative analysis of pharmacophore screening tools. Journal of Chemical Information and Modeling 52 (6), 1607–1620.

Sastry, G.M., Adzhigirey, M., Day, T., Annabhimoju, R., Sherman, W., 2013. Protein and ligand preparation: parameters, protocols, and influence on virtual screening enrichments. Journal of Computer-Aided Molecular Design 27 (3), 221–234.

Schrödinger, 2016. LLC, New York, NY.

Schweiker, S.S., Loughlin, W.A., Lohning, A.S., Petersson, M.J., Jenkins, I.D., 2014. Synthesis, screening and docking of small heterocycles as Glycogen phosphorylase inhibitors. European Journal of Medicinal Chemistry 84, 584–594.

Smith, G.F., 2011. Designing drugs to avoid toxicity. Progress in Medicinal Chemistry 50, 1–47.

Somsák, L., Czifrák, K., Tóth, M., Bokor, E., Chrysina, E.D., Alexacou, K.-M., Hayes, J.M., Lazoura, E., Leonidas, D.D., Zographos, S.E., Oikonomakos, N.G., 2008. New inhibitors of glycogen phosphorylase as potential antidiabetic agents. Current Medicinal Chemistry 15 (28), 2933–2983.

Somsak, L., 2011. Glucose derived inhibitors of glycogen phosphorylase. Comptes Rendus Chimie 14 (2–3), 211–223.

Suárez, D., Diaz, N., 2015. Direct methods for computing single-molecule entropies from molecular simulations. Computational Molecular Science 5 (1), 1–26.

Tetko, I.V., Tanchuk, V.Y., 2002. Application of associative neural networks for prediction of lipophilicity in ALOGPS 2.1 program. Journal of Chemical Information and Computer Sciences 42 (5), 1136–1145.

Tetko, I.V., Gasteiger, J., Todeschini, R., Mauri, M.A., Livingstone, D., Ertl, P., Palyulin, V.A., Radchenko, E.V., Zefirov, N.S., Makarenko, A.S., Tanchuk, V.Y., Prokopenko, V.V., 2005. Virtual computational chemistry laboratory - design and description. Journal of Computer-Aided Molecular Design 19 (6), 453–463.

Thomson, S.A., Banker, P., Bickett, D.M., Boucheron, J.A., Carter, H.L., Clancy, D.C., Cooper, J.P., Dickerson, S.H., Garrido, D.M., Nolte, R.T., Peat, A.J., Sheckler, L.R., Sparks, S.M., Tavares, F.X., Wang, L., Wang, T.Y., Weiel, J.E., 2009. Anthranilimide based glycogen phosphorylase inhibitors for the treatment of type 2 diabetes. Part 3: X-ray crystallographic characterization, core and urea optimization and in vivo efficacy. Biorganic and Medicinal Chemistry Letters 19 (4), 1177–1182.

Tiraidis, C., Alexacou, K.M., Zographos, S.E., Leonidas, D.D., Gimisis, T., Oikonomakos, N.G., 2007. FR258900, a potential anti-hyperglycemic drug, binds at the allosteric site of glycogen phosphorylase. Protein Science 16 (8), 1773–1782.

Trott, O., Olson, A.J., 2010. Autodock vina: improving the speed and accuracy of docking with a new scoring function, efficient optimization and multithreading. Journal of Computational Chemistry 31 (2), 455–461.

Tsirkone, V.G., Tsoukala, E., Lamprakis, C., Manta, S., Hayes, J.M., Skamnaki, V.T., Drakou, C., Zographos, S.E., Komiotis, D., Leonidas, D.D., 2010. 1-(3-deoxy-3-fluoro-β-D-glucopyranosyl) pyrimidine derivatives as inhibitors of glycogen phosphorylase b: kinetic, crystallographic and modeling studies. Biorganic and Medicinal Chemistry 18 (10), 3413–3425.

Tsitsanou, K.E., Skamnaki, V.T., Oikonomakos, N.G., 2000. Structural basis of the synergistic inhibition of glycogen phosphorylase a by caffeine and a potential antidiabetic drug. Archives of Biochemistry and Biophysics 384 (2), 245–254.

Tsitsanou, K.E., Hayes, J.M., Keramioti, M., Mamais, M., Oikonomakos, N.G., Kato, A., Leonidas, D.D., Zographos, S.E., 2013. Sourcing the affinity of flavonoids for the glycogen phosphorylase inhibitor site via crystallography, kinetics and QM/MM-PBSA binding studies: comparison of chrysin and flavopiridol. Food and Chemical Toxicology 61, 14–27.

Venkatarangan, P., Hopfinger, A.J., 1999. Prediction of ligand-receptor binding thermodynamics by free energy force field three dimensional quantitative structure-activity relationship analysis: applications to a set of glucose analogue inhibitors of glycogen phosphorylase. Journal of Medicinal Chemistry 42 (12), 2169–2179.

Verdonk, M.L., Cole, J.C., Hartshorn, M.J., Murray, C.W., Taylor, R.D., 2003. Improved protein-ligand docking using GOLD. Proteins: Structure, Function and Bioinformatics 52 (4), 609–623.

Villoutreix, B.O., Renault, N., Lagorce, D., Sperandio, O., Montes, M., Miteva, M.A., 2007. Free resources to assist structure-based virtual screening experiments. Current Protein and Peptide Science 8 (4), 381–411.

Wen, X., Sun, H., Liu, J., Cheng, K., Zhang, P., Zhang, L., Hao, J., Ni, P., Zographos, S.E., Leonidas, D.D., Alexacou, K.M., Gimisis, T., Hayes, J.M., Oikonomakos, N.G., 2008. Naturally occurring pentacyclic triterpenes as inhibitors of glycogen phosphorylase: synthesis, structure-activity relationships and X-ray crystallographic studies. Journal of Medicinal Chemistry 51 (12), 3540–3554.

Williamson, G., 2013. Possible effects of dietary polyphenols on sugar absorption and digestion. Molecular Nutrition and Food Research 57 (1), 48–57.

Wolber, G., Langer, T., 2005. LigandScout: 3-D pharmacophores derived from protein-bound ligands and their use as virtual screening filters. Journal of Chemical Information and Modeling 45 (1), 160–169.

Xu, W., Lucke, A.J., Fairlie, D.P., 2015. Comparing sixteen scoring functions for predicting biological activities of ligands for protein targets. Journal of Molecular Graphics and Modeling 57, 76–88.

Zhang, L., Li, H., Zhu, Q., Liu, J., Chen, L., Leng, Y., Jiang, H., Liu, H., 2009. Benzamide derivatives as dual-action hypoglycemic agents that inhibit glycogen phosphorylase and activate glucokinase. Bioorganic and Medicinal Chemistry 17 (20), 7301–7312.

Zhou, R.H., Friesner, R.A., Ghosh, A., Rizzo, R.C., Jorgensen, W.L., Levy, R.M., 2001. New linear interaction method for binding affinity calculations using a continuum solvent model. Journal of Physical Chemistry B 105, 10388–10397.

Zographos, S.E., Oikonomakos, N.G., Tsitsanou, K.E., Leonidas, D.D., Chrysina, E.D., Skamnaki, V.T., Bischoff, H., Goldman, S., Watson, K.A., Johnson, L.A., 1997. The structure of glycogen phosphorylase b with an alkyl-dihydropyridine-dicarboxylic acid compound, a novel and potent inhibitor. Structure 5 (11), 1413–1425.

Chapter 3

Identification and Extraction of Antidiabetic Antioxidants From Natural Sources

K. Rashid, P.C. Sil
Bose Institute, Calcutta, West Bengal, India

1. INTRODUCTION

Diabetes mellitus is considered as an incurable endocrine degenerative and chronic metabolic disorder. No report depicted complete recovery of any patients from this chronic clinical state (Malviya et al., 2010, p. 113; Li et al., 2004a, p. 1). On the other hand, the rapid increase in the global incidence of diabetes has already recognized it as a life-threatening disease to the people from the different regions of the world. In 2013, millions of cases of diabetes have been recorded and in 2035 a huge rise is expected in the occurrence of this disease as a result of its continuous annual increment (Mohammed et al., 2015, p. 78). It occurs due to the relative or absolute deficiency of insulin secretion and/or insulin action that causes chronic hyperglycemia and impaired metabolism of different biomolecules (proteins, lipids, carbohydrate, and nucleic acids) (Modak et al., 2007, p. 164). The hyperglycemic state is a precondition for the initiation and progression of diabetic complications and is accompanied by an increase in the production of free radicals (Baynes, 1991, p. 405; Oberley, 1988, p. 113). Diabetic pathophysiological state arises as a result of an imbalance between free radical formation and neutralization of the body using inbuilt natural antioxidants (Rashid et al., 2013b, p. 585). The present modalities available for the treatment of diabetes and its associated complications include insulin therapy, transplantations, and various medications (in the form of sulfonylurea, biguanide, α-glucosidase inhibitors, etc.). These therapeutic strategies are restricted due to their limited actions, pharmacokinetics property, secondary failure rate, serious side effects, and being expensive to most of the people (Asche et al., 2008, p. 638; Jacobsen et al., 2009, p. 145). Moreover, World Health Organization (WHO) has depicted that only protection from the diabetic state is not the major concern for future but maintenance of individual's good health is also essential

Discovery and Development of Antidiabetic Agents From Natural Products.
http://dx.doi.org/10.1016/B978-0-12-809450-1.00003-X

(Malviya et al., 2010, p. 113; Demain and Sanchez, 2009, p. 5). Therefore, during the last few years, considerable attention has been given on the identification and isolation of some of the traditional formulations from plants as they contain various bioactive components with good antidiabetic property having negligible side effects compared to other hypoglycemic agents used in regular clinical practice. Hence, these components exhibit a bright future in the therapy and management of diabetes. Keeping the fact in mind that plants are biodegradable, safe, and cost-effective with limited side effects, the necessity of screening different medicinal plants with hypoglycemic effects in diabetes gains momentum (Rizvi and Mishra, 2013, p. 1). Using the antioxidant property of most of these plants may lower blood glucose level, mitigate metabolic abnormalities, and delay the development of diabetes-related complications. In this book chapter, we would like to present the role of some active ingredients [e.g., curcumin, mangiferin, and arjunolic acid (AA)] of well-regarded medicinal plants that have antidiabetic and antioxidant potential on the current status of diabetes. Along with their extraction and identification from natural sources, potential molecular mechanism of action of these active components has also been discussed elaborately herein. Antioxidant potential of these molecules was assessed using some assays that include ferric-reducing antioxidant power (FRAP), 2,2-diphenyl-1-picryl hydrazyl (DPPH), nitric oxide (•NO), and superoxide (O_2^{-}) scavenging and lipid peroxidation inhibition assays; whereas antidiabetic activity was investigated by α-amylase and α-glucosidase inhibition tests.

2. CURCUMIN: EXTRACTION, IDENTIFICATION, AND ANTIDIABETIC–ANTIOXIDANT POTENTIAL

The gold-colored spice turmeric is obtained from the rhizome of the plant *Curcuma longa* L. (Zingiberaceae) cultivated in the tropical climate (Aggarwal et al., 2007, p. 1; Zhang et al., 2013, p. 1). It is consumed as a food ingredient in Iran; also used in Indian, Chinese, Polynesian, Malaysian, and Thai curries, and in the preparation of western mustard (Agrawal and Chakrabarti, 2010, p. 30; Ghorbani et al., 2014, p. e18081; Gupta et al., 2013, p. 1510). Since 1900 BC, this component has been applied for therapeutic purposes for more than 2500 years for the treatment of various inflammatory and oxidative stress-mediated disorders in the Asian countries (Gupta et al., 2013, p. 1510). The use of turmeric in Western countries to prevent and treat arthritis has also been adopted due to its traditional applications against a number of diseases. In Ayurvedic and traditional medicinal practices, this component has also been applied to treat diabetes and its associated complications for thousands of years (Aggarwal et al., 2007, p. 1). The most important chemical components of turmeric include curcumin (diferuloylmethane), desmethoxycurcumin (DMC), and bisdesmethoxycurcumin (BDMC), altogether which are called curcuminoids. Curcumin, the best-studied active component of turmeric, constitutes almost

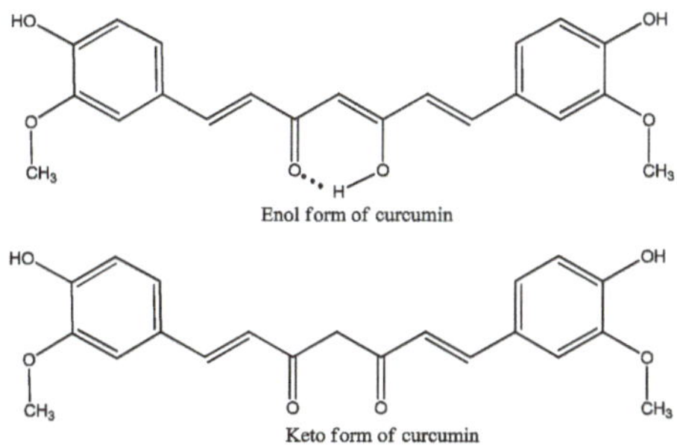

FIGURE 3.1 Enol and keto form of curcumin

3.14% of powdered turmeric. This polyphenolic compound is responsible for the yellow color of turmeric and is quite stable in the acidic pH of stomach (Hatchera et al., 2008, p. 1631; Jurenka, 2009, p. 141). It was first isolated almost two centuries ago, and its structure as diferuloylmethane was determined in 1910 by Lampc and Milobedeska (Jurenka, 2009, p. 141). Extensive research within the last half century has proven that most of the therapeutic activities assigned to turmeric are associated with the presence of curcumin. Therefore, this principal curcuminoid of turmeric is considered as its most active constituent (Perez-Torreset., 2013, p. 25). It can exist in its keto–enol tautomeric forms (Fig. 3.1), and the enol form is more energetically stable both in the solid and solution phase. This pharmacologically active component has been shown to exhibit a wide spectrum of biological activities, which include its antiinflammatory, antioxidant, anticarcinogenic, antimutagenic, anticoagulant, antidiabetic, antibacterial, antifungal, antiprotozoal, antiviral, antivenom, antiulcer, anti-Alzheimer's, and hypocholesterolemic activity; thus, it has a potential against various malignant diseases, diabetes, allergies, arthritis, Alzheimer's disease, and other chronic illnesses (Chattopadhyay et al., 2004, p. 44; Ono et al., 2004, p. 742; Shehzad and Lee, 2010, p. 113). On the other hand, being an inexpensive and relatively safe drug, curcumin has attracted the attention of the scientific world for the treatment of diabetic pathophysiology (Goel et al., 2008, p. 787; Shehzad et al., 2011, p. 151; Chuengsamarn et al., 2012, p. 2121). Evidence implicates hyperglycemia-derived oxygen free radicals as a mediator of diabetic complications (Kowluru et al., 2001, p. 1938; Aldebasi et al., 2013, p. 194). The inflammatory damage that characterizes type 1 diabetes is mediated at least in part through islets reactive oxygen species (ROS), and in type 2 diabetes, the high nutrient flux and consequent ROS production appear to mediate loss of β-cell function. In insulin-sensitive tissues (including muscle, liver, and heart), high fatty acid flux results in oxidative damage whereas non-insulin-sensitive tissues (including the eye, kidney, and nervous system) undergo damage on exposure to both high circulating glucose and fatty acid levels (Sivitz et al., 2010, p. 552). ROS-induced damages consequently result in diabetic complications and damage to different organs. A number of molecular pathways, like increased intracellular advanced glycation end products (AGEs) formation, protein kinase C (PKC) activation, enhanced activity of hexosamine pathway, increased polyol pathway flux, etc., are emphasized as being major contributors of hyperglycemia-induced ROS production (Bierhaus et al., 1998, p. 586; King et al., 1997, p. S77; Brownlee, 2005, p. 1616; Williamson et al., 1993, p. 801). The antidiabetic activity of curcumin is attributed to its antioxidant and antiinflammatory properties. Being a potent antioxidant, curcumin could scavenge free radicals, which are considered as a crucial factor for the management of diabetes. The paper describing the blood glucose lowering effect of curcumin in a diabetic patient was first published in 1972 (Srinivasan, 1972, p. 269). Curcumin has since been extensively studied in experimental animal models of diabetes and found to be effective in reducing hyperglycemia and hyperlipidemia. It is also applied in a few clinical trials of type 2 diabetes to treat its complications (Sahebkar, 2013, p. 197). Hypoglycemic effect of this molecule is mediated through the regulation of various signaling pathways, which includes transcription factors, growth factors, inflammatory cytokines, protein kinases, and other enzymes.

2.1 Extraction and Estimation of Curcumin

2.1.1 Extraction of Curcuminoid

The main coloring substance in *C. longa* is curcumin. A combination of curcumin itself and two other related compounds, namely, DMC and BDMC is commonly known as curcuminoid. Researchers from various groups conducted different experiments for the extraction, isolation, and purification of curcuminoids from *C. longa*. Curcumin was purified from curcuminoid applying a number of suitable chromatographic processes and finally its purity and identification were carried out using different physicochemical techniques (Waghmare et al., 2015, p. 111; Revathy et al., 2011, p. 22). In this part of the book chapter, we would like to discuss some commonly accepted procedures for the extraction, purification, identification, and quantitation of curcumin.

2.1.1.1 Solvent Extraction

A common step for the extraction of curcumin is the washing of the fresh *C. longa* rhizomes with water (deionized preferred), followed by slicing and drying in the sun for a long time (approximately 1 week or so). Additional final drying is done in a hot air oven for a few hours at approximately 50°C. These dried slices are then cut into small pieces, followed by powdering (in an electronic mill) and subjected to solvent extraction. Based on the varying polarity, different solvents like chloroform, ethyl acetate, methanol, acetone, etc., are used for the extraction of curcuminoids from turmeric rhizomes. Here, it should be mentioned that the procedure is set up with various solvents in such a way so that the processes change from nonpolar to polar. After completion of the extraction, a dark brown extract is obtained. It is cooled and further evaporation is carried out to obtain a high concentration of the sample by using a rotary evaporator. The bath temperatures are usually maintained at 40, 50, 60, and 70°C and the rotation speed is maintained at 90 rpm. Sometimes one more step is followed by vacuum suction to enhance its concentration further. The crude dried extract, so obtained from the evaporator, is usually black orange in color. After completion of all these steps for each extract, the total yield is determined and percentage yield of curcuminoid in the extract is calculated by spectrophotometric analysis using a suitable spectrophotometer at 420 nm. If needed, the extract can also be subjected to the thin layer chromatography (TLC) to have the respective R_f values for curcumin, DMC, and BDMC (Waghmare et al., 2015, p. 113).

2.1.1.2 Maceration Extraction

In this extraction procedure, the dried powder (prepared as mentioned in the solvent extraction) is added with ethanol (70%) and is allowed to be mixed thoroughly on a shaker with 210 rpm at room temperature for a long time (2 days). The extract so obtained is filtered through Whatman no. 1 filter paper. The extraction procedure is repeated by adding the other portions of the solvent to the solids until a clear solution is obtained. This process helps to remove the powder, but the filtrate does not become pure enough as it also contains the other impurities and ethanol. The combined filtrates are concentrated using a rotary evaporator under reduced pressure. The crude extract so obtained is heated on a boiling water bath. This procedure is allowed to continue until a constant weight of the extract is obtained. It was then analyzed and curcuminoid is collected (Waghmare et al., 2015, p. 115).

2.1.1.3 Microwave-Assisted Extraction

For microwave-assisted extraction (MAE) of the curcuminoid, the powdered samples are mixed thoroughly with a suitable modifier as needed for the specific experiment. Sufficient time is allowed for the powder to absorb the modifier and get saturated. The saturated powder is then placed into the extraction vessel, and an appropriate amount of the extracting solvent is added. Different time of irradiation with the microwave extractor operating at different power levels is needed for MAE. The sample is treated in an intermittent way, i.e., irradiation–cooling–irradiation under microwave maintaining a particular ratio of irradiation and cooling time and cycles. The samples are then centrifuged; the supernatant is filtered, concentrated under vacuum, dissolved in a solvent (usually methanol), and quantified by high-performance thin layer chromatography (HPTLC) (Waghmare et al., 2015, p. 116).

2.1.1.4 Sonication Extraction

For this extraction procedure, an ultrasonic cleaning bath and a double jacket cell are usually used. The cell is maintained at a constant temperature (usually 25°C) by circulating water from a thermostat bath so that the rising temperature caused by ultrasonic exposure is avoided. The samples are filtered, washed with the solvent, evaporated to dryness using a rotary evaporator, and diluted appropriately to inject to high-performance liquid chromatography (HPLC) for further analysis. To obtain total curcuminoids, the optical density of the samples at 420 nm was measured (after certain dilution) using a suitable spectrophotometer (Waghmare et al., 2015, p. 116).

2.1.1.5 Slurry Extraction

For this extraction procedure, turmeric powder is mixed with acetone and stirred for 2 days. The solution is then filtered on a silica filter and the filtrate is concentrated. The extract so obtained is then subjected to a column chromatography for separation and further purification of the components (Waghmare et al., 2015, p. 117).

2.1.2 Estimation of Curcuminoid

2.1.2.1 Spectrophotometric Analysis

A certain amount of pure curcumin was dissolved into methanol and water such that an appropriate concentration is reached and then absorbance is taken at 420 nm. A standard graph is constructed with the known concentrations against respected absorbance. To find out the concentration of extracted samples by using the same spectrophotometer, fixed amount of the samples is dissolved in the same solvent (used as a standard solution) and absorbance is determined at 420 nm (as all three components of curcuminoid have λ_{max} at 420 nm). The concentration of each preparation of curcuminoid is then calculated from the standard graph.

2.1.2.2 Separation of Curcuminoids by Thin Layer Chromatography

For this separation technique, usually methanol extracts are tested on TLC for the presence of different curcuminoid. Thin layer chromatographic plates are made up of silica gel. Plates are presaturated with mobile phase in a glass tank and are developed up to a certain height of the solvent front. After development, plates are removed from the tank, dried, and the spots are analyzed. If the resolution of separation is achieved using chloroform and methanol at the ratio of 95:5 as the mobile phase, the R_f values of curcuminoids were reported to be 0.75, 0.55, and 0.27, for curcumin, DMC, and BDMC, respectively, and this better resolution of R_f value suggests that for the separation of compounds in column chromatography, chloroform and methanol can be used as a suitable solvent mixture (Revathy et al., 2011, p. 24).

2.1.3 Separation of Individual Components of Curcuminoid

Usually, TLC, HPTLC, and column chromatography (CC) are used to separate the individual components of curcuminoid pigments. The stationary phase most commonly used for this separation purpose is silica gel although the mobile phase varies depending on the necessity of chromatographic separation. Different solvent systems consisting of the mobile phase include ethanol, chloroform, acetic acid, hexane, benzene, ethyl acetate, methanol, etc. Among these techniques, HPLC method is found to be highly sensitive, precise, and accurate for detection and quantification of individual components of curcuminoid. Here, separation is mostly done using reverse phase principle in a suitable column employing mixtures of water, acetonitrile, ethanol, and methanol as separating mobile phase. A chromatographic analysis is done on HPLC system consisting of a suitable injector with a matching loop. The sample is injected and an ultraviolet detector is used at 425 nm, for detection of curcumin. Then the chromatograms are processed by a suitable chromatography management system, and the amount of curcumin in different samples is quantified using the computer software. The curcumin content in different varieties of turmeric is determined by using a comparison between the peak areas of experimental samples and the area of standard curcumin.

2.2 Curcumin as an Antioxidant

An antioxidant should possess two fundamental characters: firstly, the compound should be present in fewer amounts as compared to the substrate to be oxidized and secondly, the species resulting from its oxidation must be stable through intramolecular hydrogen bonding (Halliwell, 1990, p. 1). Curcumin, the active lipid soluble principle of turmeric, contains a heptadiene chain with two terminal benzene rings; both of the rings possess a symmetrically attached hydroxyl and methoxy group. Curcumin is considered as a phenolic antioxidant due to an important role of the benzene ring-attached hydroxyl groups in the antioxidant activity (Agnihotri and Mishra, 2011, p. 14221). Earlier reports suggest that the remarkable antioxidant property of curcumin is due to its ability to scavenge ROS (includes free radical oxidants and molecular oxidants) and to inhibit lipid peroxidation. Curcumin has three reactive functional groups: one diketone moiety and two phenolic groups. The free radical scavenging activity is correlated with either phenolic group or —CH_2 group of β-diketone moiety (Nimse and Pal, 2015, p. 27996). Free radicals depending upon the pH of the medium are able to undergo electron transfer or abstract H atom from either of these two groups. Using laser flash and pulse radiolysis techniques, a group of scientists demonstrated that during antioxidant mechanism curcumin involves H-atom abstraction mainly from the central —CH_2 group (in the chain of the keto form) whereas only 15% of H-abstraction takes place from the phenolic group in the whole reaction process (Jovanovic et al., 2001, p. 3064). In contrast, the study of another group revealed the involvement of phenolic group that donates H-atom for the neutralization of peroxyl radical (Barclay et al., 2000, p. 2841). Lately, a study showed that the antioxidant activity of lipophilic curcumin increased in the polar media (cytoplasm) though the mechanism is not clearly known (Litwinienko and Ingold, 2005, p. 8982). However, based on the theoretical study, Sun et al. (2002) suggested that the antioxidant property of curcumin is attributed to the H-atom abstraction from the phenolic group rather than the central —CH_2 group (Sun et al., 2002, p. 2909).

The mechanism of autooxidation of curcumin has been shown when H-atom is abstracted by the free radicals either from the methylenic moiety or one of the phenolic hydroxyl groups of curcumin (Fig. 3.2A and B, respectively).

The antioxidant potential of curcumin was evaluated by applying various in vitro assays that include, ferric-reducing power assay, free radical scavenging activity by DPPH (a powerful free radical and used to detect the electron-donating capacity of any antioxidant), hydroxyl radical scavenging activity, nitric oxide radical scavenging activity (excess concentration of NO is associated with several diseases), etc. Curcumin scavenges free radicals in a concentration-dependent manner and inhibits the production of hydroxyl and nitric oxide free radicals (Satapathy and Panda, 2013, p. 2808). Curcumin also inhibits superoxide radical by mimicking superoxide dismutase (Mishra et al., 2004, p. 355).

In addition, curcumin is also able to react with molecular oxidants (peroxynitrite, hydrogen peroxide, etc.). In several biological systems, curcumin has been observed to protect cells from excessive production of these oxidants. Some reports in the literature depicted the direct reaction of curcumin with peroxynitrite (Kim et al., 2003, p. 481; Iwunze and McEwan, 2004, p. 749). Curcumin shows its powerful antioxidant activity against peroxynitrite-induced oxidative stress-mediated damage in diabetes by preventing nitrotyrosine formation.

2.3 Curcumin as an Antidiabetic Agent

Diabetes, characterized by abnormally high plasma glucose levels, results in the cellular stress-mediated major complications. The most effective therapeutic approaches for the treatment of diabetes and its associated complications are the inhibition of digestion of carbohydrates and absorption of carbohydrates in the small intestine (Najafian, 2015, p. 29). That can happen in two manners either by attenuating the activity of carbohydrate-hydrolyzing enzymes (α-amylase and α-glucosidase) in the digestive organs or by inhibiting the absorption of glucose via the sugar transporter systems, sodium/glucose cotransporters (SGLT1 and SGLT2) (Najafian et al., 2010, p. 553, 2011, p. 1; Kumar et al., 2011, p. 19). α-Amylase catalyzes the hydrolysis of α-(1,4)-D-glycosidic linkages of starch and other glucose polymers (glycogen) to yield glucose and maltose. In addition, an exotype membrane-bound carbohydrase α-glucosidase (α-D-glucosideglucohydrolase) is widely distributed in animal tissue (Kimura et al., 2004, p. 1035). It is mainly present in the epithelium of small intestine where it facilitates the absorption of glucose from the small intestine. It catalyzes the hydrolytic cleavage of oligosaccharides from the nonreducing end to produce absorbable α-glucose (Irfan, 2002, p. 1). Inhibition of these two enzymes helps to spread the carbohydrates to the lower part of the small intestine that in turn results in the reduction of the rate of digestion and absorption of carbohydrates. This mechanism lowers the postprandial rise of blood glucose level and helps in avoiding diabetic complication-associated damages (Irfan, 2002, p. 1). Natural

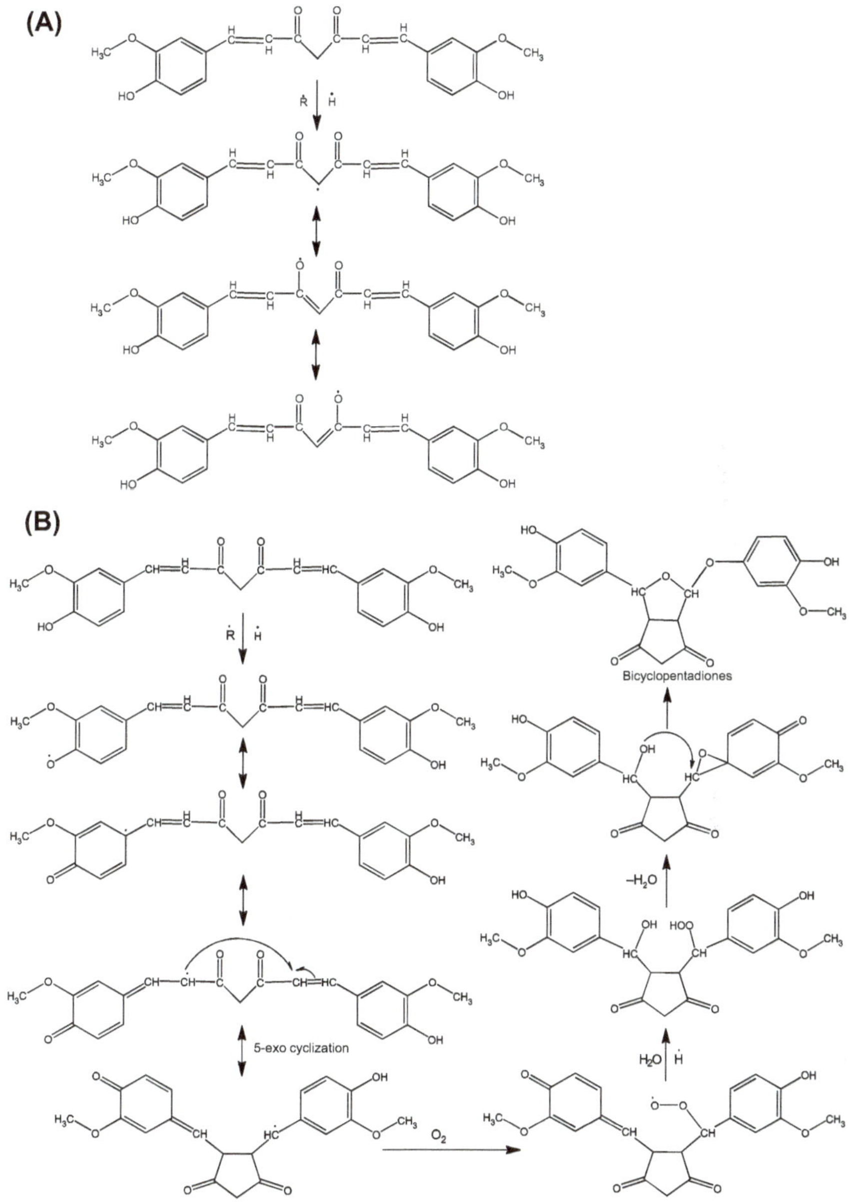

FIGURE 3.2 A) Mechanism of free radical scavenging activity of curcumin initiated by methylenic moiety, (B) mechanism of free radical scavenging activity of curcumin initiated by phenolic moiety.

curcumin acts as a competitive inhibitor of α-amylase and α-glucosidase with IC_{50} values about 51.32 and 23.0 μM, respectively. Curcumin reduces blood glucose level by interacting with these enzymes and also elevates insulin concentration in serum. Therefore, curcumin could be used in the treatment of diabetes.

2.4 Curcumin as an Antiinflammatory Agent

Earlier evidence suggests that persistent oxidative stress leads to chronic inflammation that in turn results in various chronic diseases including diabetes and its related complications (Reuter et al., 2010, p. 1603). Nuclear factor kappa B (NF-κB) and tumor necrosis factor α (TNF-α) pathways are located at the center of the inflammatory response. ROS plays an important role in the activation and enhancement of the inflammatory response by acting on both upstream and downstream molecules of these signaling pathways. The three loops that are involved in the amplification of reactive species formation includes NF-κB–TNF-α positive feedback loop, redox-sensing loop by ROS–NF-κB–TNF-α, and ROS-modified proteins generated loop. The final loop among the three promotes the autoimmune response by feeding back into loops 1 and 2 (Ishibashi, 2013, p. 6375; Debnath et al., 2013, p. 7253). Hydroxyl radical shows the most damaging effect to cellular biomolecules and is responsible for the activation of all these inflammatory loops. Curcumin is able to block the function of these different looping either by scavenging hydroxyl radical directly or by NF-κB pathways. Curcumin reduces the production of TNF-α, and on binding with this molecule it blocks TNF-α-mediated signaling associated with a number of expressions of various inflammatory cytokines in different cells (Chang et al., 1997, p. 779; Anthwal et al., 2014, p. 1; Gupta et al., 2014, p. 91). These cytokines have a causal association with high fasting glucose, decrease insulin sensitivity, insulin resistance (IR), type 2 diabetes, and cardiovascular disease (Asghari et al., 2012, p. 368). The antiinflammatory agent, curcumin is also able to repress the release of different interleukins and interferons (e.g., IL-1β, IFN-γ, etc.). Several reports depicted the reciprocal regulation of nuclear erythroid 2-related factor 2 (Nrf-2) and NF-κB in inflammation. Thus, Nrf-2 also plays a vital role in inflammatory response through the regulation of different molecules [mitogen-activated protein kinase (MAPK), PKC, phosphatidylinositol 3-kinase (PI3K), etc.] (Cordero et al., 2015, p. 597; Lin et al., 2015, p. 203). Upon binding with Nrf-2, curcumin stimulates the expression of HO-1, which is considered as one the mechanism of Nrf-2's antioxidant-dependent antiinflammatory activity. Moreover, it modulates signaling through the JAK/STAT pathway that is involved in the negative regulation of inflammatory interleukins (IL-1, IL-2, IL-6, IL-8, and IL-12) and cytokines [TNF-α and monocyte chemoattractant protein-1 (MCP-1)]. Furthermore, this active constituent, downregulates the expression of several inflammatory signal-responsive enzymes [cyclooxygenase-2 (COX-2), lipoxygenase, xanthine oxidase, inducible nitric oxide synthase (iNOS), etc.], that in turn results in attenuation of STAT-3 phosphorylation and its subsequent translocation to nucleus (Grynkiewicz and Slifirski, 2012, p. 206; Shishodia et al., 2005, p. 209; Sahebkar, 2013, p. 201; Soetikno et al., 2013, p. 760). Some studies have also demonstrated that the beneficial antiinflammatory effect of curcumin is mediated through the upregulation and the activation of peroxisome proliferator-activated receptor γ (PPAR-γ) activation (Jacob et al., 2007, p. 1).

From the in vitro study with human monocytes (THP-1 cells), Yun et al. (2011) showed that curcumin suppresses the secretion of IL-6, MCP-1, and TNF-α in response to high glucose (Yun et al., 2011, p. 450). In addition, they demonstrated that curcumin helps in the attenuation of acetylation of histone proteins by increasing the activity of histone deacetylases (HDACs), decreasing the activity of histone acetyltransferase (HAT), and reducing the expression of p300 and acetylated CBP/p300. High glucose accelerates the production of cytokines via epigenetic changes and being an epigenetic modification of histone acetylation is regulated through the opposing actions of HATs and HDACs. Curcumin protects from diabetic complications by contributing to epigenetic modifications through the regulation of HATs and HDACs (Pham and Lee, 2012, p. 1868). In consistence with the previous data, Jain et al. have found that curcumin can reduce the secretion of cytokines in in vitro study with U937 monocytes along with IL-8, which was supported by their in vivo study with diabetic streptozotocin (STZ)-induced Sprague–Dawley rats. They observed significantly diminished levels of IL-6, MCP-1, TNF-α, glucose, and oxidative stress in the experimental models (Jain et al., 2009, p. 241).

2.5 Mechanism of Curcumin-Induced Protective Action in Diabetes

Diabetic patients have an increased risk of developing various serious health problems. Unlike physiological glucose concentration, chronic superphysiological glucose concentration in diabetes adversely affects the structure and function of a large number of organs (e.g., pancreas, kidneys, heart, nerves, liver, spleen, feet, eyes, testes, etc.) and tissues via cellular stress (oxidative stress, endoplasmic reticulum (ER) stress, nitrosative stress, and inflammatory responses). In this section, we would like to discuss the adverse effect of cellular stress on pancreas, liver, and testes. In addition, we have offered emphasis on the molecular and cellular mechanism of action of curcumin to provide protection to these organs in diabetes.

2.5.1 Protective Action on Pancreatic Damage

Type 1 and type 2 both kinds of diabetes are associated with a reduction in the pancreatic β-cell mass, which is responsible for the secretion of insulin. Glucose enters into the β-cell via glucose transporter 2 (GLUT-2) and undergoes oxidation to produce ATP. Increased ATP/ADP ratio is responsible for the closure of ATP-sensitive potassium channel (Serrano-Martín et al., 2006, p. 4214). The closing of potassium channel is associated with membrane depolarization and opens voltage-gated Ca^{2+} channel. Ca^{2+} enters inside the pancreatic β-cells through this channel to stimulate insulin secretion. Oxidative and ER stresses have been observed as crucial factors for the damage of pancreatic β-cells (Marhfour et al., 2012, p. 2417). Cellular stress (oxidative and ER stress) is able to exert ill effects on β-cells due to low-antioxidant defense machinery and exposition to the toxic substances in daily life (Lenzen et al., 1996, p. 463). β-Cells of pancreatic islets contain active ER for the folding, export, and processing of newly synthesized insulin. Excessive Ca^{2+} release from this active organelle, hyperglycemia, oxidative stress, and/or misfolded mutant insulin proteins are responsible for the loss of ER homeostasis, which results in an adaptive unfolded protein response (UPR). UPR helps ER to mitigate stress and increase its folding capacity (Ozcan et al., 2004, p. 457). Protective action of curcumin on pancreatic β-cells has been extensively evaluated. Curcumin helps in the increment of islets viability and reduction of the reactive species formation. Some in vitro as well as in vivo studies showed that curcumin exerts its protective effects on islets by increasing the activation of HSP-70 and upregulation of Nrf-2-dependent HO-1 expression (Kanitkar and Bhonde, 2008, p. 182; Rashid and Sil, 2015b, p. 297). Protection of curcumin mediated through antioxidant defense mechanism (HO-1, glutathione, NAD(P)H:quinoneoxidoreductase 1) helps it and its analogues to increase basal insulin secretion in human islets and improves the outcome of islets transplantation (Balamurugan et al., 2009, p. 454). It suppresses the inflammation in the pancreas by inhibiting cytokine (TNF-α, IL-1β, and IFN-γ)-induced NF-κB translocation into the nucleus and activation of downstream signaling molecules (NO and iNOS) without harming normal islet function(Rashid and Sil, 2015b, p. 297). This active component helps in the translocation of GLUT-2 to the membrane and normalizes NO-dependent release of Ca^{2+} from the ER (Kanitkar et al., 2008, p. 702). In addition, Rashid and Sil (2015b) showed that by downregulating the concentration of intracellular Ca^{2+} and the expression of caspase-12, curcumin attenuates ER-dependent cell death (Rashid and Sil, 2015b, p. 304). It has also been found that administration of curcumin results in diminished infiltration of lymphocytes to islets (Chanpoo et al., 2010, p. S152). Pancreatic islets regeneration and insulin secretion from β-cells were enhanced due to the inclusion of curcumin in transplantation of bone marrow (El-Azab et al., 2011, p. 41). Curcumin also takes part in the opening and activation of anion channels. Curcumin-mediated depolarization of the membrane helps in the generation of electronic activity and subsequently release of insulin (Best et al., 2007, p. 1768). Moreover, curcumin reduces type 2 diabetes-associated human islet amyloid polypeptide (h-IAPP) fibril formation and aggregation by involving two mechanisms. First mechanism is the modulation of self-assembly of h-IAPP by unfolding α-helix and second mechanism is attenuation of MCP-1-dependent amylin mRNA expression (Daval et al., 2010, p. 118; Sparks et al., 2012, p. 551; Cai et al., 2011, p. 1). Antiapoptotic action of curcumin also helps it to rescue the viability of β-cells containing pancreatic islets in diabetes (Meghana et al., 2007, p. 183). Protection exerted by curcumin on pancreatic β-cells in hyperglycemia-mediated oxidative state has been briefly summarized in Fig. 3.3. Altogether, protective action of curcumin ameliorates hyperglycemic state in diabetic subjects.

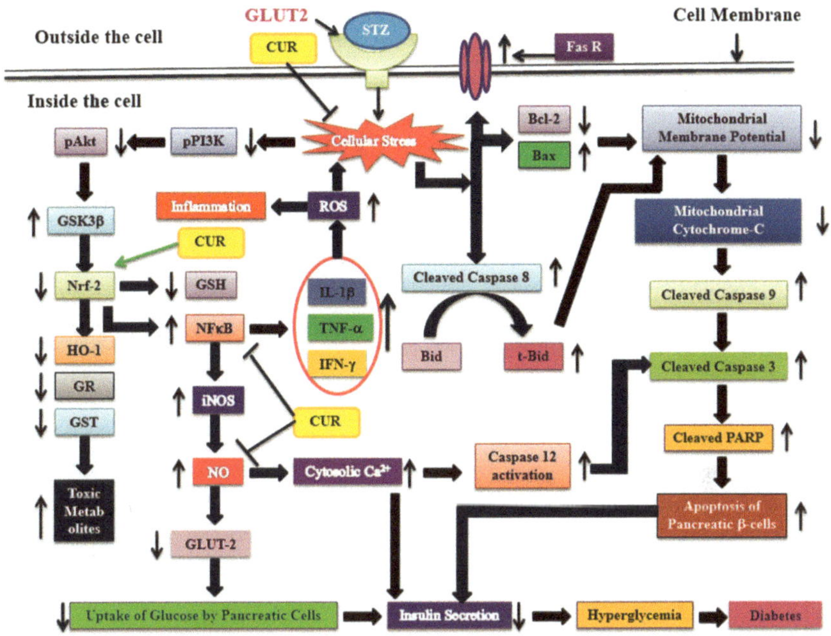

FIGURE 3.3 Cell signaling mechanism of cellular stress-mediated pancreatic damage in streptozotocin (STZ)-induced diabetic rats and protective action of curcumin. STZ increases cellular stress by accelerating the production of reactive oxygen and nitrogenous species that in turn modulates the expression of different proteins and their downstream signaling molecules to induce an inflammatory response as well as inhibiting intracellular antioxidant defense mechanisms. Enhanced cellular stress also alters the expression of pro- and antiapoptotic proteins, which promote mitochondrial pore formation and release of apoptotic factors through this pore to the cytosol. These events finally lead to apoptotic pancreatic β-cell death via extrinsic as well as intrinsic pathways to reduce insulin secretion and resulting hyperglycemia. Treatment with curcumin effectively alters these adverse effects to protect pancreatic β-cells from cellular stress-mediated damage. *GLUT-2*, glucose transporter 2; *GST*, glutathione S-transferase; *iNOS*, inducible nitric oxide synthase; *Nrf-2*, nuclear erythroid 2-related factor 2; *ROS*, reactive oxygen species; *TNF-β*, tumor necrosis factor-β. *Reprinted from Rashid, K., Sil, P.C., 2015. Curcumin enhances recovery of pancreatic islets from cellular stress induced inflammation and apoptosis in diabetic rats. Toxicology and Applied Pharmacology 282 (3), 297–310, with the permission from Elsevier.*

2.5.2 Protective Action on Liver Damage

Several liver disorders including fatty liver are found in most of the diabetic individuals (Babu and Srinivasan, 1995, p. 13). A different group of scientists reported that treatment with curcumin for 8 weeks diminished the level of excreted albumin, creatine, and inorganic phosphates in STZ-exposed diabetic animals (Babu and Srinivasan, 1995, p. 13; Ghosh et al., 2015, p. 368). Curcumin helps in the reduction of liver weight, the increment in body weight, inhibition of the release of the hepatic damage marker enzymes (ALT and ALP), diminution of oxidative stress-related markers (lipid peroxidation and protein carbonylation), and NO production in the hepatic tissue and plasma in diabetic rats.

Other group of investigators also depicted that curcumin showed hypolipidemic effect in diabetic subjects by regulating the activity of cholesterol-7α-hydroxylase (Babu and Srinivasan, 1997, p. 169; Patumraj et al., 2006, p. 481; Mahesh et al., 2004, p. 639; Gutierres et al., 2012, p. 440). Supplementation with curcumin improved lipidemia by the reduction of lipid load (includes total lipid, cholesterol, triglyceride, and low density lipoprotein cholesterol) and induction of the activity of adipogenesis-associated enzyme PPAR-γ (Yousef et al., 2008, 3506; Nishiyama et al., 2005, p. 959; Kuroda et al., 2005, p. 937; Deng et al., 2011, p. 30723). Amelioration of lipidemia is correlated with restoration of enzymes related to lipid peroxidation, glucose metabolism, redox balance (includes superoxide dismutase, catalase, glutathione peroxidase, etc.), glucose regulation in liver (glucose-6-phosphatase, phosphoenolpyruvate carboxykinase, etc.), and lipid regulation in hepatic tissue (fatty acid synthase, 3-hydroxy-3-methylglutaryl coenzyme reductase, and acyl-CoA:cholesterol acyltransferase) (Mahesh et al., 2004, p. 639; Gutierres et al., 2012, p. 440; Seo et al., 2008, p. 995). Increased expression of glucose-6-phosphatase and phosphoenolpyruvate carboxykinase (enzymes involved in hepatic gluconeogenesis) in type 2 diabetes showed an adverse impact on patients. Curcumin attenuates these two enzymes and increases the phosphorylation of AMP-activated protein kinase (AMPK) in H4IIE rat hepatoma and Hep3B human hepatoma cells to control glucose homeostasis of the body through liver (Franckhauser et al., 2006, p. 273; Kim et al., 2009, p. 377; Fujiwara et al., 2008, p. 185). AMPK helps in the glucose homeostasis by increasing glucose uptake through the liver and suppressing gluconeogenesis in that tissue in type 2 diabetic patients (Schultze et al., 2012, p. 11). Type 2 diabetes-associated hyperleptinemia results in hepatic fibrosis-mediated activation of hepatic stellate cells (HSCs). Curcumin activates AMPK, which is involved in fatty acid oxidation and lipogenesis. This molecule serves as a sensor of cellular energy homeostasis and results in the inactivation of HSCs due to the decreased stimulation by leptin, insulin, hyperglycemia, AGEs, and oxidized low density lipoprotein (ox-LDL) (Tang and Chen, 2010, p. 4168; Lin and Chen, 2011, p. 160; Lin et al., 2012, p. 2212; Kang and Chen, 2009, p. 1275). The antiinflammatory and antilipolytic properties of curcumin are responsible for its lipid accumulation lowering property in liver tissue in high fat diet (HFD)-induced experimental animals. It helps in the suppression of TNF-α, plasma free fatty acid (FFA), and serum fetuin-A (involves in insulin resistance and fatty liver) (El-Moselhy et al., 2011, p. 1129; Xie et al., 2012, p. 3; Oner-Iyidogan et al., 2013, p. 677; Haukeland et., 2012, p. 503; Stefan et al., 2006, p. 853). In addition, curcumin has also found to protect hepatic cells from oxidative stress-induced cell death in diabetes by attenuating the expression of p53, stress-induced MAPKs, and mitochondria-dependent apoptosis-inducing proteins (Ghosh et al., 2015, p. 365). Acute coronary syndrome patients have been observed to possess a low level of total cholesterol and LDL cholesterol after oral consumption of curcumin at a dose of 45 mg/day for 2 months (Alwiet al., 2008, p. 201).

2.5.3 Protective Action on Testicular Damage

Earlier reports suggest that in diabetic patients, prolonged hyperglycemic state resulting oxidative and ER stress leads to infertility/subfertility and causes damage to the reproductive organ (Kanter et al., 2012, p. 719; La Vignera et al., 2009, p. 1; Khaneshi et al., 2013, p. 347; Wang et al., 2014, p. E14; Hajizadeh et al., 2014, p. 123; Liu et al., 2013, p. 78). Oxidation of high amount of polyunsaturated fatty acid in sperm cells results in infertility by affecting the motility, counts, and structure of the sperm cells in mammals (Aitken et al., 1989, p. 183). This pathogenesis involves hormonal deregulation in the process of spermatogenesis, an anomalous behavior of sperm cells, germ cell death, and varying the degree of the testicular lesions (Cameron et al., 1990, p. E881; Ballester et al., 2004, p. 706; Scarano et al., 2006, p. 482). Kanter et al. (2013) observed that intragastric administration of curcumin results in the amelioration of lipid peroxidation, activities of antioxidant enzymes (superoxide dismutase and glutathione peroxidase), serum testosterone level, and severe oxidative stress-mediated germ cell death in STZ-induced diabetic animals (Kanter et al., 2013, p. 1578). For the regulation of spermatogenesis and maintenance of the normal physiological state of seminiferous tubules, upregulation in testosterone level is required (Sharpeet al., 1992, p. 172). Consistent with the previous study, in 2015 Rashid and Sil showed that curcumin is also able to scavenge the ROS in testicular tissue in diabetic state and restore the intracellular redox balance by elevating GSH/GSSG ratio (Rashid and Sil, 2015a, p. 70). This active antioxidant molecule also increases the activity of different enzymes required for the normal function of testes like sorbitol dehydrogenase (provides energy to the sperm cells to perform metabolic activities), glucose-6-phosphate dehydrogenase (facilitates the hydroxylation of steroid hormones by supplying reducing equivalents), 3β-hydroxysteroid dehydrogenase and 17β-hydroxysteroid dehydrogenase (both the enzymes catalyze testicular androgenesis and help in the synthesis of testosterone) (Prasad et al., 1995, p. 484; Jana et al., 2006, p. 9). Curcumin helps in the enhancement of sperm count, sperm motility, and diminution of sperm abnormality. Additionally, histological assessment showed that treatment with curcumin at a dose of 100 mg/kg body weight daily for 8 weeks mitigates disruption of the seminiferous tubule, loss of sertoli cells, diminution of leydig cells, and absence of centrally located spermatozoa in STZ-administered rats (Fig. 3.4). This active component provides protection to testicular cells from hyperglycemia-induced apoptotic death in diabetic condition by regulating a number of signaling pathways. It suppresses the release of inflammatory cytokines (TNF-α and IL-1β), inhibits the translocation of NF-κB into nucleus and activation of downstream COX-2 protein, and attenuates subsequent inflammatory response. Curcumin also regulates the expression of antioxidant defense protein Nrf-2 and other antioxidant enzymes. In addition, it downregulates the phosphorylation of stress-responsive proteins like c-jun-NH2-terminal protein kinase (JNK) and p38. Curcumin-mediated anti-apoptotic effect is associated with the attenuation of ER-dependent apoptosis

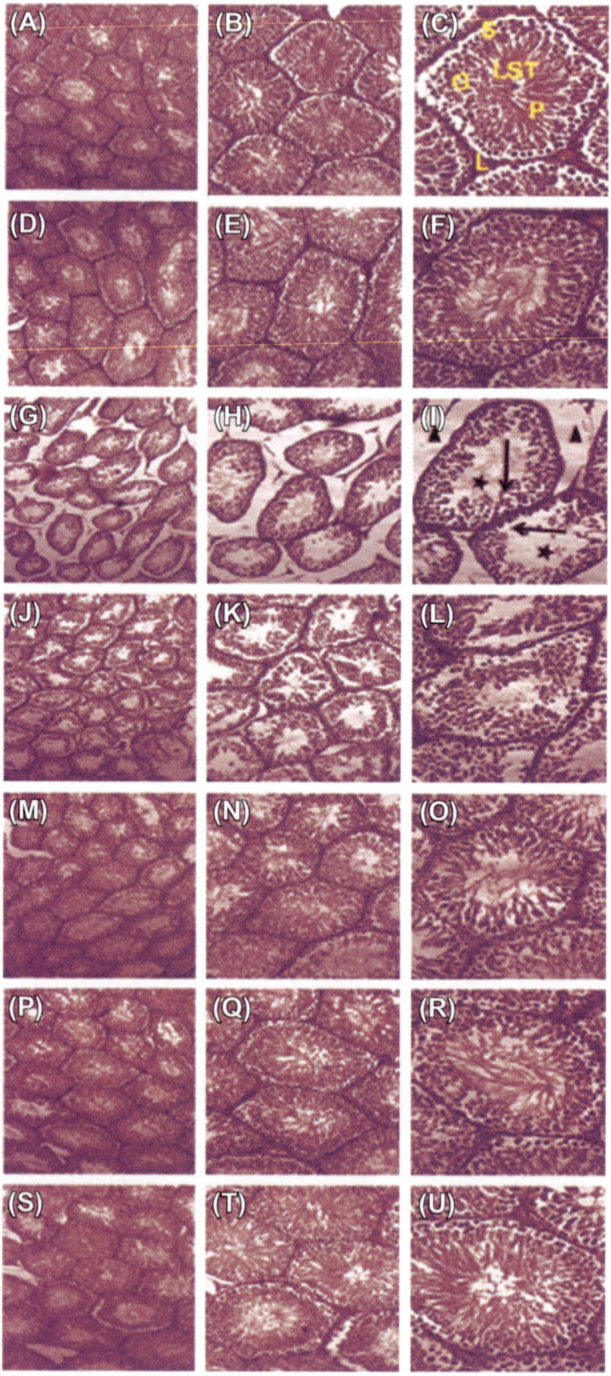
(A)
(B)
(C)
S
LST
G
P
L
(D)
(E)
(F)
(G)
(H)
(I)
(J)
(K)
(L)
(M)
(N)
(O)
(P)
(Q)
(R)
(S)
(T)
(U)

(includes increased intracellular calcium level, enhanced expression of calpain and caspase-12) and mitochondria-dependent apoptosis (involves reduction of mitochondria membrane potential, reduced expression of antiapoptotic proteins and increased expression of cleaved caspases) (Rashid and Sil, 2015a, p. 81) to provide protection to the testicular cells in diabetes.

2.6 Clinical Impact of Curcumin on Human

Administration of fermented turmeric powder in patients of 20 years or above at a dose of 3 g/day for 12 weeks results in a diminution of ALT and ALP level with normal glucose level without imparting any toxic effects (Kim et al., 2013, p. 6). Treatment of obese people with curcuminoids (300 mg/day for 3 months) results in the lowering of fasting blood glucose level, insulin resistance index, serum free fatty acids, and serum triglycerides due to the oxidation and subsequent utilization of free fatty acids (Na et al., 2013, p. 1569). In addition, curcumin supplementation in the form of capsules (six capsules each day and each of the capsule contains 250 mg curcuminoids) for 9 months to the prediabetic patients helps in reducing the number of patients who eventually developed type 2 diabetes. This effect of curcumin was due to amelioration of the function of β-cells with least side effects (Chuengsamarn et al., 2012, p. 2125). To check the impact of curcumin on 21–38 years old people with normal fasting blood glucose and cholesterol level, they were ingested with turmeric at a dose of 2.8 g/day for 4 weeks. In these people, turmeric has not shown any blood glucose and lipid-altering activity suggesting its nontoxic effect on normal individuals (Tang et al., 2008, p. 1265). Another study showed that ingestion of *C. longa* (6 g) increased the secretion of insulin without affecting glucose level in normal individuals. However, compared to different serious side effects of synthetic antidiabetic drugs, curcumin could be recommended for future clinical trials for diabetic pathophysiology.

FIGURE 3.4 Histopathological changes in testicular tissue (stained with hematoxylin and eosin dye) associated with diabetes and dose-dependent protective action of curcumin showing optimum protection at a dose of 100 mg/kg body weight. (A–C) Testicular section from control (untreated) rat testis, A (×100), B (×200), C (×400); (D–F) testicular section from animals treated with CUR-100 only, D (×100), E (×200), F (×400); (G–I) testicular section from the diabetic group, G (×100), H (×200), I (×400); (J–U) testicular section from the animals treated with CUR at a dose of 40 (J–L), J (×100), K (×200), L (×400); 70 (M–O), M (×100), N (×200), O (×400); 100 (P–R), P (×100), Q (×200), R (×400) and 130 (S–U) mg/kg body weight after diabetic induction, S (×100), T (×200), U (×400). Normal control shows spermatogonia (G), the cytoplasm of Sertoli cells (S), Leydig cells (L), lumen of seminiferous tubule (LST), and spermatocytes (P). In the STZ-intoxicated group, *triangles* indicate loss of Leydig cells, *arrows* indicate loss of Sertoli cells, and *asterisks* indicate loss of centrally located spermatozoa, whereas CUR treatment shows normal morphology. *Reprinted from Rashid, K., Sil, P.C., 2015. Curcumin ameliorates testicular damage in diabetic rats by suppressing cellular stress mediated mitochondria and endoplasmic reticulum-dependent apoptotic death. Biochimica et Biophysica Acta Molecular Basis of Disease 1852 (1), 70–82.*

2.7 Bioavailability of Curcumin

Shoba et al. (1998) and Anand et al. (2007) showed that the bioavailability of curcumin is very low, which in turn decreases its clinical application (Shoba et al., 1998, p. 353; Anand et al., 2007, p. 807). The reason behind the reduced bioavailability of curcumin includes less absorption, rapid metabolism, and quick elimination (Anand et al., 2007, p. 807). Recently, pronounced attention has been paid to increase the delivery of curcumin in cells by discovering suitable delivering system and its therapeutic efficacy in human clinical trials. Limited application of curcumin is due to its hydrophobic nature (poor solubility in water) and quick degradation in the presence of UV light (Iwunze, 2004, p. 164; Canamares et al., 2006, 1386). Till now, different delivery systems of curcumin have been discovered to increase the bioavailability of this active component that includes piperine, liposomal curcumin, nanoparticles, and complexion with phospholipids (Shaikh et al., 2009, p. 223; Kakkar et al., 2011, p. 502; Mohanty et al., 2012, p. 1347). In addition, synthesis of structural analogues has also been observed for increasing pharmacokinetics and pharmacological properties of curcumin (Bisht and Maitra, 2009, p. 192).

2.8 Toxicity of Curcumin

From ancient time, curcumin has been used in the diet and traditional medicine in the form of turmeric. Negligible reports are associated with the toxic effect of this molecule even at high doses (Chainani-Wu, 2003, p. 161). According to FAO (The Food and Agriculture Organization of the United Nations), WHO (The World Health Organization), and EFSA (European Food Safety Authority), ADI (acceptable daily intake) value of curcumin is about 3 mg/kg body weight per day (Trujillo et al., 2013, p. 450; Nabavi et al., 2015, p. 2448). Few reports suggest a high dose of curcumin exerts a teratogenic effect, embryonic toxicity, chromosome aberrations, and astrocyte cell death (Wu et al., 2007, p. 1336). Physicians do not recommend curcumin-based capsule in case of gallstone patients as it leads to the contraction of human gallbladder (Rasyid et al., 2002, p. 314). Therefore, investigations need to be carried out to find out the extent of the toxic effect of curcumin.

3. MANGIFERIN: EXTRACTION, IDENTIFICATION, AND ANTIDIABETIC–ANTIOXIDANT POTENTIAL

Investigation on effective antidiabetic agents from natural sources has gained momentum after the approval of World Health Organization on diabetes mellitus (WHO, 1980, p. 45; Rashid et al., 2012, p. 240, 2013a, p. 317). A second polyphenolic herbal medicine which is used in the management of diabetes includes a *C*-glucosidic xanthone, named "mangiferin" (2-β-D-glucopyranosyl-1,3,6,7-tetrahydroxyxanthone) (Telang et al., 2013, p. 1561). This molecule is made of highly condensed aromatic ring system, which is attached to a glucose molecule

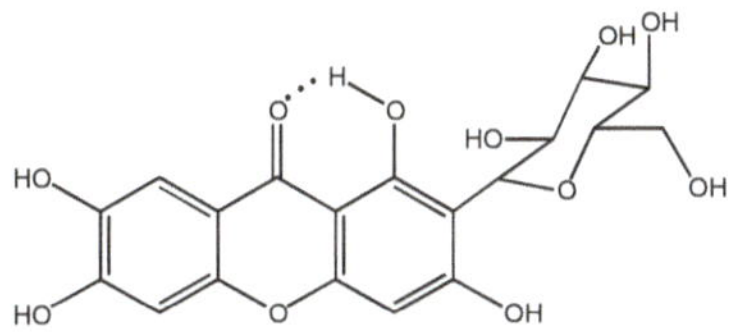

FIGURE 3.5 Chemical structure of Mangiferin.

using C—C bond (Fig. 3.5) (Raihanet al., 2013, p. 74). Two different aromatization pathways are involved in the formation of mangiferin: the shikimate and the ketate pathways (Wauthoz et al., 2007, p. 114). The unique structure of this molecule helps it to possess Lipinski's drug like properties that include molecular weight lower than 500 Da, less than 5 donor functions for hydrogen bonds, less than 10 acceptor functions for hydrogen bonds and favorable octanol/water partition coefficient, i.e., logP mangiferin = +2.73 (Lipinski et al., 1997, p. 3; Nuñez-Selles, 2005, p. 704). The chemical formula of mangiferin is $C_{19}H_{18}O_{11}$ and it is resistant to hydrolysis due to the presence of C-glucosides. The stable light yellow colored crystalline compound, mangiferin, is sufficiently soluble in water as well as other polar solvents, moderately soluble in medium-grade polar solvents and insoluble in lipophilic media (Smejkal, 2014, p. 64). This compound is relatively abundant in nature and reacts with less number of oxidizing agents. It has been reported that mangiferin traditionally used by native citizens of different countries (includes Bolivia, Southern Guiana, Antilles, Columbia, Philippines, and India) in the treatment of various diseases including diabetes as it possess a broad range of pharmacological effects such as antidiabetic, antitumor, antiviral, immunomodulatory, antimicrobial, antioxidant, antiparasitic, antiseptic, and antiasthmatic activities (Periyar Selvam et al., 2012, p. S1583; Li et al., 2013, p. 296; Guha et al., 1996, p. 443; Yoosook et al., 2000, p. 411; Leiro et al., 2004, p. 763; Stoilova et al., 2005, p. 37). Along with its drug like properties, it shows least or no harmful effects in clinical practice and bears relatively low costs as compared to oral antidiabetic synthetic drugs, which increase its applications in diabetic therapeutic strategy (Periyar Selvam et al., 2012, p. S1586; Sellamuthu et al., 2013, p. 2). In Sri Lanka, it is also present in the brand name "Salaretin" for the management of diabetes. This phytochemical commonly appears with three different components like isomangiferin (4-*C*-β-D-glucopyranosyl-1,3,6,7-tetrahydroxyxanthone), homomangiferin (2-*C*-β-D-glucopyranosyl-1,6,7-tetrahydroxy-3-methoxyxanthone), and neomangiferin (mangiferin-7-*O*-β-D-glucopyranoside) (Smejkal, 2014, p. 64; Matkowski et al., 2013, p. 439). Mangiferin is found in various families, mainly in Anacardiaceae and Gentianaceae families. Other sources include rhizome of *Anemarrhena asphodeloides, Aphloia theiformis, Iris unguicularis, Swertia chirata, Salacia reticulata, Cyclopia* spp., *Fabaceae, Mangifera indica*, etc. (Denisova et al., 1980, p. 76; Islam et al., 2008, p. 1271). Extraction of mangiferin from different parts of *M. indica* (commonly known as mango tree) such as leave, fruits, stems bark, heartwood and roots, etc., is considered as the most common and convenient method to get significant amount of this active component (Qin et al., 2008, 955; Desai et al., 1966, p. 457; El Ansari et al., 1971, p. 2239; El Ansari et al., 1967; p. 1363; Jash and Brahmachari, 2015, p. 1; Ramanathan and Seshadri, 1960, p. 131; Nigam and Mitra,1964, 378).

3.1 Extraction and Identification of Mangiferin

3.1.1 Extraction and Isolation of Mangiferin

The plant, *M. indica*, is the main source of mangiferin and can be extracted from various parts of this plant. Besides, it is also present in a considerable (1–7% dry mass) amount in rhizomes of *A. asphodeloides* (TCM drug Zhimu), *Iris domestica* (syn. *Belamcandachinensis* TCM drug Shegan) and some other *Iris* sp., roots of *Salacia* sp. Celastraceae (Ayurvedic medicine), in *Gentianaceae* family members—*Gentiana* and *Swertia* ssp., and in *Hypericum* ssp. Besides, mangiferin can be found in Bignoniaceae, Moraceae, Thymelaceae, Aphloiaceae, and Malvaceae families. A recent report showed that sufficient amount of this molecule could be isolated from *Coffea pseudozanguebariae* (Rubiaceae). The report also claimed that the presence of the huge amount (exceeding 6% of dry mass) makes this plant one of the richest natural sources of mangiferin (Matkowski et al., 2013, p. 439). As mentioned earlier, mangiferin can be extracted and isolated from various sources and different groups of researchers described various procedures for its extraction. Some common extraction methods of mangiferin from *M. indica* are briefly presented here:

3.1.1.1 Extraction From the Bark of *Mangifera indica*

Ghosh et al. (2012) described its extraction from the bark of *M. indica* (Ghosh et al., 2012, p. 117). Briefly, the bark of the *M. indica* was collected from the plant, cut into small parts and dried, ground into powder, and thoroughly extracted with ethanol (95%) in a Soxhlet apparatus for 56 h. The different parts of the extract were combined and concentrated under reduced pressure until a yellow amorphous powder was obtained. The dried alcoholic extract was allowed to adsorb on silica gel (60–120 mesh) and subjected to chromatography on silica gel column packed in petroleum ether (60–80°C). Mangiferin was obtained in the eluent as a pale yellow solution after elution of the column with chloroform: methanol (1:1). Evaporation of the solvent provides it as a yellow amorphous solid and that on crystallization with the ethanol gave pale yellow needle-shaped mangiferin crystals.

3.1.1.2 Extraction From the Flowering Buds of *Mangifera indica*

Biswas et al. (2015) described its extraction from the flowering buds of *M. indica.* The *buds* were cut into small pieces, dried at room temperature, and coarsely powdered. It was then extracted exhaustively in Soxhlet apparatus with petroleum ether (60–80°C) for 56h to remove fat. Defatted powder was extracted by using required quantity of ethanol (95%) as a solvent for 21h and concentrated under reduced pressure to yield a semisolid mass. The entire process was repeated several times and finally dissolved in ethanol at room temperature and was concentrated under reduced pressure to get yellow amorphous powder (Biswas et al., 2015, p. 51).

3.1.1.3 Extraction from the Fruits of *Mangifera indica*

Extraction of mangiferin from the fruits has recently been reported by Luo et al. (2012) in which the fruits were allowed to ripen at room temperature until fully ripen according to firmness, color, and formation of the aroma of mango characteristics (Luo et al., 2012, p. 11,263). Fruits were manually separated into peel, pulp, and seed kernel; cut into small pieces, freeze-dried in liquid nitrogen followed by grounding into a fine powder; and extracted with ethanol–methanol–acetone, 1:1:1 solvent mixture followed by sonication in an ultrasonic bath. The solution was then centrifuged at ambient temperature. The same extraction procedure of the sample was carried out twice; the supernatants were combined and used for the mangiferin content analysis. Results showed that mango peel and seed kernel are good sources of mangiferin with high contents of the same.

3.1.1.4 Extraction from the Leaves of *Mangifera indica*

For this extraction, the leaves were cleaned by rinsing with tap water, dried at 40–50°C in an oven, ground to a powder with a cutting mill, and subjected to maceration extraction with three different solvents including methanol, ethanol, and 70% acetone. The maceration was made up at room temperature and the solvents were changed at a particular interval (every week for 3 weeks). Each extract was filtered and evaporated under vacuum until dryness to obtain the thickening mass of the three solvent extracts. The dried methanol extract was resuspended in methanol and partitioned with dichloromethane. The procedure was repeated for four times and the aqueous methanolic phase was hydrolyzed by refluxing with sulfuric acid at low pH with continuous stirring. It was cooled and partitioned with ethyl acetate four times. The combined ethyl acetate layer was dried, dissolved in methanol, and the precipitate formed was isolated by filtration. Finally, for crystallization, the precipitate was dissolved in 70% aqueous methanolic solution and left in a refrigerator when the pale yellow needle-shaped crystals of mangiferin were formed. These were dried and collected. The dried ethanol and the dried acetone extracts were resuspended in 50% ethanol and 50% acetone, respectively, and processed following by the same techniques as described earlier. The isolated mangiferin from each extract was further characterized using various techniques as described here (Jutiviboonsuk and Sardsaengjun, 2010, p. 122).

3.1.2 *Identification of the Isolated Mangiferin*

A number of techniques such as TLC, HPLC), NMR (^{1}H, ^{13}C) spectroscopy, etc., are being applied to the analysis of mangiferin (obtained from various extraction procedures as mentioned earlier) at the final step of isolation to get pure molecule.

3.1.2.1 Thin Layer Chromatography

For the identification of the isolated mangiferin obtained from different sources, an isolation method is performed by TLC analysis and that is usually compared to standard reference of mangiferin. Silica gel (a particular variety) is used as an adsorbent for normal phase and two solutions [e.g., ethyl acetate:acetone:formic acid:water (8:2:1:1) and ethyl acetate:methanol:formic acid:water (8:2:1:1)], which are used as the developing solvents. However, if cellulose is used as an adsorbent for reverse phase, the solution of ethyl acetate:formic acid:water (67:13:20) is used as a developing solvent. The spots on the plates are detected by spraying sulfuric acid (in methanol) and heating at a high temperature (110°C).

3.1.2.2 High-Performance Liquid Chromatography

The identification of mangiferin is best achieved by reversed-phase high performance liquid chromatography (RP-HPLC). An analytical determination is conducted by RP-HPLC with ultraviolet detection at a specific wave length (278 nm). The chromatographic separation was made on a reverse phase (C18) analytical column using a different solvent mixture (chloroform:acetone:formic acid as 8:1.5:0.5) as the mobile phase, at a flow rate (1 mL/min) at room temperature. The amount of mangiferin in the sample was quantified from the standard curve initially constructed by applying pure standard mangiferin under identical conditions.

3.2 Mangiferin as an Antioxidant

Being an inhibitor of lipid autooxidation, the knowledge about the antioxidant property of mangiferin would spread out the scope of its application to the food industry as well as clinical implementation for the treatment of diabetes. An investigation by using pulse radiolysis techniques, it has been observed that this molecule efficiently reacts with different oxidizing and reducing radicals ($\cdot OH$, $N_3\cdot$, and $CCl_3O_2\cdot$), and inhibits them (Stoilovaet al., 2008, p. 2707; Wauthoz et al., 2007, p. 115). This molecule is also found to attenuate lipid peroxidation, which was initiated either enzymatically by reduced nicotinamide adenine dinucleotide phosphate (NADPH) or nonenzymatically by ascorbic acid or Fenton's reagent ($H_2O_2+Fe^{2+}$) in hepatic microsomes of rats. Mangiferin rapidly scavenges DPPH (1,1-diphenyl-2-picrylhydrazyl) free radical suggesting its strong antioxidant activity as compared with a common medically applied antioxidant rutin (Dar et al., 2005, p. 596). The concentration of mangiferin causing 50%

inhibition of the free radical (IC_{50}) was assayed to be 7.5 ± 0.0075 μg/mL. Mangiferin is unable to terminate radical chain reaction during the process of lipid peroxidation either in the enzymatic system or the linoleic acid hydroperoxide-induced peroxidation system. These observations proposed that the antioxidant activity of mangiferin is due to its ability to scavenge free radicals involved in lipid peroxidation initiation process (Sato et al., 1992, p. 721). This polyphenolic molecule shows its antioxidative activity due to the presence of free hydroxyl groups and catechol moiety. The catechol moiety is essential for the formation of stable mangiferin–Fe^{2+}/Fe^{3+} complexes in mitochondria and prevention of Fenton reaction (Andreu et al., 2005, p. 47). By scavenging free radicals, it protects mitochondria from swelling and loss of its membrane potential in different oxidative damages. Mangiferin has four phenolic H-atoms, upon reaction with free radicals (e.g., ROS) two of which could easily be abstracted to form two phenoxyl radicals (λ_{max}, 390 and 470 nm, respectively), which undergo radical–radical reactions and stabilized by resonance (Fig. 3.6). The second radical could react with ascorbate for the regeneration of mangiferin (Mishra et al., 2006, p. 70; Wauthoz et al., 2007, p. 115). This antioxidant compound helps in maintaining the balance of the enzymes, which play a pivotal role in cellular defense (superoxide dismutase, catalase, and the glutathione system) against oxidative stress (Rajendran et al., 2008, p. 137). The contribution of mangiferin as an effective antioxidant has been confirmed in cellular stress-mediated in vitro as well as in vivo conditions. Protection of mangiferin was observed against oxidative stress in different cell lines (HepG2, hepatocytes, myocytes, and mesangial cells) using various oxidative stress inducing agents

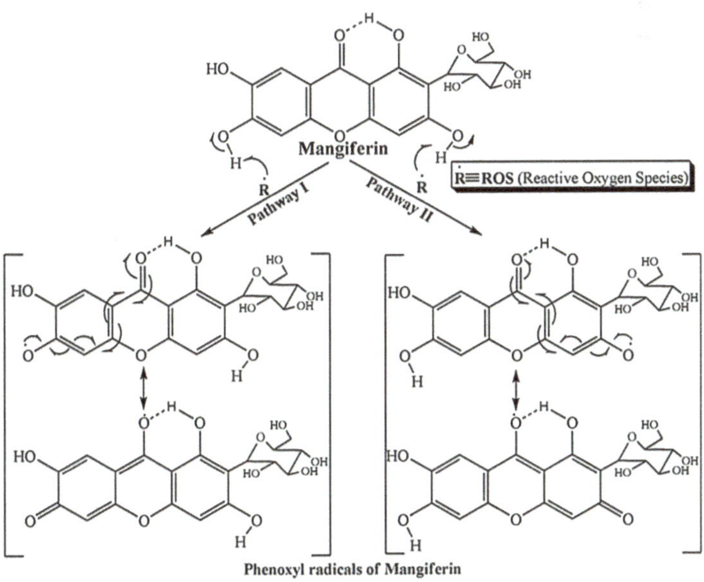

FIGURE 3.6 Reaction of mangiferin with reactive oxygen species (free radicals) and resonating canonical forms of two phenoxyl radicals.

($HgCl_2$, CCl_4, tertbutyl hydroperoxide, isoproterenol, and hyperglycemia, respectively) (Dar et al., 2005, p. 596; Agarwala et al., 2012, p. 117; Rodeiro et al., 2008, p. 1242; Prabhu et al., 2006, p. 126; Li et al., 2010, p. 893). In addition, mangiferin is able to ameliorate 12-*O*-tetradecanoylphorbol-13-acetate (TPA, an oxidative stress-stimulating agent)-induced oxidative damage in in vivo system (in OF1 mice) (Sanchez et al., 2000, p. 565).

3.3 Mangiferin as an Antidiabetic Agent

The most important feature of diabetes is hyperglycemia (postprandial and fasting). The hyperglycemic state arises mainly due to three reasons: an overload of carbohydrate, inadequate utilization, and excess formation of glucose. Overloaded carbohydrates result in excess production of glucose by the help of α-amylase and α-glucosidase. We have described earlier that these two enzymes are involved in the metabolism of carbohydrates. α-Amylase degrades complex dietary carbohydrates to oligosaccharides and disaccharides, which, in turn, is transformed into monosaccharide by the action of α-glucosidase. Gut (gastrointestinal tract) then absorbs the liberated glucose and results in postprandial hyperglycemia. Inadequate utilization of glucose occurs as a result of low insulin secretion from pancreatic β-cells, resistance to insulin action, and reduced activity of glycolytic enzymes (glucose-utilizing enzymes). Excess formation of glucose caused due to the increased activity of the enzymes involved in gluconeogenesis pathway (e.g., glucose-6-phosphatase, fructose-1,6-bisphosphatase, glucose-6-phosphate dehydrogenase, etc.). In STZ (a diabetogenic agent)-induced diabetic animals and KK-Ay diabetic mice (a model of type 2 diabetes), mangiferin has been found to decrease plasma glucose level and increase insulin sensitivity and glucose tolerance (Muruganandan et al., 2005, p. 497; Miura et al., 2001, p. 85; Sellamuthu et al., 2009, p. 206). The mechanisms, using which mangiferin shows antidiabetic activity, involve appreciable inhibition of α-amylase and α-glucosidase enzymes as compared with acarbose (a standard drug for diabetes) (Dineshkumar et al., 2010, p. 75). This molecule enhances utilization of glucose by accelerating the translocation of the GLUT-2 transporter to the membrane in rat myotubes and subsequently increasing glucose uptake via activation of AMPK (Girónet al., 2009, p. 565). Mangiferin also stimulates the action of enzymes in glycolysis (hexokinase and pyruvate kinase), and the enzymes in glycogenesis (glycogen phosphorylase and glycogen synthase). On the other hand, mangiferin has been reported to reduce the activity of glucose-6-phosphatase and fructose-1,6-bisphosphatase to slow down gluconeogenesis (Sellamuthu et al., 2009, p. 206). Therefore, mangiferin could be useful in the management of hyperglycemia in diabetes.

3.4 Utilization of Mangiferin and Its Derivatives in Diabetes

Lately, in the management of diabetes, mangiferin or its derivatives (e.g., norathyriol, mangiferin calcium salt, halogenated hydrocarbons of mangiferin, etc.)

either alone or combination with another herbal therapeutic agents are found to be effective. It has been claimed that mangiferin and neomangiferin are considered as the new formulation of antidiabetic medicine. Acting as an insulin sensitizer, calcium salt of mangiferin helps in the reduction of plasma glucose level and lipid. Blood sugar level could also be reduced by the action of an aglycone of mangiferin, norathyriol. In a rat model, few derivatives of mangiferin aglycone (propionate derivatives) act more efficiently to reduce hyperglycemia than mangiferin and its aglycone do lonely by themselves. The potential antidiabetic agent foliamangiferoside acts by inhibiting α-glucosidase. One of the active derivatives of mangiferin is used to treat type 2 diabetes by attenuating the protein tyrosine phosphatase 1B activity. Momangiferin, a carbonic acid glucoside of tetrahydroxy pyridine, has been used in the medication of type 2 diabetes and acts by modifying the expression of uncoupling protein-2. Another study claimed the composition of mangiferin and berberine is useful for the prevention of different complications associated with type 2 diabetes (Telang et al., 2013, p. 1562).

3.5 Mechanism of Mangiferin-Induced Protective Action in Diabetes

Evidence represents hyperglycemia as a well-established pathogenic factor of chronic complications in diabetes mellitus. This pathogenic state not only accelerates the formation of free radicals but also affects antioxidant machinery through glycation of the antioxidant enzymes (Bonnefont-Rousselot et al., 2004, p. 147). Excess formation of ROS results in oxidative damage to different organs mainly of kidney and heart (Rolo and Palmeira, 2006, p. 167).

3.5.1 Protective Action on Kidney

Diabetic nephropathy is considered as one of the serious microvascular diabetic complications. Patients with both kinds of diabetes suffer from this disease. It is the most common cause of end stage renal failure (Yan et al., 2007, p. 787). It results from the damage of small capillaries in the kidneys' glomeruli and that in turn functions less efficiently or finally could not function properly (Ibrahim and Hostetter, 1997, p. 489). Characterization of diabetic nephropathy includes glomerular hypertrophy, glomerular hyperfiltration, increased urinary albumin secretion, increased basement membrane thickness, and mesangial expansion with the accumulation of extracellular matrix proteins (ECM, type IV collagen) (Zelmanovitz et al., 2009, p. 1). In renal fibrosis, increased deposition of collagen has been observed in the glomerular region. Therefore for determining the extent of fibrosis, content of hydroxyproline has been considered. Recently, Pal et al. (2014) showed that mangiferin effectively prevents renal fibrosis in diabetes by reducing the content of hydroxyproline (Pal et al., 2014, p. 1). In addition, they have also observed that mangiferin could effectively decrease the plasma level of blood urea nitrogen (BUN), creatinine, uric acid, and urinary albumin in diabetic rats. Activation of polyol pathways; AGEs; and its precursor methylglyoxal, autoxidation of glucose, xanthine oxidase activity, mitochondrial respiratory chain deficiencies, NAD(P)H oxidase, and NOS are considered as the probable sources of hyperglycemia induced ROS-mediated diabetic nephropathy (Forbes et al., 2008, p. 1446; Brownlee, 2001, p. 813). Extensive research suggests that the cytoprotective effect of mangiferin in diabetes is attributed to its antioxidant activity. Administration of this antioxidant molecule effectively reduced the expression of AGEs and its receptor in the protein as well as mRNA level (Liu et al., 2013, p. 35). Moreover, this antioxidant molecule increased the activity and expression of glyoxalase-1 (a detoxifying enzyme of methylglyoxal) in both protein and mRNA level (Liu et al., 2013, p. 355). Pal et al. (2014) also suggest that hyperglycemia-induced ROS activates MAPKs pathway (p38, JNK, and ERK1/2) and different isoforms of PKC pathway (PKCα, PKCβ, and PKCε) to worsen the condition of diabetic nephropathy (Pal et al., 2014, p. 1). Activation of MAPKs pathway promotes apoptotic death in the diabetic state. However, treatment with mangiferin significantly decreased the expression of MAPKs and PKC pathways-related proteins. NF-κB, a transcription factor, plays an important role in the inflammatory response, cell survival, and death. Phosphorylation of IKKα leads to the dissociation of IκBα from NF-κB and subsequent phosphorylation of p65 subunit of the NF-κB molecule. This event results in translocation of NF-κB into the nucleus. NF-κB results in the upregulation of COX-2 and transforming growth factor (TGF)-β1 in the renal cortex of diabetic rats (Zhuet al., 2015, p. 295; Shah et al., 2007, p. 20). COX-2 involves in the synthesis of prostaglandin and results in inflammatory response. TGF-β1 is a crucial factor in diabetic nephropathy as it causes accumulation of ECM and develops renal hypertrophy (Wolf and Ziyadeh, 1999, p. 393; Park et al., 1997, p. 473). Mangiferin supplementation decreased the phosphorylation of IKKα, elevated the expression of IκBα, reduced the translocation of NF-κB into the nucleus, and downregulated the expression of COX-2 and TGF-β1 in the diabetic kidney. Another study showed that mangiferin provides protection by scavenging free radicals and activating various antioxidant enzymes (SOD, CAT, GPx, etc.) in this renal dysfunction (Li et al., 2010, p. 893). Attenuation of different oxidative stress markers (protein carbonyl and malondialdehyde) and elevation of GSH/GSSG ratio were also observed in mangiferin-administered group in diabetic nephropathy (Pal et al., 2014, p. 1).

Inhibition of TNF-α-mediated extrinsic apoptotic pathway and mitochondrial-dependent intrinsic apoptotic pathway were also found in the kidney of diabetic rats fed with mangiferin as shown in Fig. 3.7.

3.5.2 Protective Action on Heart

Most of the diabetic people suffer from a disorder of heart muscle known as diabetic cardiomyopathy (DCM). In this state, heart becomes unable to circulate efficiently blood throughout the body and results in accumulation of fluid in the lungs and legs. Additionally, the heart in DCM shows a tendency of slow increase in fibrosis. Severe cardiac fibrosis leads to the myocardial stiffness

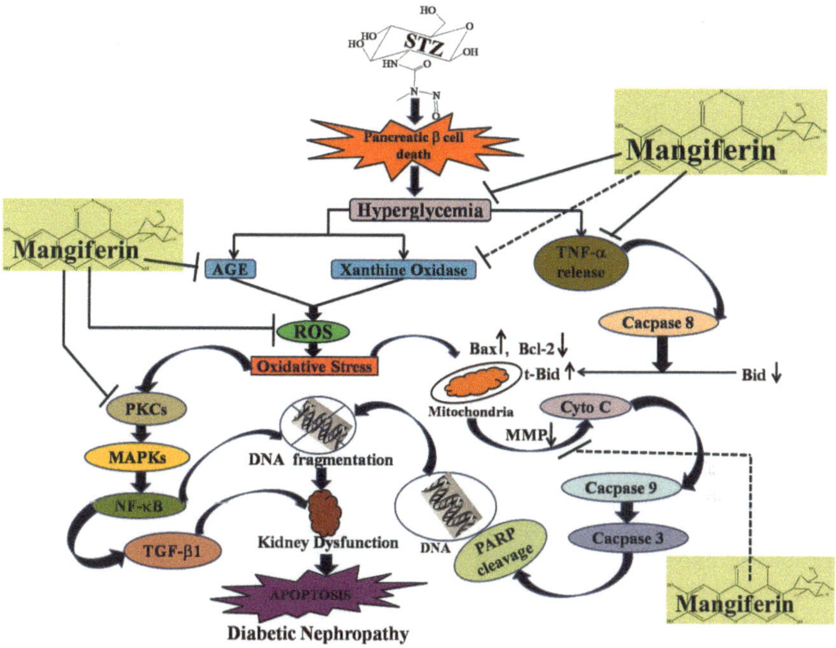

FIGURE 3.7 Mechanism of protection of mangiferin in streptozotocin (STZ)-induced diabetic nephropathy. *AGE*, advanced glycation end product; *NF-κB*, nuclear factor kappa B; *PARP*, poly(ADP-ribose) polymerase; *PKC*, protein kinase C; *MAPK*, mitogen-activated protein kinase; *ROS*, reactive oxygen species; *TGF-β1*, transforming growth factor β1; *TNF-α*, tumor necrosis factor α.

that in turn results in ventricular dysfunction. This pathophysiological state finally causes heart failure in the patients. Postprandial hyperglycemia plays a key function in the development and progression of type 2 diabetes and cardiovascular complications in the diabetic state. Mangiferin successfully prevented the accumulation of myocardial collagen to mitigate cardiac fibrosis in DCM (Hou et al., 2013, p. 759). This active antioxidant molecule is also able to protect Zucker diabetic fatty (ZDF) rats from cardiac hypertrophy by decreasing weights of the hearts and left ventricles, reducing cardiomyocyte cross-sectional areas, and inhibiting upregulation of cardiac AT-1 (angiotensin receptor 1). In response to high blood volume, atrial natriuretic peptide (ANP) is released by atrial myocytes (upper chamber of heart muscle cells). It is involved in the homeostatic control of the body (water, different ions, and fat). ANP acts to reduce the water, sodium, and adipose loads on the circulatory system, thereby reducing blood pressure. Administration of mangiferin suppressed angiotensin II-induced ANP mRNA overexpression and protein synthesis in H9c2 cells (a cell line derived from embryonic BD1X rat heart tissue) (Huang et al., 2008, p. 574). This molecule exerts its protective action by markedly attenuating the overexpression of mRNAs encoding TGF-β1 and TGF-β3 in the obese Zucker rats (OZRs) heart (Li et al., 2004b, p. 1735). Supplementation with mangiferin exhibited significant reduction in plasma glucose level, glycosylated hemoglobin, and serum creatine phosphokinase (CPK) levels along with the amelioration of oxidative stress (alters the activity of different antioxidant enzymes, e.g., SOD, CAT, etc.) in the cardiac tissue of the diabetic rats (Muruganandan et al., 2002, p. 165). Finally, mangiferin significantly improves cardiac function by modulating some molecules and could act as a therapeutic agent in the treatment of DCM.

3.6 Clinical Impact and Practical Use of Mangiferin

No authentic implementation of mangiferin in the clinical study has yet been reported. To evaluate the effectiveness of mangiferin in humans, a study was conducted by Pardo-Andreu et al. (2006) on elderly people aged over 65 years (Pardo-Andreu et al., 2006, p. 158). They were supplemented with Vimang (mangiferin-rich extract from *M. indica*). Enhanced activity of extracellular SOD, upregulation of antioxidant status in serum (2,2′-azino-bis(3-ethylbenzothiazoline-6-sulfonic acid (ABTS) test), and reduced level of oxidized glutathione and lipid peroxidation were detected in the Vimang-administered group. Antioxidant status was evaluated using ABTS test, and lipid peroxidation was measured using thiobarbituric acid-reactive substance (TBARS) assay. The presence of other phenols may also contribute to the protective action of Vimang against oxidative stress. Mangiferin possesses very low toxicity. Safe dose of this antioxidant molecule in mice is 25 g/kg, whereas LD_{50} in rats after intraperitoneal ingestion was found as 365 mg/kg body weight (Bhattacharya et al., 1972, p. 651; Niu et al., 2012, p. 1177).

3.7 Bioavailability of Mangiferin

The structure of mangiferin makes it a probable candidate to fulfill the requisite for well oral bioavailability. The structure of this molecule satisfies all the rules made by Lipinski to acquire drug like properties. Mangiferin and its metabolites have been detected in the serum and urine sample within 1 h of single oral administration in rats due to its fast absorption (Liu et al., 2012, p. 2116, 2010, p. 3349). This molecule is able to cross the blood–brain barrier in gerbils to exert its effect (Martínez Sánchez et al., 2001, p. 465). It has been reported that concentration of mangiferin was found in nanomolar (plasmatic concentration) range after oral ingestion. This result indicated low bioavailability of this molecule as it is absorbed quickly and underwent rapid metabolism in the liver to generate different metabolites. Participation of these metabolites in the biological activity of mangiferin is also appreciable.

3.8 Toxicity of Mangiferin

A first criterion for applying any substance as a pharmacological agent is the determination of the toxicity of that compound. Mangiferin administration in rats at a dose of 25–500 mg/kg body weight per day for 30 days did not exert any toxic effect on the animals. Mangiferin-supplemented animals not even showed any respiratory distress, hair loss, restlessness, convulsions, laxative, coma, weight loss, etc. (Sellamuthu et al., 2009, p. 208). Though fewer studies have been demonstrated the nontoxic effect of mangiferin in people but more studies are required to confirm its state of toxicity in human trials.

4. ARJUNOLIC ACID: EXTRACTION, IDENTIFICATION, AND ANTIDIABETIC–ANTIOXIDANT POTENTIAL

In the medical literature of ancient India (Charaka, Samhita, and Ashtanga Hridayam), the cardioprotective effect of a deciduous tree *Terminalia arjuna* has clearly been discussed (Hemalatha et al., 2010, p. 238). *Terminalia arjuna* belongs to Combretaceae family and has also been able to cure hepatic, urogenital, and viral diseases (Kumar and Prabhakar, 1987, p. 173). It possesses a number of biological activities due to the presence of various active constituents in its bark and AA is considered as one of them (Ghosh and Sil, 2013, p. 1098; Anonymous Monograph, 1999, p. 436). AA, a pentacyclic triterpenoid saponin, is used as an effective cardioprotective tonic (Fig. 3.8). Though the first isolating source of AA is the bark of *T. arjuna*, later on, it has been isolated from various other plants which have therapeutic values (e.g., *Combretum nelsonii*, *Campsis grandiflora*, *Leandra chaeton*, *Cochlospermum tinctorium*, *Cornus capitata*, *Syzygium guineense*, etc.) (Masoko et al., 2008, p. 1074; Kim et al., 2005, p. 550; Zang et al., 2003, p. 582; Diallo et al., 1989, p. 879; Bhakuni et al., 2002, p. S68; Djoukeng et al., 2005, p. 283). Extensive research showed that cardioprotective action of AA is resulting from its preventive effect on myocardial necrosis, lowering of blood pressure, platelet aggregation and coagulation, heart rate, and cholesterol levels (Hemalatha et al., 2010, p. 238). Secondary metabolites help the plants to grow and develop but do not take part in its survival. Triterpenoids are derived from C30 precursors and considered as a vital secondary metabolite of plant (Xu et al., 2004, p. 261). The chiral triterpenic acid has a rigid pentacyclic backbone with two equatorial hydroxyl groups and one equatorial hydroxymethyl group. The carboxyl group is attached at the ring junction of the *cis*-fused two rings (Ajees and Balakrishna, 2002, p. 682; Hemalatha et al., 2010, p. 239).

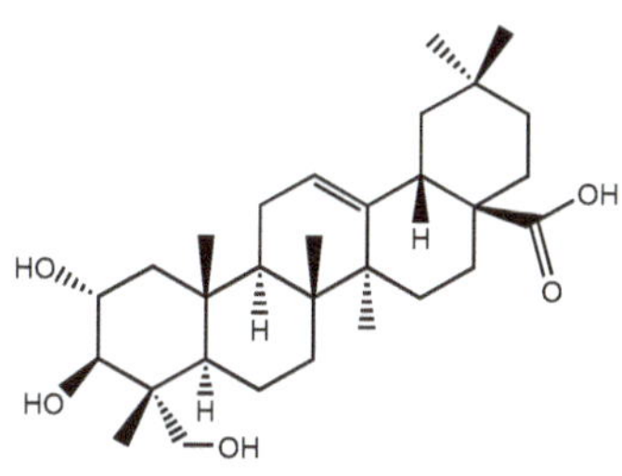

FIGURE 3.8 Chemical structure of arjunolic acid (AA).

Besides its cardioprotective effect, AA is also known for its wide range of biological properties, which includes wound healing, antioxidant, antidiabetic, antiinflammatory, antitumor, anticholinesterase, antiasthmatic, and antimicrobial activities (Sindambiwe et al., 1998, p. 585).

4.1 Extraction and Identification of Arjunolic Acid

Various methods for the extraction of AA were described by a number of authors. Some common methods for its extraction have been described here:

Extraction of arjunolic acid from the bark

AA is usually extracted from the barks of *T. arjuna*. Barks are cut into small pieces, dried, ground into powder, and extracted with petroleum ether followed by diethyl ether. The crude material is filtered off and the mother liquor is concentrated under reduced pressure when a white amorphous solid is obtained which on TLC shows two spots. The material was then subjected to separation on silica gel column for further separation. Homogeneity of AA after the final step of purification was confirmed by using reverse phase HPLC, NMR (^{1}H, ^{13}C), mass spectroscopy (MS), and optical rotation studies (Ghosh et al., 2011, p. 4858).

Microwave-assisted extraction from stem bark of Terminalia arjuna

Recently, a new extraction procedure for the AA was reported by Verma et al. (2012) in which the authors included the microwave in the extraction process (Verma et al., 2012, p. 1627). The stem bark of *T. arjuna* was cut into small pieces, dried, and ground to powder. The powder was put in conical flasks and several solvents (ranging from nonpolar to polar) were added separately. The suspension was preleached for 10 min and extracted at a certain temperature (below the boiling point of the solvent) by applying microwave irradiation at 6000 W for 5 min. An intermittent process (irradiation–cooling–irradiation in which irradiation was applied for 1 min and the mixture was kept for another 1 min for cooling between two irradiations) was then followed. The cooled extract was filtered on Whatman filter paper; solvents were removed under vacuum pressure and lyophilized to get dry mass. HPLC was finally used to identify AA.

4.2 Arjunolic Acid as an Antioxidant

AA is well recognized for the utilization of its antioxidant properties to provide protection against oxidative stress (imbalance between the production and elimination of ROS). Its antioxidant property is associated with its free radical scavenging potential and metal-chelating ability. In cell-free system, free radical scavenging activity of this molecule was assessed using DPPH and its in vivo antioxidant potential was measured using FRAP assay (Manna et al., 2008, p. 137). AA-containing carboxylic hydrogen is responsible for its free radical scavenging activity as any free radical can easily abstract hydrogen from this group and get reduced. Being a good metal chelator, AA forms a

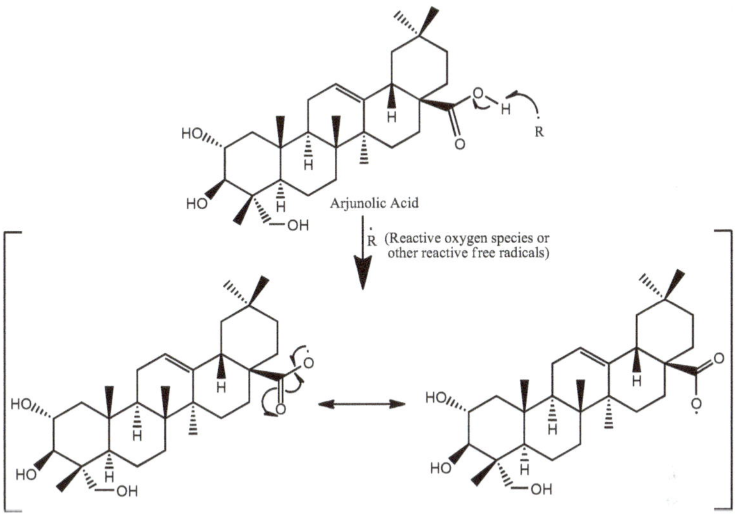

FIGURE 3.9 Free radical scavenging activity of arjunolic acid: two equivalent resonating canonical forms.

five-membered chelate complex with arsenic using its two equatorial hydroxyl groups and provides protection to the cardiac tissue against oxidative stress-mediated damage. Upon chelation, free toxin can easily be removed from the system and blocks toxin mediated further oxidative damage of the tissue. A research group demonstrated that the protective action exhibited by AA is similar to the protective role of vitamin C (Manna et al., 2007, p. 187). AA possesses one primary and two secondary hydroxyl groups. Therefore, it can be assumed that like vitamin C, AA could also be oxidized by ROS and other reactive species to form analogous keto derivative. Besides arsenic, this molecule exerts its protective action against a number of toxin and drug-induced tissue damages that include sodium arsenite, sodium fluoride, acetaminophen, etc. (Hemalatha et al., 2010, p. 242). Free radical scavenging activity of AA has been presented in Fig. 3.9.

4.3 Arjunolic Acid as an Antidiabetic Agent

We have discussed earlier that two enzymes, α-glucosidase and α-amylase, play a pivotal role (mainly in type 2 diabetes) in the reduction of excess dietary carbohydrate and formation of glucose from those molecules. Therefore, inhibition of these two enzymes could be helpful in the treatment of diabetic pathophysiology and its associated complications. In the case of previous two antioxidant molecules (curcumin and mangiferin), we have found significant inhibitory action against these two carbohydrate-metabolizing enzymes whereas AA is not able to do so to a considerable extent. It has been reported that AA isolated from the leaves of *Lagerstroemia speciosa* attenuates the α-glucosidase to some extent but possesses insignificant inhibitory activity against α-amylase (Hou et al., 2009, p. 614). Though this antioxidant molecule is unable to suppress the activity of aforementioned two enzymes, a number of research studies suggest its protective action in diabetes. It can be depicted that the antidiabetic action of this component is attributed to its antioxidant activity.

Protective role of AA on pancreatic tissue in STZ-induced type 1 diabetic rat model (Manna et al., 2009a, p. 53) was described by Manna et al. The intoxication of STZ leads to the formation of reactive oxygen and nitrogen species in the β-cell containing pancreatic tissue of the experimental group. Formation of excessive reactive species overwhelmed the intracellular antioxidant defense system and enhanced the peroxidation of membrane lipids as well as carbonylation of different proteins. Increased serum glucose and proinflammatory cytokine TNF-α were also found in the STZ-administered group. However, treatment with AA is able to ameliorate the condition by the efficient scavenging of free reactive species and reducing the expression of stress-induced MAPKs (phospho ERK1/2 and phospho p38), transcription factor NF-κB and regulating the molecules involved in mitochondrial-dependent apoptotic pathways.

This group also demonstrated the efficacy of AA against STZ-induced spleen damage under hyperglycemic state (Manna et al., 2010, p. 114). Most of the primary immune responses are initiated in the spleen and cytokine production by spleen cells is probably considered as one of the reasons for the β-cell destruction in type 1 diabetes (Thorvaldson et al., 2003, p. 1609). AA has been found to regulate the expression of different cytokines (IL-2, IFN-γ, and TNF-α) and maintain intracellular redox environment to protect spleen from oxidative stress-induced inflammatory response-mediated damage. In addition, the histopathological study showed that AA ameliorates splenic white pulp (responsible for immunological function) and decreased red pulp to enhance the immune response in type 1 diabetes (Fig. 3.10).

Efficacy of AA against STZ-induced diabetic nephropathy was also studied by Manna et al (Manna et al., 2009b, p. 297). They have shown that the diabetic renal damage is associated with increased kidney weight to body weight ratio, glomerular area and volume, blood glucose level, serum creatinine level, and BUN. Alteration in kidney-related and other parameters indicates renal dysfunction that in turn increases the generation of reactive species, enhances oxidative stress-related markers, and reduces intracellular antioxidant defense in the renal tissue of diabetic group. Treatment with AA successfully ameliorates diabetes-induced oxidative stress-mediated renal damage by reducing oxidative as well as nitrosative stress and deactivating the polyol pathways of the kidney. The histological pattern showed multiple foci of hemorrhagic necrosis and cloudy swelling of tubules in the diabetic renal tissue (Fig. 3.11). On treatment with AA amelioration of histopathological state was observed. This result suggests

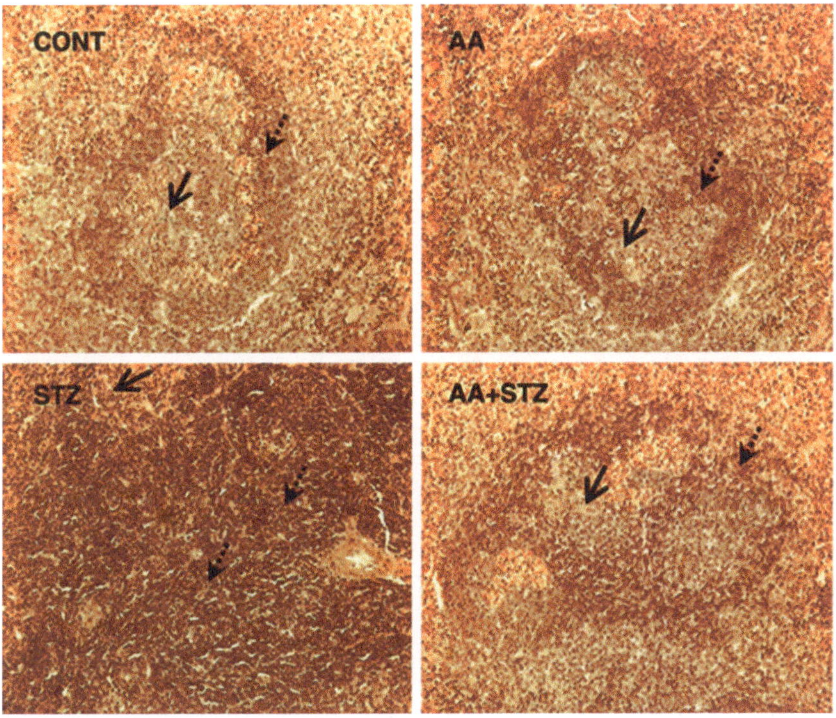

FIGURE 3.10 Hematoxylin- and eosin-stained spleen section of (Cont) normal rats spleen (×100); (AA) spleen from animals treated with AA; (STZ) STZ-exposed group, showing white pulp (marked with *solid arrow*) diminished in size and the red pulp (marked with *dotted arrow*) tended to enlarge (×100); (AA+STZ) AA pretreated spleen section before STZ administration (×100) showing almost normal morphology. *Reprinted from Manna, P., Ghosh, J., Das, J., Sil, PC., 2010. Streptozotocin induced activation of oxidative stress responsive splenic cell signaling pathways: protective role of arjunolic acid. Toxicology and Applied Pharmacology 244 (2), 114–129, with the permission from Elsevier.*

that AA could act as a protective agent against the renal oxidative stress-induced damage in diabetes.

5. CONCLUDING REMARKS

This chapter offers an overview of the protective role of various naturally occurring active principles such as curcumin, mangiferin, and AA on the oxidative stress-mediated diabetic pathophysiology. These natural biomolecules showed their efficacy against diabetes mainly due to their potential antioxidant property. Some of them also possess the ability to reverse insulin resistance in diabetic state. Although they have a number of beneficial outcomes, some side effects and low bioavailability of these molecules make their application limited in the treatment of this pandemic disease. Little amount of ROS is required for the

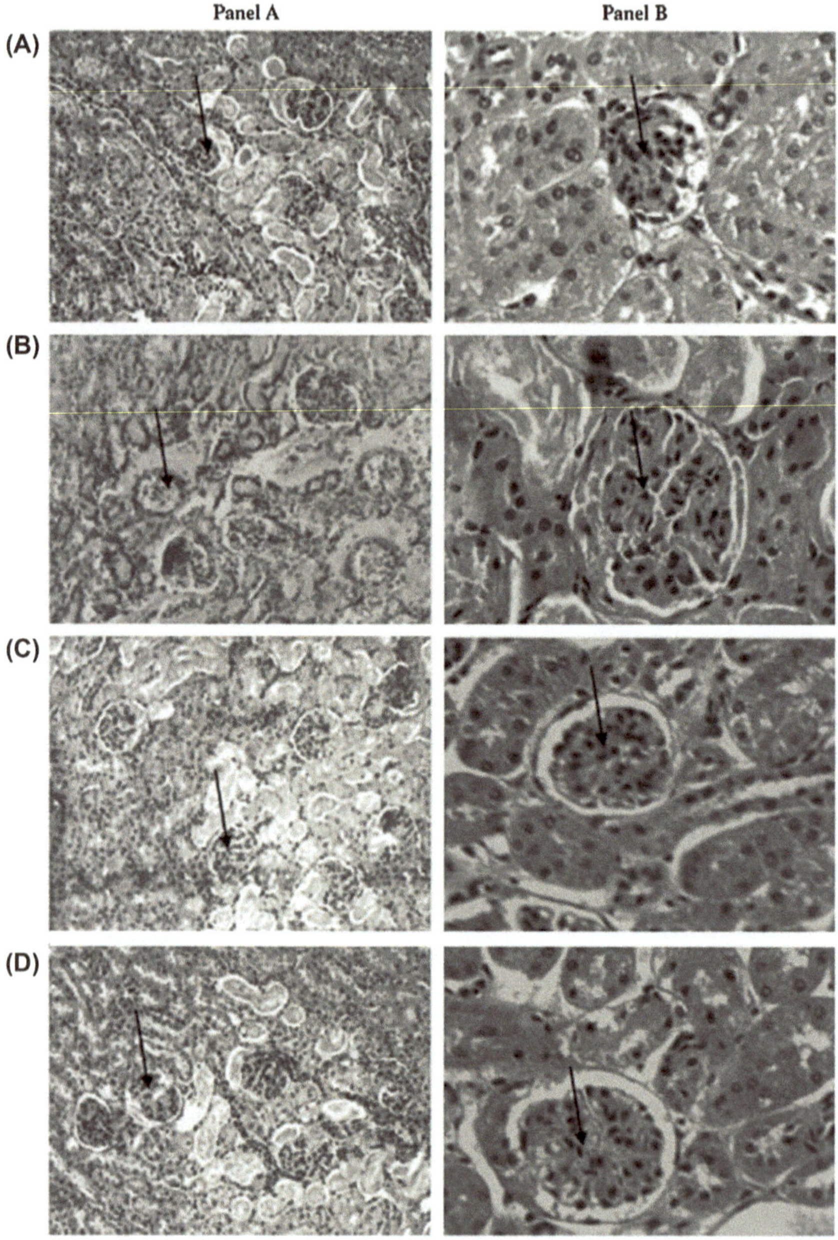

FIGURE 3.11 Hematoxylin- and eosin-stained kidney section. (A) Kidney section from normal healthy animals showing normal appearance of glomeruli (marked with *arrows*); (B) STZ-exposed animals, showing multiple foci of hemorrhage, necrosis, and cloudy swelling of tubules; (C) kidney section from AA-pretreated animals, showing almost normal appearance of glomeruli (marked with *arrows*) and tubules in kidney; and (D) kidney section from AA posttreated. Panel A represents microscopic magnification (100×) and panel B represents microscopic magnification (1000×). *Reprinted from Manna, P., Sinha, M., Sil, PC., 2009. Prophylactic role of arjunolic acid in response to streptozotocin mediated diabetic renal injury: activation of polyol pathway and oxidative stress responsive signaling cascades. Chemico-Biological Interactions 181 (3), 297–308, with the permission from Elsevier.*

normal physiological activities of the body. Certain dose of these antioxidant molecules eliminates excessive amount of the ROS and shows harmful effects. Additionally, people taking other medications should be careful about the intake of these antioxidants as they badly influence the effect of some medicines. To increase the implementation of these molecules, researchers are working on the increment of the bioavailability of these molecules and determining their effective dose which will be favorable without generating any complications. Scientists are also engaged to identify some other potent drugs to which these molecules interact to affect their activities and to aware the consumers about the adverse side of these compounds. Besides, we cannot ignore the enormous advantageous role of these magic molecules as compared to their dark side. Finally, we can conclude that the pleiotropic beneficial activities and the clinical efficacy of these molecules would propose their application for the discovery of cost-effective drugs in future for the management of diabetes.

ABBREVIATIONS

AA Arjunolic acid
ABTS 2,2′-Azino-bis(3-ethylbenzothiazoline-6-sulfonic acid)
ADI Acceptable daily intake
AGEs Advanced glycation end products
ALT Alanine transaminase
ALP Alkaline phosphatase
ALX Alloxan monohydrate
AMPK AMP-activated protein kinase
ANP Atrial natriuretic peptide
AT-1 Angiotensin receptor 1
ATP Adenosine triphosphate
CAT Catalase
COX-2 Cyclooxygenase-2
DCM Diabetic cardiomyopathy
DPPH2 2-Diphenyl-1-picrylhydrazyl
DSL D-Saccharic acid-1,4-lactone
EFSA European Food Safety Authority
ER Endoplasmic reticulum
ERK Extracellular signal-regulated kinases
FAO The Food and Agriculture Organization of the United Nations
FFA Free fatty acids
FRAP Ferric-reducing antioxidant power
GLUT-2 Glucose transporter 2
GSH Reduced glutathione
GST Glutathione S-transferase
GPx Glutathione peroxidase
HAT Histone acetyltransferase
HDACs Histone deacetylases
HDL High density lipoprotein

HFD High fat diet
h-IAPP Human islet amyloid polypeptide
HO-1 Heme oxygenase-1
HSCs Hepatic stellate cells
HSP-70 70 kDa heat shock proteins
IFN-γ Interferon γ
IL-1β Interleukin-1β
IR Insulin resistance
JNK c-Jun-NH2-terminal protein kinase
LDH Lactate dehydrogenase
LDL Low density lipoprotein
MAPKs Mitogen-activated protein kinases
MCP-1 Monocyte chemoattractant protein-1
MGF Mangiferin
MPT Mitochondrial membrane permeability transition
NADPH Reduced nicotinamide adenine dinucleotide phosphate
NF-κB Nuclear factor kappa B
NO Nitric oxide
iNOS Inducible nitric oxide synthase
Nrf-2 Nuclear erythroid 2-related factor 2
PARP Poly(ADP-ribose) polymerase
PI3K Phosphatidylinositol 3-kinases
PKC Protein kinase C
PPAR-γ Peroxisome proliferator-activated receptor γ
ROS Reactive oxygen species
SGLT Sodium–glucose-linked transporter
SOD Superoxide dismutase
STAT3 Signal transducer and activator of transcription 3
STZ Streptozotocin
TA *Terminalia arjuna*
TBARS Thiobarbituric acid-reactive substances
TGF-β1 Transforming growth factor β1
TNF-α Tumor necrosis factor α
UPR Unfolded protein response
WHO World Health Organization
ZDF Zucker diabetic fatty rats

REFERENCES

Agarwala, S., NR, B., Mudholkar, K., Bhuwania, R., Satish Rao, B.S., 2012. Mangiferin, a dietary xanthone protects against mercury-induced toxicity in HepG2 cells. Environmental Toxicology 27 (2), 117–127.

Aggarwal, B.B., Sundaram, C., Malani, N., Ichikawa, H., 2007. Curcumin: the Indian solid gold. Advances in Experimental Medicine and Biology 595, 1–75.

Agnihotri, N., Mishra, P.C., 2011. Scavenging mechanism of curcumin toward the hydroxyl radical: a theoretical study of reactions producing ferulic acid and vanillin. The Journal of Physical Chemistry A 115 (49), 14221–14232.

Agrawal, S., Chakrabarti, A., 2010. Potential nutraceutical ingredients from plant origin. In: Pathak, Y. (Ed.), Handbook of Nutraceuticals Editor. CRC Press, pp. 27–68.

Aitken, R.J., Clarkson, J.S., Fishel, S., 1989. Generation of reactive oxygen species, lipid peroxidation, and human sperm function. Biology of Reproduction 41, 183–197.

Ajees, A.A., Balakrishna, K., 2002. Arjunolic acid. Acta Crystallographica Section E 58, 682–684.

Aldebasi, Y.H., Aly, S.M., Rahmani, A.H., 2013. Therapeutic implications of curcumin in the prevention of diabetic retinopathy via modulation of antioxidantactivity and genetic pathways. International Journal of Physiology, Pathophysiology and Pharmacology 5 (4), 194–202.

Alwi, I., Santoso, T., Suyono, S., Sutrisna, B., Suyatna, F.D., Kresno, S.B., Ernie, S., 2008. The effect of curcumin on lipid level in patients with acute coronary syndrome. Acta Medica Indonesiana 40 (4), 201–210.

Anand, P., Kunnumakkara, A.B., Newman, R.A., Aggarwal, B.B., 2007. Bioavailability of curcumin: problems and promises. Molecular Pharmaceutics 4 (6), 807–818.

Andreu, G.P., Delgado, R., Velho, J.A., Curti, C., Vercesi, A.E., 2005. Iron complexing activity of mangiferin, a naturally occurring glucosylxanthone, inhibits mitochondrial lipid peroxidation induced by Fe^{2+}-citrate. European Journal of Pharmacology 513 (1–2), 47–55.

Anonymous Monograph, 1999. Terminalia arjuna. Alternative Medicine Review 4, 436–437.

Anthwal, A., Thakur, B.K., Rawat, M.S., Rawat, D.S., Tyagi, A.K., Aggarwal, B.B., 2014. Synthesis, characterization and in vitro anticancer activity of C-5 curcumin analogues with potential to inhibit TNF-α-induced NF-κB activation. Biomed Research International 2014 (1–10), 524161.

Asche, C.V., McAdam-Marx, C., Shane-McWhorter, L., Sheng, X., Plauschinat, C.A., 2008. Association between oral antidiabetic use, adverse events and outcomes in patients with type 2 diabetes. Diabetes Obesity and Metabolism 10, 638–645.

Asghari, G., Sheikholeslami, S., Mirmiran, P., Chary, A., Hedayati, M., Shafiee, A., Azizi, F., 2012. Effect of pomegranate seed oil on serum TNF-alpha level in dyslipidemic patients. International Journal of Food Sciences and Nutrition 63 (3), 368–371.

Babu, P.S., Srinivasan, K., 1995. Influence of dietary curcumin and cholesterol on the progression of experimentally induced diabetes in albino rat. Molecular and Cellular Biochemistry 152 (1), 13–21.

Babu, P.S., Srinivasan, K., 1997. Hypolipidemic action of curcumin, the active principle of turmeric (*Curcuma longa*) in streptozotocin induced diabetic rats. Molecular and Cellular Biochemistry 166 (1–2), 169–175.

Balamurugan, A.N., Akhov, L., Selvaraj, G., Pugazhenthi, S., 2009. Induction of antioxidant enzymes by curcumin and its analogues in human islets: implications in transplantation. Pancreas 38 (4), 454–460.

Ballester, J., Munoz, M.C., Dominguez, J., Rigau, T., Guinovart, J.J., Rodriguez-Gil, J.E., 2004. Insulin-dependent diabetes affects testicular function by FSH- and LH-linked mechanisms. Journal of Andrology 25, 706–719.

Barclay, L.R.C., Vinqvist, M.R., Mukai, K., Goto, H., Hashimoto, Y., Tokunaga, A., Uno, H., 2000. On the antioxidant mechanism of curcumin: classical methods are needed to determine antioxidant mechanism and activity. Organic Letters 2 (18), 2841–2843.

Baynes, J.W., 1991. Role of oxidative stress in development of complication of diabetes. Diabetes 40, 405–412.

Best, L., Elliott, A.C., Brown, P.D., 2007. Curcumin induces electrical activity in rat pancreatic β-cells by activating the volume-regulated anion channel. Biochemical Pharmacology 73 (11), 1768–1775.

Bhakuni, R.S., Shukla, Y.N., Tripathi, A.K., Prajapati, V., Kumar, S., 2002. Insect growth inhibitor activity of arjunolic acid isolated from *Cornus capitata*. Phytotherapy Research 16 (1), S68–S70.

Bhattacharya, S., Sanyal, A.K., Ghosal, S., 1972. Monoamine oxidase-inhibiting activity of mangiferin isolated from Canscora-Decussata. Naturwissenschaften 59, 651.

Bierhaus, A., Hofmann, M.A., Ziegler, R., Nawroth, P., 1998. AGE and other interaction with AGE-receptors in vascular disease and diabetes-I. The AGE concept. Cardiovascular Research 37, 586–590.

Bisht, S., Maitra, A., 2009. Systemic delivery of curcumin: 21st century solutions for an ancient conundrum. Current Drug Discovery Technology 6 (3), 192–199.

Biswas, T., Sen, A., Roy, R., Maji, S., Maji, H.S., 2015. Isolation of mangiferin from flowering buds of *Mangifera indica* L and its evaluation of in vitro antibacterial activity. Research & Reviews: Journal of Pharmaceutical Analysis 4, 49–56.

Bonnefont-Rousselot, D., Beaudeux, J.L., Thérond, P., Peynet, J., Legrand, A., Delattre, J., 2004. Diabetes mellitus, oxidative stress and advanced glycation endproducts. Annales Pharmaceutiques Francaises 62, 147–157.

Brownlee, M., 2001. Biochemistry and molecular cell biology of diabetic complications. Nature 414, 813–820.

Brownlee, M., 2005. The pathobiology of diabetic complications: a unifying mechanism. Diabetes 54, 1615–1625.

Cai, K., Qi, D., Hou, X., Wang, O., Chen, J., Deng, B., Qian, L., Liu, X., Le, Y., 2011. MCP-1 upregulates amylin expression in murine pancreatic-β cells through ERK/JNK-AP1 and NF-κB related signaling pathways independent of CCR2. PLoS One 6 (5), e19559.

Cameron, D.F., Rountree, J., Schultz, R.E., Repetta, D., Murray, F.T., 1990. Sustained hyperglycemia results in testicular dysfunction and reduced fertility potential in BBWOR diabetic rats. American Journal of Physiology 259, E881–E889.

Canamares, M., Garcia-Ramos, J., Sanchez-Cortes, S., 2006. Degradation of curcumin dye in aqueous solution and on Ag nanoparticles studied by ultraviolet–visible absorption and surface-enhanced Raman spectroscopy. Applied Spectroscopy 60 (12), 1386–1391.

Chainani-Wu, N., 2003. Safety and anti-inflammatory activity of curcumin: a component of tumeric (*Curcuma longa*). Journal of Alternative and Complementary Medicine 9 (1), 161–168.

Chang, N.S., Joki, N., Mattison, J., Dinh, T., John, S., 1997. Characterization of serum adhesive proteins that block tumor necrosis factor-mediated cell death. Cell Death and Differentiation 4, 779–786.

Chanpoo, M., Petchpiboonthai, H., Panyarachun, B., Anupunpisit, V., 2010. Effect of curcumin in the amelioration of pancreatic islets in streptozotocin-induced diabetic mice. Journal of the Medical Association of Thailand Chotmaihet Thangphaet 93, S152–S159.

Chattopadhyay, I., Biswas, K., Uday, B., et al., 2004. Turmeric and curcumin: biological actions and medicinal applications. Current Science India 87 (1), 44–53.

Chuengsamarn, S., Rattanamongkolgul, S., Luechapudiporn, R., Phisalaphong, C., Jirawatnotai, S., 2012. Curcumin extract for prevention of Type 2 diabetes. Diabetes Care 35 (11), 2121–2127.

Cordero, H.I., Martín, M.A., Goya, L., Ramos, S., 2015. Cocoa flavonoids protect hepatic cells against high-glucose-induced oxidative stress: relevance of MAPKs. Molecular Nutrition and Food Research 59, 597–609.

Dar, A., Faizi, S., Naqvi, S., Roome, T., Zikr-ur-Rehman, S., Ali, M., Firdous, S., Moin, S.T., 2005. Analgesic and antioxidant activity of mangiferin and its derivatives: the structure activity relationship. Biological and Pharmaceutical Bulletin 28 (4), 596–600.

Daval, M., Bedrood, S., Gurlo, T., Huang, C.J., Costes, S., Butler, P.C., Langen, R., 2010. The effect of curcumin on human islet amyloid polypeptide misfolding and toxicity. Amyloid 17 (3–4), 118–128.

Debnath, T., Kim, H., Lim, B.O., 2013. Natural products as a source of anti-inflammatory agents associated with inflammatory bowel disease. Molecules 18, 7253–7270.

Demain, A.L., Sanchez, S., 2009. Microbial drug discovery: 80 years of progress. Journal of Antibiotics (Tokyo) 62, 5–16.

Deng, T., Sieglaff, D.H., Zhang, A., Lyon, C.J., Ayers, S.D., Cvoro, A., Gupte, A.A., Xia, X., Baxter, J.D., Webb, P., Hsueh, W.A., 2011. A peroxisome proliferator-activated receptor γ (PPARγ)/PPARγ coactivator 1β autoregulatory loop in adipocyte mitochondrial function. The Journal of Biological Chemistry 286 (35), 30723–30731.

Denisova, O.A., Glyzin, V.I., Patudin, A.V., Gavrilenko, B.D., 1980. Determination of the content of the xanthone glycoside mangiferin in iris, gentiana and hedysarum Plants. Khimiko-Farmatsevticheskli Zhurnal 14, 76–77.

Desai, P.D., Ganguly, A.K., Govindachari, T.R., Joshi, B.S., Kamat, V.N., Manmade, A.H., Mohamed, P.A., Nagle, S.K., Nayak, R.H., Saksena, A.K., Sathe, S.S., Vishwanathan, N., 1966. Chemical investigation of some Indian plants: Part II. Indian Journal of Chemistry 4, 457–549.

Diallo, B., Vanhaelen, M., Vanhaelen-Fastré, R., Konoshima, T., Kozuka, M., Tokuda, H., 1989. Studies on inhibitors of skin-tumor promotion. Inhibitory effects of triterpenes from *Cochlospermum tinctorium* on Epstein-Barr virus activation. Journal of Natural Products 52 (4), 879–881.

Dineshkumar, B., Mitra, A., Manjunatha, M., 2010. Studies on the anti-diabetic and hypolipidemic potentials of mangiferin (xanthone glucoside) in streptozotocin-induced Type 1 and Type 2 diabetic model rats. International Journal of Advances in Pharmaceutical Sciences 1, 75–85.

Djoukeng, J.D., Abou-Mansour, E., Tabacchi, R., Tapondjou, A.L., Bouda, H., Lontsi, D., 2005. Antibacterial triterpenes from *Syzygium guineense* (Myrtaceae). Journal of Ethnopharmacology 101 (1–3), 283–286.

El Ansari, M.A., Rajadurai, S., Nayudamma, Y., 1967. Studies on the polyphenols of *Mangifera indica* bark. Leather Science. 14, 247–251.

El Ansari, M.A., Reddy, K.K., Sastry, K.N.S., Nayudamma, Y., 1971. Dicotyledonae, anacardiaceae polyphenols of *Mangifera indica*. Phytochemistry 10, 2239–2241.

El-Azab, M.F., Attia, F.M., El-Mowafy, A.M., 2011. Novel role of curcumin combined with bone marrow transplantation in reversing experimental diabetes: effects on pancreatic islet regeneration, oxidative stress, and inflammatory cytokines. European Journal of Pharmacology 658 (1), 41–48.

El-Moselhy, M.A., Taye, A., Sharkawi, S.S., El-Sisi, S.F.I., Ahmed, A.F., 2011. The antihyperglycemic effect of curcumin in high fat diet fed rats: role of TNF-α and free fatty acids. Food and Chemical Toxicology 49 (5), 1129–1140.

Forbes, J.M., Coughlan, M.T., Cooper, M.E., 2008. Oxidative stress as a major culprit in kidney disease indiabetes. Diabetes 57, 1446–1454.

Franckhauser, S., Muñoz, S., Elias, I., Ferre, T., Bosch, F., 2006. Adipose overexpression of phosphoenolpyruvate carboxykinase leads to high susceptibility to diet-induced insulin resistance and obesity. Diabetes 55 (2), 273–280.

Fujiwara, H., Hosokawa, M., Zhou, X., et al., 2008. Curcumin inhibits glucose production in isolated mice hepatocytes. Diabetes Research and Clinical Practice 80 (2), 185–191.

Ghorbani, Z., Hekmatdoost, A., Mirmiran, P., 2014. Anti hyperglycemic and insulin sensitizer effects of turmeric and its principle constituent curcumin. International Journal of Endocrinology and Metabolism 12 (4), e18081.

Ghosh, J., Sil, P.C., 2013. Arjunolic acid: a new multifunctional therapeutic promise of alternative medicine. Biochimie 95 (6), 1098–1109.

Ghosh, J., Das, J., Manna, P., Sil, P.C., 2011. The protective role of arjunolic acid against doxorubicin induced intracellular ROS dependent JNK-p38 and p53-mediated cardiac apoptosis. Biomaterials 32 (21), 4857–4866.

Ghosh, M., Das, J., Sil, P.C., 2012. D(+) galactosamine induced oxidative and nitrosative stress-mediated renal damage in rats via NF-κB and inducible nitric oxide synthase (iNOS) pathways is ameliorated by a ployphenol xanthone, mangiferin. Free Radical Research 46, 116–132.

Ghosh, S., Bhattacharyya, S., Rashid, K., Sil, P.C., 2015. Curcumin protects rat liver from streptozotocin-induced diabetic pathophysiology by counteracting reactive oxygen species and inhibiting the activation of p53 and MAPKs mediated stress response pathways. Toxicology Reports 2, 365–376.

Girón, M.D., Sevillano, N., Salto, R., Haidour, A., Manzano, M., Jiménez,ML., et al., 2009. *Salacia oblonga* extract increases glucose transporter 4-mediated glucose uptake in L6 rat myotubes: role of mangiferin. Clinical Nutrition 28 (5), 565–574.

Goel, A., Kunnumakkara, A.B., Aggarwal, B.B., 2008. Curcumin as "Curecumin": from kitchen to clinic. Biochemical Pharmacology 75 (4), 787–809.

Grynkiewicz, G., Slifirski, P., 2012. Curcumin and curcuminoids in quest for medicinal status. Acta Biochimica Polonica 59 (2), 201–212.

Guha, S., Ghosal, S., Chattopadhyay, U., 1996. Antitumor, immunomodulatory and anti-HIV effect of mangiferin, a naturally occurring glucosyl xanthone. Chemotherapy 42 (6), 443–445.

Gupta, S.C., Sung, B., Kim, J.H., Prasad, S., Li, S., Aggarwal, B.B., 2013. Multitargeting by turmeric, the golden spice: from kitchen to clinic. Molecular Nutrition and Food Research 57 (9), 1510–1528.

Gupta, S.C., Tyagi, A.K., Deshmukh-Taskar, P., Hinojosa, M., Prasad, S., Aggarwal, B.B., 2014. Downregulation of tumor necrosis factor and other proinflammatory biomarkers by polyphenols. Archives of Biochemistry and Biophysics 559, 91–99.

Gutierres, V.O., Pinheiro, C.M., Assis, R.P., Vendramini, R.C., Pepato, M.T., Brunetti, I.L., 2012. Curcumin-supplemented yoghurt improves physiological and biochemical markers of experimental diabetes. The British Journal of Nutrition 108 (3), 440–448.

Hajizadeh, M.R., Eftekhar, E., Zal, F., Jafarian, A., Mostafavi-Pour, Z., 2014. Mulberry leaf extract attenuates oxidative stress-mediated testosterone depletion in streptozotocin induced diabetic rats. Iranian Journal of Medical Sciences 39 (2), 123–129.

Halliwell, B., 1990. How to characterise a biological antioxidant. Free Radical Research Communications 9, 1–32.

Hatchera, H., Planalp, R., Chob, J., et al., 2008. Curcumin: from ancient medicine to current clinical trials. Cellular and Molecular Life Sciences 65 (11), 1631–1652.

Haukeland, J.W., Dahl, T.B., Yndestad, A., Gladhaug, I.P., Løberg, E.M., Haaland, T., Konopski, Z., Wium, C., Aasheim, E.T., Johansen, O.E., Aukrust, P., Halvorsen, B., Birkeland, K.I., 2012. Fetuin A in nonalcoholic fatty liver disease: in vivo and in vitro studies. European Journal of Endocrinology 166 (3), 503–510.

Hemalatha, T., Pulavendran, S., Balachandran, C., Manohar, B.M., Puvanakrishnan, R., 2010. Arjunolic acid: a novel phytomedicine with multifunctional therapeutic applications. Indian Journal of Experimental Biology 48 (3), 238–247.

Hou, W., Li, Y., Zhang, Q., Wei, X., Peng, A., Chen, L., Wei, Y., 2009. Triterpene acids isolated from *Lagerstroemia speciosa* leaves as alpha-glucosidase inhibitors. Phytotherapy Research 23 (5), 614–618.

Hou, J., Zheng, D., Zhong, G., Hu, Y., 2013. Mangiferin mitigates diabetic cardiomyopathy in streptozotocin-diabetic rats. Canadian Journal of Physiology and Pharmacology 91 (9), 759–763.

Huang, T.H., He, L., Qin, Q., Yang, Q., Peng, G., Harada, M., Qi, Y., Yamahara, J., Roufogalis, B.D., Li, Y., 2008. *Salacia oblonga* root decreases cardiac hypertrophy in Zucker diabetic fatty rats: inhibition of cardiac expression of angiotensin II type 1 receptor. Diabetes Obesity and Metabolism 10 (7), 574–585.

Ibrahim, H.N., Hostetter, T.H., 1997. Diabetic nephropathy. Journal of the American Society of Nephrology 8, 487–493.

Irfan, B., 2002. Phytochemical Studies on *Ferula mongolica* and Other Mongolian Medicinal Plants. Ph.D. thesis). International Centre for Chemical Sciences, University of Karachi.

Ishibashi, T., 2013. Molecular hydrogen, new antioxidant and anti-inflammatory therapy for rheumatoid arthritis and related diseases. Current Pharmaceutical Design 19, 6375–6381.

Islam, M.N., Yoo, H.H., Lee, J., Nam, J.W., Seo, E.K., Jin, C., Kim, D.H., 2008. Simultaneous determination of bioactive xanthone glycosides and norlignans from ethanolic extract of *Anemarrhena asphodeloides* by liquid chromatography. Journal of AOAC International 91, 1271–1277.

Iwunze, M.O., McEwan, D., 2004. Peroxynitrite interaction with curcumin solubilised in ethanol. Cellular and Molecular Biology 50, 749–752.

Iwunze, M.O., 2004. Binding and distribution characteristics of curcumin solubilized in CTAB micelle. Journal of Molecular Liquids 111 (1), 161–165.

Jacob, A., Wu, R., Zhou, M., Wang, P., 2007. Mechanism of the anti-inflammatory effect of curcumin: PPAR-gamma activation. PPAR Research 2007, 89369.

Jacobsen, I.B., Henriksen, J.E., Beck-Nielsen, H., 2009. The effect of metformin in overweight patients with type 1 diabetes and poor metabolic control. Basic and Clinical Pharmacology and Toxicology 105, 145–149.

Jana, K., Jana, S., Samanta, P.K., 2006. Effects of chronic exposure to sodium arsenite on hypothalamo-pituitary-testicular activities in adult rats: possible an estrogenic mode of action. Reproductive Biology and Endocrinology 4, 9–22.

Jain, S.K., Rains, J., Croad, J., Larson, B., Jones, K., 2009. Curcumin supplementation lowers TNF-alpha, IL-6, IL-8, and MCP-1 secretion in high glucose-treated cultured monocytes and blood levels of TNF-alpha, IL-6, MCP-1, glucose, and glycosylated hemoglobin in diabetic rats. Antioxidants and Redox Signaling 11 (2), 241–249.

Jash, S.K., Brahmachari, G., 2015. Chemical profile and health benefits of fruit mango—an emerging functional food: an update. Signpost Open Access Journal of Organic and Biomolecular Chemistry 3, 1–27.

Jovanovic, S.V., Boone, C.W., Steenken, S., Trinoga, M., Kaskey, R.B., 2001. How curcumin works preferentially with water soluble antioxidants. Journal of the American Chemical Society 123 (13), 3064–3068.

Jurenka, J.S., 2009. Anti-inflammatory properties of curcumin, a major constituent of *Curcuma longa*: a review of preclinical and clinical research. Alternative Medicine Review 14 (2), 141–153.

Jutiviboonsuk, A., Sardsaengjun, C., 2010. Mangiferin in leaves of three Thai mango (*Mangifera indica* L.) varieties. Isan Journal of Pharmaceutical Science 6 (3), 122–129.

Kakkar, V., Singh, S., Singla, D., Kaur, I.P., 2011. Exploring solid lipid nanoparticles to enhance the oral bioavailability of curcumin. Molecular Nutrition and Food Research 55 (3), 495–503.

Kang, Q., Chen, A., 2009. Curcumin eliminates oxidized LDL roles in activating hepatic stellate cells by suppressing gene expression of lectin-like oxidized LDL receptor-1. Laboratory Investigation 89 (11), 1275–1290.

Kanitkar, M., Bhonde, R.R., 2008. Curcumin treatment enhances islet recovery by induction of heat shock response proteins, Hsp70 and heme oxygenase-1, during cryopreservation. Life Sciences 82 (3–4), 182–189.

Kanitkar, M., Gokhale, K., Galande, S., Bhonde, R.R., 2008. Novel role of curcumin in the prevention of cytokine-induced islet death in vitro and diabetogenesis in vivo. British Journal of Pharmacology 155 (5), 702–713.

Kanter, M., Aktas, C., Erboga, M., 2012. Protective effects of quercetin against apoptosis and oxidative stress in streptozotocin-induced diabetic rat testis. Food and Chemical Toxicology 50, 719–725.

Kanter, M., Aktas, C., Erboga, M., 2013. Curcumin attenuates testicular damage, apoptotic germ cell death, and oxidative stress in streptozotocin-induced diabetic rats. Molecular Nutrition and Food Research 57, 1578–1585.

Khaneshi, F., Nasrolahi, O., Azizi, S., Nejati, V., 2013. Sesame effects on testicular damage in streptozotocin-induced diabetes rats. Avicenna Journal of Phytomedicine 3 (4), 347–355.

Kim, J.E., Kim, A.R., Chung, H.Y., Han, S.Y., Kim, B.S., Choi, J.S., 2003. In vitro peroxynitrite scavenging activity of diarylheptanoids from *Curcuma longa*. Phytotherapy Research 17 , 481–484.

Kim, D.H., Han, K.M., Chung, I.S., Kim, D.K., Kim, S.H., Kwon, B.M., Jeong, T.S., Park, M.H., Ahn, E.M., Baek, N.I., 2005. Triterpenoids from the flower of *Campsis grandiflora* K. Schum. as human acyl-CoA: cholesterol acyltransferase inhibitors. Archives of Pharmacal Research 28 (5), 550–556.

Kim, T., Davis, J., Zhang, A.J., He, X., Mathews, S.T., 2009. Curcumin activates AMPK and suppresses gluconeogenic gene expression in hepatoma cells. Biochemical and Biophysical Research Communications 388 (2), 377–382.

Kim, S.W., Ha, K.C., Choi, E.K., Jung, S.Y., Kim, M.G., Kwon, D.Y., Yang, H.J., Kim, M.J., Kang, H.J., Back, H.I., Kim, S.Y., Park, S.H., Baek, H.Y., Kim, Y.J., Lee, J.Y., Chae, S.W., 2013. The effectiveness of fermented turmeric powder in subjects with elevated alanine transaminase levels: a randomised controlled study. BMC Complementary and Alternative Medicine 13, 58–64.

Kimura, K., Lee, J.H., Lee, I.S., Lee, H.S., Park, K.H., Chiba, S., 2004. Two potent competitive inhibitors discriminating alpha-glucosidase family I from family II. Carbohydrate Research 339, 1035–1040.

King, G.L., Ishii, H., Koya, D., 1997. Diabetic vascular dysfunction: a model of excessive activation of protein kinase C. Kidney International 60, S77–S85.

Kowluru, R.A., Tang, J., Kern, T.S., 2001. Abnormalities of retinal metabolism in diabetes and experimental galactosemia. VII. Effect of long-term administration of antioxidants on the development of retinopathy. Diabetes 50, 1938–1942.

Kumar, D.S., Prabhakar, Y.S., 1987. On the ethnomedical significance of the Arjun tree, *Terminalia arjuna* (Roxb.) Wight & Arnot. Journal of Ethnopharmacology 20 (2), 173–190.

Kumar, S., Narwal, S., Kumar, V., Prakash, O., 2011. α-glucosidase inhibitors from plants: a natural approach to treat diabetes. Pharmacognosy Reviews 5 (9), 19–29.

Kuroda, M., Mimaki, Y., Nishiyama, T., Mae, T., Kishida, H., Tsukagawa, M., Takahashi, K., Kawada, T., Nakagawa, K., Kitahara, M., 2005. Hypoglycemic effects of turmeric (*Curcuma longa* L. rhizomes) on genetically diabetic KK-Ay mice. Biological and Pharmaceutical Bulletin 28 (5), 937–939.

La Vignera, S., Calogero, A.E., Condorelli, R., Lanzafame, F., Giammusso, B., Vicari, E., 2009. Andrological characterization of the patient with diabetes mellitus. Minerva Endocrinologica 34, 1–9.

Leiro, J., Arranz, J.A., Yáñez, M., Ubeira, F.M., Sanmartín, M.L., Orallo, F., 2004. Expression profiles of genes involved in the mouse nuclear factor-kappa B signal transduction pathway are modulated by mangiferin. International Immunopharmacology 4 (6), 763–778.

Lenzen, S., Drinkgern, J., Tiedge, M., 1996. Low antioxidant enzyme gene expression in pancreatic islets compared with various other mouse tissues. Free Radical Biology and Medicine 20, 463–466.

Li, W.L., Zheng, H.C., Bukuru, J., De Kimpe, N., 2004a. Natural medicines used in the traditional Chinese medical system for therapy of diabetes mellitus. Journal of Ethnopharmacology 92 (1), 1–21.

Li, Y., Peng, G., Li, Q., Wen, S., Huang, T.H., Roufogalis, B.D., Yamahara, J., 2004b. *Salacia oblonga* improves cardiac fibrosis and inhibits postprandial hyperglycemia in obese Zucker rats. Life Sciences 75 (14), 1735–1746.

Li, X., Cui, X., Sun, X., Li, X., Zhu, Q., Li, W., 2010. Mangiferin prevents diabetic nephropathy progression in streptozotocin-induced diabetic rats. Phytotherapy Research 24, 893–899.

Li, X.J., Du, Z.C., Huang, Y., et al., 2013. Synthesis and hypoglycemic activity of esterified-derivatives of mangiferin. Chinese Journal Natural Medicine 11, 296–301.

Lin, J., Chen, A., 2011. Curcumin diminishes the impacts of hyperglycemia on the activation of hepatic stellate cells by suppressing membrane translocation and gene expression of glucose transporter-2. Molecular and Cellular Endocrinology 333 (2), 160–171.

Lin, J., Tang, Y., Kang, Q., Feng, Y., Chen, A., 2012. Curcumin inhibits gene expression of receptor for advanced glycation endproducts (RAGE) in hepatic stellate cells in vitro by elevating PPAR-gamma activity and attenuating oxidative stress. British Journal of Pharmacology 166 (8), 2212–2227.

Lin, M., Zhai, X., Wang, G., Tian, X., Gao, D., Shi, L., Wu, H., Fan, Q., Peng, J., Liu, K., Yao, J., 2015. Salvianolic acid B protects against acetaminophen hepatotoxicity by inducing Nrf2 and phase II detoxification gene expression via activation of the PI3K and PKC signaling pathways. Journal of Pharmacological Sciences 127, 203–210.

Lipinski, C.A., Lombardo, F., Dominy, B.W., Feeney, P.J., 1997. Experimental and computational approaches to estimate solubility and permeability in drug discovery and development setting (review). Advanced Drug Delivery Reviews 23, 3–25.

Litwinienko, G., Ingold, K.U., 2005. Abnormal solvent effects on hydrogen atom abstraction. 3. Novel kinetics in sequential proton loss electron transfer chemistry. The Journal of Organic Chemistry 70 (22), 8982–8990.

Liu, Y., Xu, F., Zeng, X., Yang, L., Deng, Y., Wu, Z., Feng, Y., Li, X., 2010. Application of a liquid chromatography/tandem mass spectrometry method to pharmacokinetic study of mangiferin in rats. Journal of Chromatography B, Analytical Technologies in the Biomedical and Life Sciences 878 (32), 3345–3350.

Liu, H., Wu, B., Pan, G., He, L., Li, Z., Fan, M., Jian, L., Chen, M., Wang, K., Huang, C., 2012. Metabolism and pharmacokinetics of mangiferin in conventional rats, pseudogerm-free rats and streptozotocin-induced diabetic rats. Drug Metabolism and Disposition 40 (11), 2109–2118.

Liu, G.L., Zhang, Y.M., Dai, D.Z., Ding, M.J., Cong, X.D., Dai, Y., 2013. Male hypogonadism induced by high fat diet and low dose streptozotocin is mediated by activated endoplasmic reticulum stress and IκBβ and attenuated by argirein and valsartan. European Journal of Pharmacology 713 (1–3), 78–88.

Liu, Y.W., Zhu, X., Zhang, L., Lu, Q., Wang, J.Y., Zhang, F., Guo, H., Yin, J.L., Yin, X.X., 2013. Up-regulation of glyoxalase 1 by mangiferin prevents diabetic nephropathy progression in streptozotocin-induced diabetic rats. European Journal of Pharmacology 721 (1–3), 355–364.

Luo, F., Lv, Q., Zhao, Y., Hu, G., Huang, G., Zhang, J., Sun, C., Li, X., Chen, K., 2012. Quantification and purification of mangiferin from Chinese mango (*Mangifera indica* L.) cultivars and its protective effect on human umbilical vein endothelial cells under H_2O_2-induced stress. International Journal of Molecular Sciences 13, 11260–11274.

Mahesh, T., Sri Balasubashini, M.M., Menon, V.P., 2004. Photoirradiated curcumin supplementation in streptozotocin induced diabetic rats: effect on lipid peroxidation. Therapie 59 (6), 639–644.
Malviya, N., Jain, S., Malviya, S., 2010. Antidiabetic potential of medicinal plants. Acta Poloniae Pharmaceutica 67 (2), 113–118.
Manna, P., Sinha, M., Pal, P., Sil, P.C., 2007. Arjunolic acid, a triperpenoid saponin, ameliorates arsenic induced cytotoxicity in hepatocytes. Chemico-Biological Interactions 170, 187–200.
Manna, P., Sinha, M., Sil, P.C., 2008. Arsenic–induced oxidative myocardial injury: protective role of arjunolic acid. Archives of Toxicology 82, 137–149.
Manna, P., Sinha, M., Sil, P.C., 2009a. Protective role of arjunolic acid in response to streptozotocin-induced type-I diabetes via the mitochondrial dependent and independent pathways. Toxicology 257, 53–63.
Manna, P., Sinha, M., Sil, P.C., 2009b. Prophylactic role of arjunolic acid in response to streptozotocin mediated diabetic renal injury: activation of polyol pathway and oxidative stress responsive signaling cascades. Chemico-Biological Interactions 181 (3), 297–308.
Manna, P., Ghosh, J., Das, J., Sil, P.C., 2010. Streptozotocin induced activation of oxidative stress responsive splenic cell signaling pathways: protective role of arjunolic acid. Toxicology and Applied Pharmacology 244 (2), 114–129.
Marhfour, I., Lopez, X.M., Lefkaditis, D., Salmon, I., Allagnat, F., Richardson, S.J., Morgan, N.G., Eizirik, D.L., 2012. Expression of endoplasmic reticulum stress markers in the islets of patients with type 1 diabetes. Diabetologia 55, 2417–2420.
Martínez Sánchez, G., Candelario-Jalil, E., Giuliani, A., León, O.S., Sam, S., Delgado, R., et al., 2001. *Mangifera indica* L. extract (QF808) reduces ischaemia-induced neuronal loss and oxidative damage in the gerbil brain. Free Radical Research 35 (5), 465–473.
Masoko, P., Mdee, L.K., Mampuru, L.J., Eloff, J.N., 2008. Biological activity of two related triterpenes isolated from *Combretum nelsonii* (Combretaceae) leaves. Natural Product Research 22 (12), 1074–1084.
Matkowski, A., KuÊ, P., Góralska, E., Woêniak, D., 2013. Mangiferin – a bioactive xanthonoid, not only from mango and not just antioxidant. Mini-Reviews in Medicinal Chemistry 13, 439–455.
Meghana, K., Sanjeev, G., Ramesh, B., 2007. Curcumin prevents streptozotocin-induced islet damage by scavenging free radicals: a prophylactic and protective role. European Journal of Pharmacology 577 (1–3), 183–191.
Mishra, B., Priyadarsini, K.I., Bhide, M.K., Kadam, R.M., Mohan, H., 2004. Reactions of superoxide radicals with curcumin: probable mechanisms by optical spectroscopy and EPR. Free Radical Research 38, 355–362.
Mishra, B., Indira Priyadarsini, K., Sudheerkumar, M., Unnikrishhnan, M.K., Mohan, H., 2006. Pulse radiolysis studies of mangiferin: a C-glycosyl xanthone isolated from *Mangifera indica*. Radiation Physics and Chemistry 75, 70–77.
Miura, T., Ichiki, H., Hashimoto, I., Iwamoto, N., Kato, M., Kubo, M., Ishihara, E., Komatsu, Y., Okada, M., Ishida, T., Tanigawa, K., 2001. Antidiabetic activity of a xanthone compound, mangiferin. Phytomedicine 8 (2), 85–87.
Modak, M., Dixit, P., Londhe, J., Ghaskadbi, S., Devasagayam, T.P., 2007. Indian herbs and herbal drugs used for the treatment of diabetes. Journal of Clinical Biochemistry and Nutrition 40, 163–173.
Mohammed, A., Kumar, D., Rizvi, S.I., 2015. Antidiabetic potential of some less commonly used plants in traditional medicinal systems of India and Nigeria. Journal of Intercultural Ethnopharmacology 4 (1), 78–85.
Mohanty, C., Das, M., Sahoo, S.K., 2012. Emerging role of nanocarriers to increase the solubility and bioavailability of curcumin. Expert Opinion on Drug Delivery 9 (11), 1347–1364.

Muruganandan, S., Gupta, S., Kataria, M., Lal, J., Gupta, P.K., 2002. Mangiferin protects the streptozotocin-induced oxidative damage to cardiac and renal tissues in rats. Toxicology 176 (3), 165–173.

Muruganandan, S., Srinivasan, K., Gupta, S., Gupta, P.K., Lal, J., 2005. Effect of mangiferin on hyperglycemia and atherogenicity in streptozotocin diabetic rats. Journal of Ethnopharmacology 97 (3), 497–501.

Na, L.X., Li, Y., Pan, H.Z., Zhou, X.L., Sun, D.J., Meng, M., Li, X.X., Sun, C.H., 2013. Curcuminoids exert glucose-lowering effect in type 2 diabetes by decreasing serum free fatty acids: a double-blind, placebo-controlled trial. Molecular Nutriton and Food Research 57 (9), 1569–1577 (Research Support, Non-U.S. Government).

Nabavi, S.F., Thiagarajan, R., Rastrelli, L., Daglia, M., Sobarzo-Sánchez, E., Alinezhad, H., Nabavi, S.M., 2015. Curcumin: a natural product for diabetes and its complications. Current Topics in Medicinal Chemistry 15 (23), 2445–2455.

Najafian, M., Ebrahim-Habibi, A., Yaghmaei, P., Parivar, K., Larijani, B., 2010. Core structure of flavonoids precursor as an antihyperglycemic and antihyperlipidemic agent: an in vivo study in rats. Acta Biochimica Polonica 57 (4), 553–560.

Najafian, M., Jahromi, M.Z., Nowroznejhad, M.J., Khajeaian, P., Kargar, M.M., Sadeghi, M., Arasteh, A., 2011. Phloridzin reduces blood glucose levels and improves lipids metabolism in streptozotocin-induced diabetic rats. Molecular Biology Reports 1 (1), 1–8.

Najafian, M., 2015. The effects of curcumin on alpha amylase in diabetics rats. Zahedan Journal of Research in Medical Sciences 17 (12) e5198. pp. 29–34.

Nigam, S.K., Mitra, C.R., 1964. Constituents of mango (*Mangifera indica*) roots. Indian Journal of Chemistry 2, 378–379.

Nimse, S.B., Pal, D.K., 2015. Free radicals, natural antioxidants and their reaction mechanism (review article). RSC Advances 5, 27986–28006.

Nishiyama, T., Mae, T., Kishida, H., Tsukagawa, M., Mimaki, Y., Kuroda, M., Sashida, Y., Takahashi, K., Kawada, T., Nakagawa, K., Kitahara, M., 2005. Curcuminoids and sesquiterpenoids in turmeric (*Curcuma longa* L.) Suppress an increase in blood glucose level in type 2 diabetic KK-Aγ mice. Journal of Agricultural and Food Chemistry 53 (4), 959–963.

Niu, Y., Lu, W., Gao, L., Lin, H., Liu, X., Li, L., 2012. Reducing effect of mangiferin on serum uric acid levels in mice. Pharmaceutical Biology 50, 1177–1182.

Nuñez-Selles, A.J., 2005. Antioxidant therapy: myth or reality? Journal of the Brazilian Chemical Society 16, 699–710.

Oberley, L.W., 1988. Free radicals and diabetes. Free Radical Biology and Medicine 5 (2), 113–124.

Oner-Iyidogan, Y., Kocak, H., Seyidhanoglu, M., Gürdöl, F., Gülçubuk, A., Yildirim, F., Çevik, A., Uysal, M., 2013. Curcumin prevents liver fat accumulation and serum fetuin-A increase in rats fed a high-fat diet. Journal of Physiology and Biochemistry 69 (4), 677–686.

Ono, K., Hasegawa, K., Naiki, H., Yamada, M., 2004. Curcumin has potent anti-amyloidogenic effects for Alzheimer's β-amyloid fibrils in vitro. Journal of Neuroscience Research 75 (6), 742–750.

Ozcan, U., Cao, Q., Yilmaz, E., Lee, A.H., Iwakoshi, N.N., Ozdelen, E., Tuncman, G., Görgün, C., Glimcher, L.H., Hotamisligil, G.S., 2004. Endoplasmic reticulum stress links obesity, insulin action, and type 2 diabetes. Science 306, 457–461.

Pal, P.B., Sinha, K., Sil, P.C., 2014. Mangiferin attenuates diabetic nephropathy by inhibiting oxidative stress mediated signaling cascade, TNFα related and mitochondrial dependent apoptotic pathways in streptozotocin-induced diabetic rats. PLoS One 9 (9), e107220.

Pardo-Andreu, G.L., Philip, S.J., Riano, A., Sanchez, C., Viada, C., Nunez-Selles, A.J., Delgado, R., 2006. *Mangifera indica* L. (Vimang) protection against serum oxidative stress in elderly humans. Archives of Medical Research 37, 158–164.

Park, I.S., Kiyomoto, H., Abboud, S.L., Abboud, H.E., 1997. Expression of transforming growth factor-b and type IV collagen in early streptozotocin-induced diabetes. Diabetes 46, 473–480.

Patumraj, S., Wongeakin, N., Sridulyakul, P., Jariyapongskul, A., Futrakul, N., Bunnag, S., 2006. Combined effects of curcumin and vitamin C to protect endothelial dysfunction in the iris tissue of STZ-induced diabetic rats. Clinical Hemorheology and Microcirculation 35 (4), 481–489.

Pérez-Torres, I., Ruiz-Ramírez, A., Baños, G., El-Hafidi, M., 2013. Hibiscus sabdariffa Linnaeus (Malvaceae), curcumin and resveratrol as alternative medicinal agents against metabolic syndrome. Cardiovascular & Hematological Agents in Medicinal Chemistry 11 (1), 25–37.

Periyar Selvam, S., Arulselvan, P., Balu, P.M., Murugesan, K., 2012. Effect of mangiferin isolated from *Salacia chinensis* regulates the kidney carbohydrate metabolism in streptozotocin-induced diabetic rats. Asian Pacific Journal of Tropical Biomedicine 2, S1583–S1587.

Pham, T.X., Lee, J., 2012. Dietary regulation of histone acetylases and deacetylases for the prevention of metabolic diseases. Nutrients 4 (12), 1868–1886.

Prabhu, S., Jainu, M., Sabitha, K.E., Devi, C.S., 2006. Role of mangiferin on biochemical alterations and anti-oxidant status in isoproterenol-induced myocardial infarction in rats. Journal of Ethnopharmacology 107, 126–133.

Prasad, A.K., Pant, N., Srivastava, S.C., Kumar, R., Srivastava, S.P., 1995. Effect of dermal application of hexachlorocyclohexane (HCH) on male reproductive system of rat. Human and Experimental Toxicology 14, 484–488.

Qin, J.P., Deng, J.G., Feng, X., Wang, Q., Wang, S.B., 2008. Quantitative RP-LC analysis of mangiferin and homomangiferin in *Mangifera indica* L. leaves and in *Mangifera persiciforma* C.Y. Wu et T.L. Ming leaves. Chromatographia 68, 955–960.

Rajendran, P., Ekambaram, G., Sakthisekaran, D., 2008. Cytoprotective effect of mangiferin on benzo(a)pyrene-induced lung carcinogenesis in Swiss albino mice. Basic and Clinical Pharmacology and toxicology 103 (2), 137–142.

Ramanathan, J.D., Seshadri, T.R., 1960. Constitution of mangiferin. Current Science 29, 131–132.

Rashid, K., Sil, P.C., 2015a. Curcumin ameliorates testicular damage in diabetic rats by suppressing cellular stress mediated mitochondria and endoplasmic reticulum-dependent apoptotic death. Biochimica et Biophysica Acta Molecular Basis of Disease 1852 (1), 70–82.

Rashid, K., Sil, P.C., 2015b. Curcumin enhances recovery of pancreatic islets from cellular stress induced inflammation and apoptosis in diabetic rats. Toxicology and Applied Pharmacology 282 (3), 297–310.

Rashid, K., Bhattacharya, S., Sil, P.C., 2012. Protective role of D-saccharic acid-1,4-lactone in alloxan induced oxidative stress in the spleen tissue of diabetic rats is mediated by suppressing mitochondria dependent apoptotic pathway. Free Radical Research 46 (3), 240–252.

Rashid, K., Das, J., Sil, P.C., 2013a. Taurine ameliorate alloxan induced oxidative stress and intrinsic apoptotic pathway in the hepatic tissue of diabetic rats. Food and Chemical Toxicology 51, 317–329.

Rashid, K., Sinha, K., Sil, P.C., 2013b. An update on oxidative stress-mediated organ pathophysiology. Food and Chemical Toxicology 62, 584–600.

Rasyid, A., Rahman, A.R.A., Jaalam, K., Lelo, A., 2002. Effect of different curcumin dosages on human gall bladder. Asian Pacific Journal of Clinical Nutrition 11 (4), 314–318.

Reuter, S., Gupta, S.C., Chaturvedi, M.M., Aggarwal, B.B., 2010. Oxidative stress, inflammation, and cancer, how are they linked? Free Radical Biology and Medicine 49, 1603–1616.

Revathy, S., Elumalai, S., Merina, B., Benny, A., 2011. Isolation, purification and identification of curcuminoids from turmeric (*Curcuma longa* L.) by column chromatography. Journal of Experimental Sciences 2 (7), 21–25.

Rizvi, S.I., Mishra, N., 2013. Traditional Indian medicines used for the management of diabetes mellitus. Journal of Diabetes Research. 712092.

Rodeiro, I., Donato, M.T., Martínez, I., Hernández, I., Garrido, G., González-Lavaut, J.A., Menéndez, R., Laguna, A., Castell, J.V., Gómez-Lechón, M.J., 2008. Potential hepatoprotective effects of new Cuban natural products in rat hepatocytes culture. Toxicology in Vitro 22, 1242–1249.

Rolo, A.P., Palmeira, C.M., 2006. Diabetes and mitochondrial function: role of hyperglycemia and oxidative stress. Toxicology and Applied Pharmacology 212, 167–178.

Sahebkar, A., 2013. Why it is necessary to translate curcumin into clinical practice for the prevention and treatment of metabolic syndrome? BioFactors 39 (2), 197–208.

Sanchez, G.M., Re, L., Giuliani, A., Nuñez-Selles, A.J., Davison, G.P., Leon-Fernandez, O.S., 2000. Protective effects of *Mangifera indica* L. extract, mangiferin and selected antioxidants against TPA-induced biomolecules oxidation and peritoneal macrophage activation in mice. Pharmacological Research 42, 565–573.

Satapathy, T., Panda, P.K., 2013. Evaluation of in vitro antioxidant, anti inflammatory and anti diabetic potential of curcumin. Indo American Journal of Pharmaceutical Research 3 (3), 2808–2818.

Sato, T., Kawamoto, A., Tamura, A., Tatsumi, Y., Fujii, T., 1992. Mechanism of antioxidant action of pueraria glycoside (PG)-1 (an isoflavonoid) and mangiferin (a xanthonoid). Chemical and Pharmaceutical Bulletin (Tokyo) 40 (3), 721–724.

Scarano, W.R., Messias, A.G., Oliva, S.U., Klinefelter, G.R., Kempinas, W.G., 2006. Sexual behaviour, sperm quantity and quality after short-term streptozotocin-induced hyperglycaemia in rats. International Journal of Andrology 29, 482–488.

Schultze, S.M., Hemmings, B.A., Niessen, M., Tschopp, O., 2012. PI3K/AKT, MAPK and AMPK signalling: protein kinases in glucose homeostasis. Expert Reviews in Molecular Medicine 14, e1.

Sellamuthu, P.S., Muniappan, B.P., Perumal, S.M., Kandasamy, M., 2009. Antihyperglycemic effect of mangiferin in streptozotocin induced diabetic rats. Journal of Health Sciences 55 (2), 206–214.

Sellamuthu, P.S., Arulselvan, P., Kamalraj, S., Fakurazi, S., Kandasamy, M., 2013. Protective nature of mangiferin on oxidative stress and antioxidant status in tissues of streptozotocin-induced diabetic rats. ISRN Pharmacology.

Seo, K.I., Choi, M.S., Jung, U.J., Kim, H.J., Yeo, J., Jeon, S.M., Lee, M.K., 2008. Effect of curcumin supplementation on blood glucose, plasma insulin, and glucose homeostasis related enzyme activities in diabetic db/db mice. Molecular Nutrition and Food Research 52 (9), 995–1004.

Serrano-Martín, X., Payares, G., Mendoza-León, A., 2006. Glibenclamide, a blocker of K^+(ATP) channels, shows antileishmanial activity in experimental murine cutaneous leishmaniasis. Antimicrobial Agents and Chemotherapy 50 (12), 4214–4216.

Shah, S.V., Baliga, R., Rajapurkar, M., Fonseca, V.A., 2007. Oxidants in chronic kidney disease. Journal of the American Society of Nephrology 18, 16–28.

Shaikh, J., Ankola, D., Beniwal, V., Singh, D., Kumar, M.R., 2009. Nanoparticle encapsulation improves oral bioavailability of curcumin by at least 9-fold when compared to curcumin administered with piperine as absorption enhancer. European Journal of Pharmaceutical Sciences 37 (3), 223–230.

Sharpe, R.M., Maddocks, S., Millar, M., Saunders, P.T.K., Kerr, J.B., McKinnell, C., 1992. Testosterone and spermatogenesis: identification of stage dependent, androgen-regulated proteins secreted by adult rat seminiferous tubules. Journal of Andrology 13, 172–184.

Shehzad, A., Lee, Y., 2010. Curcumin: multiple molecular targets mediate multiple pharmacological actions: a review. Drugs of the Future 35 (2), 113–120.

Shehzad, A., Ha, T., Subhan, F., Lee, Y.S., 2011. New mechanisms and the anti-inflammatory role of curcumin in obesity and obesity-related metabolic diseases. European Journal of Nutrition 50 ((3), 151–161.

Shishodia, S., Sethi, G., Aggarwal, B.B., 2005. Curcumin: getting back to the roots. Annals of the New York Academy of Sciences 1056, 206–217.
Shoba, G., Joy, D., Joseph, T., Majeed, M., Rajendran, R., Srinivas, P., 1998. Influence of piperine on the pharmacokinetics of curcumin in animals and human volunteers. Planta Medica 64 (4), 353–356.
Sindambiwe, J.B., Calomme, M., Geerts, S., Pieters, L., Vlietinck, A.J., Van den Berghe, D.A., 1998. Evaluation of biological activities of triterpenoid saponins from *Maesa lanceolata*. Journal of Natural Products 61, 585–590.
Sivitz, W.I., Yorek, M.A., 2010. Mitochondrial dysfunction in diabetes: from molecular mechanisms to functional significance and therapeutic opportunities. Antioxidants and Redox Signaling 12, 537–577.
Smejkal, K., 2014. Selected natural phenolic compounds potential treatment for peripheral neuropathy? Ceská a Slovenská Farmacie 63 (2), 55–70.
Soetikno, V., Suzuki, K., Veeraveedu, P.T., Arumugam, S., Lakshmanan, A.P., Sone, H., Watanabe, K., 2013. Molecular understanding of curcumin in diabetic nephropathy. Drug Discovery Today 18 (15–16), 756–763.
Sparks, S., Liu, G., Robbins, K.J., Lazo, N.D., 2012. Curcumin modulates the self-assembly of the islet amyloid polypeptide by disassembling alpha-helix. Biochemical and Biophysical Research Communications 422 (4), 551–555.
Srinivasan, M., 1972. Effect of curcumin on blood sugar as seen in a diabetic subject. Indian Journal of Medical Sciences 26 (4), 269–270.
Stefan, N., Hennige, A.M., Staiger, H., Machann, J., Schick, F., Kröber, S.M., Machicao, F., Fritsche, A., Häring, H.U., 2006. α2-Heremans- Schmid glycoprotein/fetuin-A is associated with insulin resistance and fat accumulation in the liver in humans. Diabetes Care 29 (4)), 853–857.
Stoilova, I., Gargova, S., Stoyanova, A., Ho, L., 2005. Antimicrobial and antioxidant activity of the polyphenol mangiferin. Herba Polonica Journal 51, 37–44.
Stoilova, I., Jirovetz, L., Stoyanova, A., Krastanov, A., Gargova, S., Ho, L., 2008. Antioxidant activity of the polyphenol mangiferin. Electronic Journal of Environmental, Agricultural and Food Chemistry 7 (13), 2706–2716.
Sun, Y.M., Zhang, H.Y., Chen, D.Z., Liu, C.B., 2002. Theoretical elucidation on the antioxidant mechanism of curcumin: a DFT study. Organic Letters 4 (17), 2909–2911.
Tang, Y., Chen, A., 2010. Curcumin protects hepatic stellate cells against leptin-induced activation in vitro by accumulating intracellular lipids. Endocrinology 151 (9), 4168–4177.
Tang, M., Larson-Meyer, D.E., Liebman, M., 2008. Effect of cinnamon and turmeric on urinary oxalate excretion, plasma lipids, and plasma glucose in healthy subjects. The American Journal of Clinical Nutrition. 87 (5), 1262–1267.
Telang, M., Dhulap, S., Mandhare, A., Hirwani, R., 2013. Therapeutic and cosmetic applications of mangiferin: a patent review. Expert Opinion on Therapeutic Patents 23 (12), 1561–1580.
Thorvaldson, L., Holstad, M., Sandler, S., 2003. Cytokine release by mueine spleen cells following multiple low dose streptozotocin-induced diabetes and treatment with a TNFα transcriptional inhibitor. International Immunopharmacology 3, 1609–1617.
Trujillo, J., Chirino, Y.I., Molina-Jijon, E., Anderica-Romero, A.C., Tapia, E., Pedraza-Chaverri, J., 2013. Renoprotective effect of the antioxidant curcumin: recent findings. Redox Biology 1 (1), 448–456.
Verma, S.C., Jain, C.L., Padhi, M.M., Devalla, R.B., 2012. Microwave extraction and rapid isolation of arjunic acid from *Terminalia arjuna* (Roxb. ex DC.) stem bark and quantification of arjunic acid and arjunolic acid using HPLC-PDA technique. Journal of Separation Science 35 (13), 1627–1633.

Waghmare, P., Patingrao, D., Kadu, P., 2015. Extraction, isolation, purification and identification of curcumin: a review article. European Journal of Biomedical and Pharmaceutical Sciences 2 (3), 108–123.

Wang, Y., Zhang, Z., Guo, W., Sun, W., Miao, X., Wu, H., Cong, X., Wintergerst, K.A., Kong, X., Cai, L., 2014. Sulforaphane reduction of testicular apoptotic cell death in diabetic mice is associated with the upregulation of Nrf2 expression and function. American Journal of Physiology. Endocrinology and Metabolism 307 (1), E14–E23.

Wauthoz, N., Balde, A., Balde, E.S., Van Damme, M., Duez, P., 2007. Ethnopharmacology of *Mangifera indica* L. Bark and Pharmacological Studies of its main c-glucosylxanthone, mangiferin. International Journal of Biomedical and Pharmaceutical Sciences 1 (2), 112–119.

WHO, 1980. Expert Committee on Diabetes Mellitus: Second Report. World Health Organization Technical Report Series, vol. 646. World Health Organization, Geneva, USA, pp. 1–80.

Williamson, J.R., Chang, K., Frangos, M., Hasan, K.S., Ido, Y., Kawamura, T., Nyengaard, J.R., Van Den Enden, M., Kilo, C., Tilton, R.G., 1993. Hyperglycemic pseudohypoxia and diabetic complications. Diabetes 42, 801–813.

Wolf, G., Ziyadeh, F.N., 1999. Molecular mechanisms of diabetic renal hypertrophy. Kidney International 56, 393–405.

Wu, J.Y., Lin, C.Y., Lin, T.W., Ken, C.F., Wen, Y.D., 2007. Curcumin affects development of zebrafish embryo. Biological and Pharmaceutical Bulletin 30 (7), 1336–1339.

Xie, X.Y., Kong, P.R., Wu, J.F., Li, Y., Li, Y.X., 2012. Curcumin attenuates lipolysis stimulated by tumor necrosis factor-alpha or isoproterenol in 3T-L1 adipocytes. Phytomedicine 20 (1), 3–8.

Xu, R., Fazio, G.C., Matsuda, S.P.T., 2004. On the origins of triterpenoid skeletal diversity. Phytochemistry 65, 261–291.

Yan, H.D., Li, X.Z., Xie, J.M., Li, M., 2007. Effects of advanced glycation end products on renal fibrosis and oxidative stress in cultured NRK-49F cells. Chinese Medical Journal 120, 787–793.

Yoosook, C., Bunyapraphatsara, N., Boonyakiat, Y., Kantasuk, C., 2000. Anti-herpes simplex virus activities of crude water extracts of Thai medicinal plants. Phytomedicine 6 (6), 411–419.

Yousef, M.I., El-Demerdash, F.M., Radwan, F.M.E., 2008. Sodium arsenite induced biochemical perturbations in rats: ameliorating effect of curcumin. Food and Chemical Toxicology 46 (11), 3506–3511.

Yun, J.M., Jialal, I., Devaraj, S., 2011. Epigenetic regulation of high glucose-induced proinflammatory cytokine production in monocytes by curcumin. Journal of Nutritional Biochemistry 22 (5), 450–458.

Zang, Q., ElSohly, H.N., Li, X.C., Walker, L.A., 2003. A new triterpene from *Liandra chaetodon*. Planta Medica 69, 582–584.

Zelmanovitz, T., Gerchman, F., Balthazar, A.P., Thomazelli, F.C., Matos, J.D., Canani, L.H., 2009. Diabetic nephropathy. Diabetology and Metabolic Syndrome 21, 1–10.

Zhang, D.W., Fu, M., Gao, S.H., Liu, J.L., 2013. Curcumin and diabetes: a systematic review. Evidence Based Complementary and Alternative Medicine 636053.

Zhu, X., Cheng, Y.Q., Du, L., Li, Y., Zhang, F., Guo, H., Liu, Y.W., Yin, X.X., 2015. Mangiferin attenuates renal fibrosis through down-regulation of osteopontin in diabetic rats. Phytotherapy Research 29 (2), 295–302.

Chapter 4

Glucose Transporter 4 Translocation Activators From Nature

K. Dev[1,2,a], E. Ramakrishna[1,a], R. Maurya[1,2]
[1]*CSIR-Central Drug Research Institute, Lucknow, India;* [2]*Academy of Scientific and Innovative Research, New Delhi, India*

1. INTRODUCTION

Natural products play a very important role on human health since ancient times as they have been used to treat several diseases. Traditionally, human beings have been using herbal medicines in the form of decoction, juice, or paste to cure or prevent different diseases. Natural products are secondary metabolites of living organisms including plants, animals, marine invertebrates, and microorganisms. These metabolites are the result of their changes or adaptations of million years of evolution to survive and to keep their species alive as long as possible. These secondary metabolites help the living organisms to survive under life-threatening situations such as extreme environmental conditions like hot, cold, and drought or to attract or to fight against their enemies. Consequently, secondary metabolites are often associated with various biological activities. Natural products have appropriate drug-like properties because they have the tendency to interact with biomolecules like receptors and enzymes. Several natural products and their analogues become a source of new drugs as well as new scaffolds with drug-like properties (Brahmachari, 2011, p. 1; Newman et al., 2000, p. 215; Newman and Cragg, 2012, p. 311). These new scaffolds or fragments play an important role in combinatorial synthesis and diversity-oriented synthesis for the generation of library of bioactive molecules (Over et al., 2013, p. 21). Natural products are also rich sources of antidiabetic drugs and hence, play a significant role in the development of new antidiabetic agents (Shapiro and Gong, 2002, p. 217; Butler, 2005, p. 162).

---

a. Authors Contributed equally.

Discovery and Development of Antidiabetic Agents From Natural Products.
http://dx.doi.org/10.1016/B978-0-12-809450-1.00004-1

Diabetes mellitus (DM) is a metabolic syndrome characterized by high blood glucose levels due to either insulin resistance, i.e., body's inability to utilize produced insulin or relative insulin deficiency, i.e., body's inability to produce insulin. The condition of insulin resistance is known as type 2 diabetes and the specific molecular cause of type 2 diabetes remains unknown, though considerable progress has been made to define the metabolic characteristics. Since, early 1980's prevalence of diabetes has increased more than 13-fold around the globe. The number of patients with diabetes is increasing at a higher rate and is predicted to reach over 430 million by 2030 (IDF, 2013). Individuals with impaired fasting glucose or metabolic syndrome are considered to have prediabetes and are at higher risk for developing diabetes.

Insulin signals to stimulate the uptake of circulating glucose into muscles and adipose tissues. The high blood glucose levels balanced via homeostatic mechanism, ie, restores glucose level by increasing insulin level in the body that leads to the suppression of hepatic glycogenolysis and stimulates insulin-dependent glucose uptake into muscle, liver, and adipose tissue. In the conditions of insulin resistance or insulin deficiency, glucose uptake diminishes and hence leads to high blood glucose levels. Conceptually, insulin-mediated glucose transport could trigger either by stimulating the activity of existing cell surface glucose transport proteins or by translocation of an intracellular transporter to the cell surface in response to insulin. Among all the effective managements of diabetes, regulation and expression of adiponectin and glucose transporter 4 (GLUT4) are of high therapeutic importance. Glucose transporter proteins are involved in transporting glucose and related hexoses; GLUT4 found in adipose tissues and striated muscle (skeletal muscle and cardiac muscle) is an important factor responsible for proper glucose metabolism (Morral, 2003, p. 169; Ross et al., 2004, p. 1255).

With the rapid advancements in medicine, novel treatments with fewer side effects have become feasible for the long-term control of this disorder. Recent scientific investigations have confirmed the efficacy of many of plant medicines on DM (Bailey and Day, 1989, p. 553; Qi et al., 2010, p. 434; Hung et al., 2012, p. 580; Salimifar et al., 2013, p. 402). Current therapeutics for diabetes is often associated with undesirable side effects. Traditional medicines derived from medicinal plants are used by 70% of the world's population as alternatives to synthetic chemical compounds. In this chapter, we are dealing with natural products, i.e., plant formulations, purified compounds, and a small portion of natural product-derived molecules (semisynthetic) that showed GLUT4 translocation activity.

2. PLANT FORMULATIONS WITH GLUCOSE TRANSPORTER 4 TRANSLOCATION ACTIVITY

With the development of civilization, human beings are using crude drugs as in the form of paste, juice, decoctions, etc. Therefore, several research groups have performed experiments to provide scientific evidence of the traditional medicines. In this context, Annapurna et al. analyzed the aqueous extract of *Cynodon dactylon* by using HPLC–ESI MS and identified the presence of apigenin (**1**), luteolin (**2**), 6-*C*-pentosyl-8-*C*-hexosyl apigenin (**3**), and 6-*C*-hexosyl-8-*C*-pentosyl luteolin (**4**) (Fig. 4.1). These compounds were docked with the GLU 270 of the GLUT4 receptor in silico. An interaction of apigenin and luteolin with Glu 270 of GLUT4 receptor showed docking score −4.23, −6.00 which was more comparable with active molecule genistein with score −5.19 (Annapurna et al., 2013, p. 87).

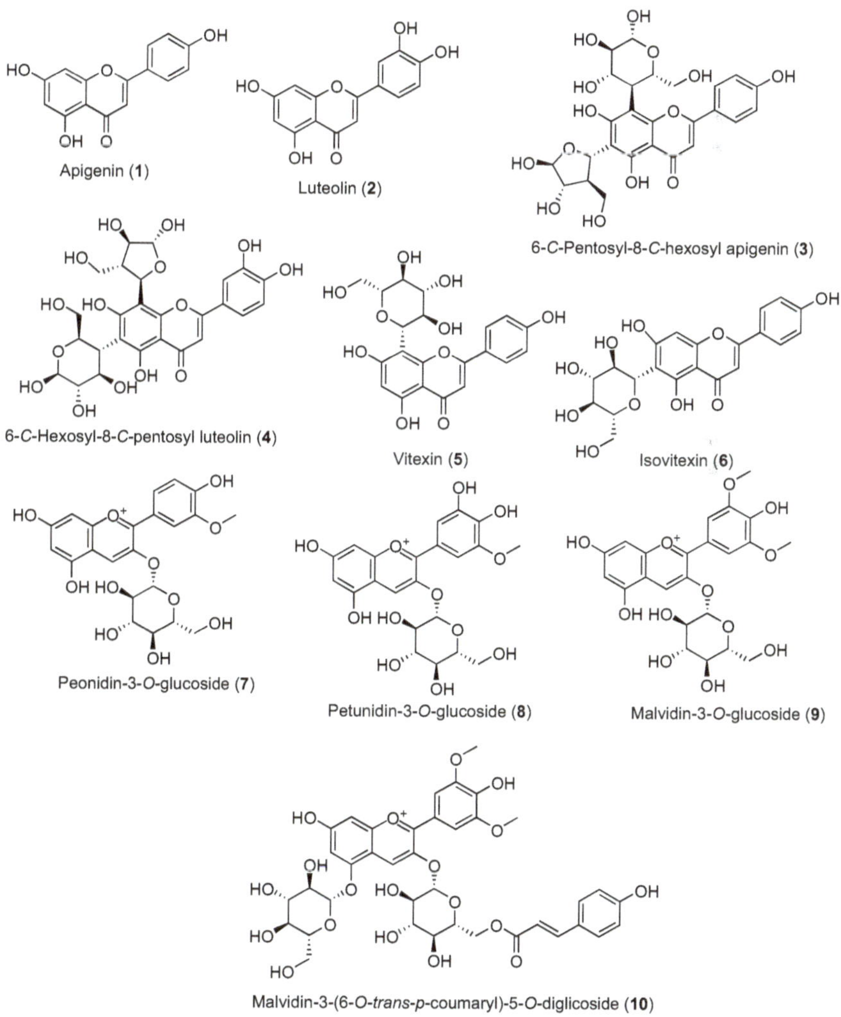

FIGURE 4.1 Chemical structure of compounds **1–10** found in the plant formulations.

HPLC analysis of *Ficus deltoidea* revealed that its methanol extract is a rich source of *C*-glycosylflavones, especially, vitexin (**5**) and isovitexin (**6**) (Fig. 4.1). The extract stimulates the hepatic GK and PPARγ (peroxisome proliferator-activated receptor γ) gene expression, which eventually lead to an increased expression of GLUT4 gene expression in skeletal muscles of streptozotocin (STZ)-induced diabetic rats (Farsi et al., 2014, p. 1). LC/UV/DAD analysis of *Vitis vinifera* grape skin extract showed the presence of four anthocyanins such as peonidin-3-*O*-glucoside (**7**), petunidin-3-*O*- glucoside (**8**), malvidin-3-*O*-glucoside (**9**), and malvidin-3-(6-*O*-*trans*-*p*-coumaryl)-5-*O*-diglicoside (**10**) (Fig. 4.1). Hyperglycemia and insulin-signaling cascade effects of *V. vinifera* grape skin extract (ACH09) were studied in alloxan-induced diabetic mice. ACH09 exerted hypoglycemic effects in control mice and antihyperglycemic effects in alloxan-induced diabetic mice. Insulin receptor (IR) content and AKT phosphorylation of alloxan+ACH09-treated group was significantly more in comparison with the alloxan-treated group. GLUT4 content in skeletal muscle suggested that ACH09 may reduce insulin resistance by inducing the activation of insulin-signaling pathways in peripheral tissues (Soares de Moura et al., 2012, p. 268). Tangzhiqing formula (TZQ-F) is a mixture of red paeony roots, mulberry leaves, lotus leaves, danshen roots, and hawthorn leaves. The GLUT4 expression of TZQ-F and its fractions were analyzed in HepG2 and L-6 cells. In HepG2 cells, TZQ-F showed no effect on GLUT4, but total flavonoid fraction of hawthorn, mulburry leaves and total saponin fraction of red peony showed upregulation effects on GLUT4 expression. All the individual fractions of TZQ-F showed enhanced GLUT4 expression by six- to eightfolds. Among them, the most notable upregulation was observed with the lotus leaf flavonoid fraction (Gao et al., 2014, p. 198/1). Pycnogenol, a procyanidin-enriched extract of *Pinus maritima* bark showed its effects on change in gene expression of GLUT4, which is involved in the glucose transport system. At concentrations 50, 100, and 200 μg/mL of pycnogenol, the expression of GLUT4 mRNA was found to be increased by approximately 20%, 30%, and 57%, respectively. The GLUT4 mRNA expression was higher at 200 μg/mL concentration of pycnogenol than that of insulin at 10 nM (Lee et al., 2010a, p. 1242).

3. NATURALLY OCCURRING COMPOUNDS WITH GLUT4 TRANSLOCATION ACTIVITY

3.1 Alkaloid and Amino Acid

Recently, Guo et al. have observed that oxymatrine (**11**) (Fig. 4.2), a quinolizidine alkaloid, exerts potent antidiabetic effect in experimental diabetic rats with a significant reduction of GLUT4 contents in muscle as compared with the normal control rats. The experimental rats were found to develop a significant elevation in GLUT4 contents in muscle, after the treatment with oxymatrine at 50, 100, and 150 mg/kg dose for 11 weeks (Guo et al., 2014, p. 807).

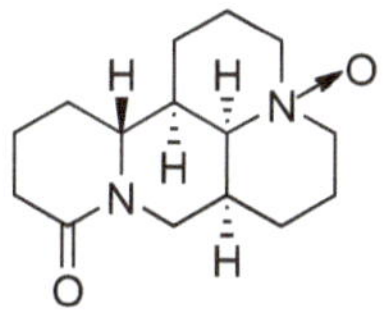

Oxymatrine (**11**)

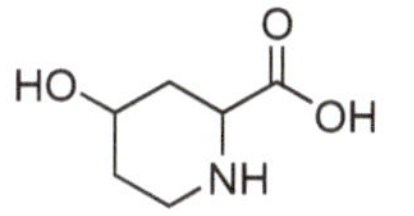

4-Hydroxypipecolic acid (**12**)

FIGURE 4.2 Oxymatrine (**11**) and 4-hydroxypipecolic acid (**12**).

On the other hand, 4-hydroxypipecolic acid (**12**) (Fig. 4.2), an amino acid isolated from *Peganum harmala*, was evaluated for increased glucose uptake activity, considerable efficacy was observed at a minimal concentration of 5 μM with the maximum activity at 25 μM (Naresh et al., 2012, p. 5648).

3.2 Fatty Acids

Naturally occurring fatty acids with a chain of an even number of carbons from 4 to 28 are preferred by skeletal muscle cells as they produce large amounts of ATP on metabolization. Sebacic acid (**13**) (Fig. 4.3), a naturally occurring fatty acid, is reported to increase insulin-dependent glucose uptake by L6 skeletal muscle cells up to 1.7-fold in a dose-dependent manner (Iaconelli et al., 2010, p. 2327). Another interesting compound phytanic acid (**14**) (Fig. 4.3), a C-20 chain fatty acid, has also been evaluated to promote the acquisition of the brown adipocyte morphology, thereby facilitating a coordinate induction of the mRNAs for GLUT4 gene markers of brown adipocyte differentiation (Schluter et al., 2002, pp. 61–69). In another study, Hu et al. demonstrated that eicosapentaenoic acid (**15**)-enriched phosphatidylcholine from sea cucumber *Cucumaria frondosa* significantly promotes the protein expression of GLUT4 in skeletal muscle cells (Hu et al., 2014b, p. 457).

3.3 Flavonoids

Flavonoids, the most abundant class of compounds occurring in large number of plant families, possess various biological activities. A number of flavonoids,

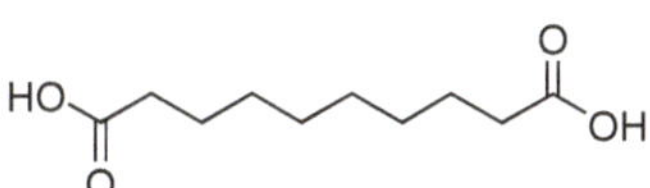

Sebacic acid (**13**)

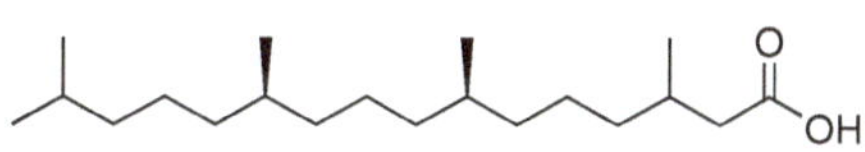

Phytanic acid (**14**)

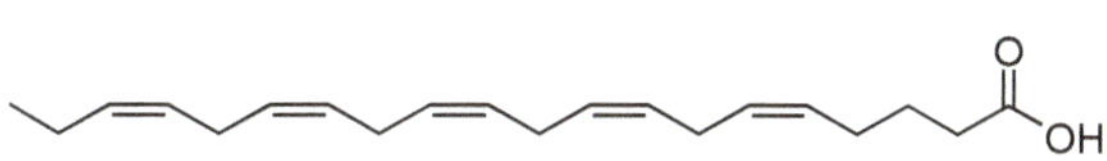

Eicosapentaenoic acid (**15**)

FIGURE 4.3 Chemical structure of fatty acids (**13–15**).

isoflavonoids, and their derivatives were screened for their antidiabetic activity by various groups and the results are available in literature (Brahmachari, 2014). (−)-Epigallocatechin gallate (**16**), the most abundant polyhydroxy flavonoid in tea, has recently been reported to exert significant inhibitory activity against the insulin-like growth factor (IGF)-increased amount of GLUT4 protein in the plasma membrane fraction and the test compound antagonized the IGF-I-decreased amount of GLUT4 protein in the low density microsome (LDM) fraction (Ku et al., 2014, p. 46). Anhe et al. reported that quercetin (**17**), a well-known flavonoid abundant in various plant sources, upregulates GLUT4 mRNA in skeletal muscle of mice and increases GLUT4 protein content by 154% and also prevents the downregulation of GLUT4 expression (Anhe et al., 2012, p. 285). Rutin (**18**), a flavonoid glycoside, induced GLUT4 translocation with increase in intracellular calcium concentration and it also induced glucose uptake via CaMKII pathway, which may result in GLUT4 translocation to the plasma membrane (Kappel et al., 2013, p. 55). In addition, two known kaempferol glycosides CO-1 (**19**) and CO-2 (**20**) (Fig. 4.4) isolated from *Cinnamomum osmophloeum* leaves exerted GLUT4 translocation activity due to enhanced adiponectin secretion, and activation of insulin signaling pathway, which involved phosphorylation of IR and activation of phosphatidylinositol-3 kinase (PI3K) (Lee et al., 2009, p. 79).

In 2010, Lee and his coworkers investigated the effect of nobiletin (**21**), a polymethoxy flavone on insulin sensitivity in obese diabetic ob/ob mice, and examined the effect of nobiletin on insulin signaling in white adipose tissue (WAT). In WAT, the GLUT4 mRNA level was increased and the expression levels in the whole plasma membrane were also significantly increased without changes in the whole cell lysates (Lee et al., 2010b, p. 1674). Treatment with kaempferitrin (**22**), a natural flavonoid glucoside, results in translocation of GLUT4 to the membrane of adipocytes. After 20 min of treatment with 15 μM kaempferitrin on differentiated cells, good results in the GLUT4 translocation were observed. In cells, when no kaempferitrin was received, the GLUT4 immunoreactivity did not colocalize with that of the plasma membrane marker Na^+/K^+ ATPase, but after treatment with 15 μM kaempferitrin the GLUT4 immunoreactivity moved to the plasma membrane and colocalized with the plasma membrane (Tzeng et al., 2009, p. 27). The effects of puerarin (**23**) on diabetic myocardial damage were investigated to explore its possible mechanism in rats. Puerarin exerted preventive and remedial effects on the diabetic myocardium, which may be related to upregulating the mRNA expressions of PPARγ and GLUT4, promoting glucose uptake and relieving damages out of oxidative stress (Chen et al., 2010, p. 650). Pectolinarin (**24**) and 5,7-dihydroxy-6,4-dimethoxyflavone (**25**), both isolated from *Cirsium japonicum*, induced PPARγ activation and enhanced adipocyte differentiation of 3T3-L1 cells in a dose-dependent manner. In addition, compounds **24** and **25** increased the expression of PPARγ target genes, like adiponectin and GLUT4, and improved the translocation of intracellular GLUT4 to the plasma membrane (Liao et al.,

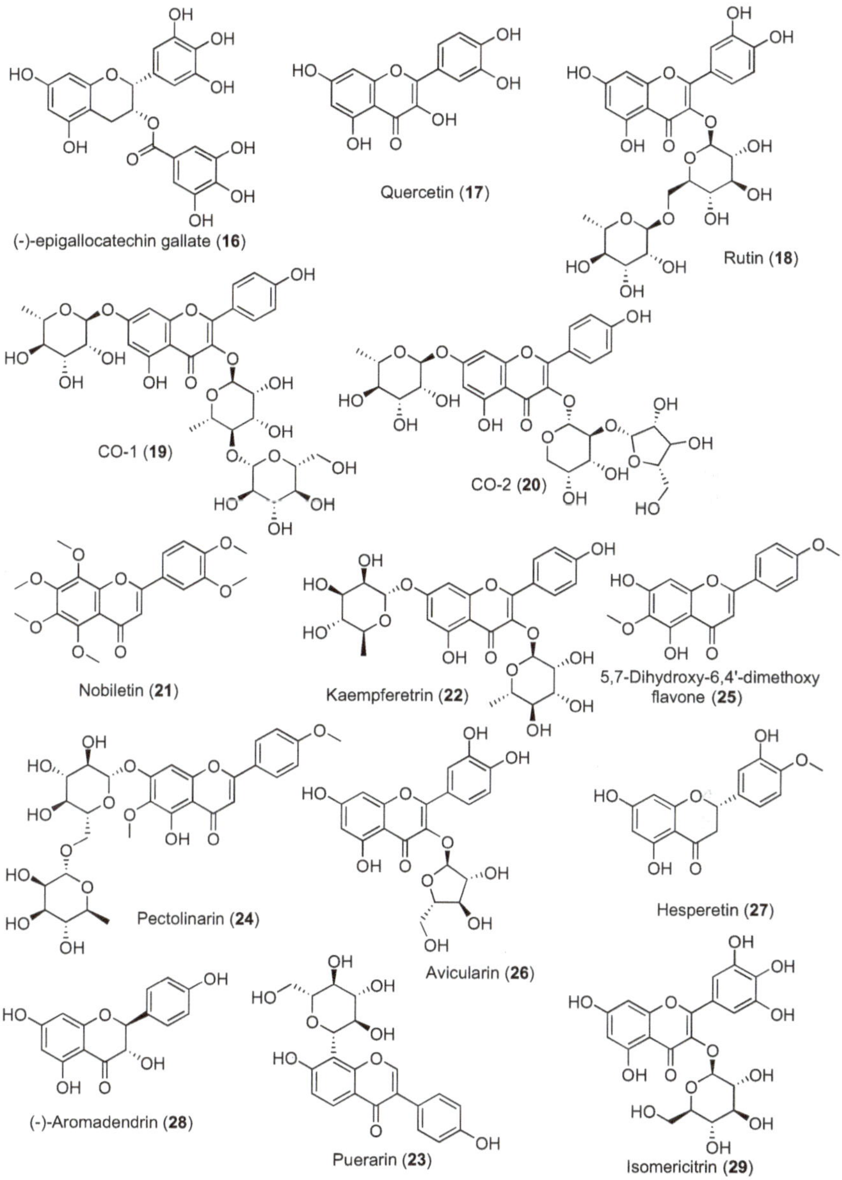

FIGURE 4.4 Chemical structure of flavonoid molecules (**16–29**).

2012, p. 855). Avicularin (**26**) exerted preventive effect on the accumulation of intracellular lipids by suppressing C/EBPα-mediated activation of GLUT4 expression and also suppressed GLUT4-mediated glucose uptake in adipocytes about 1.6-fold in 3T3-L1 cells (Fujimori and Shibano, 2013, p. 5139). Hesperetin (**27**), a dihydroflavonoid, was found to downregulate GLUT4

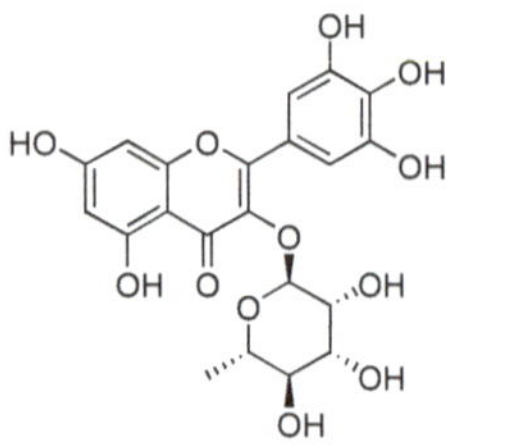

Myricetin-3-*O*-rhamnoside (**30**)

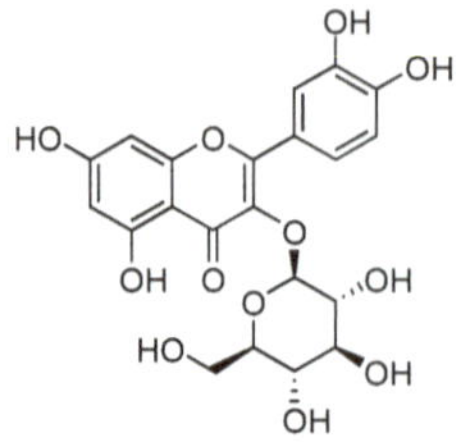

Quercetin 3-O-glucopyranoside (**31**)

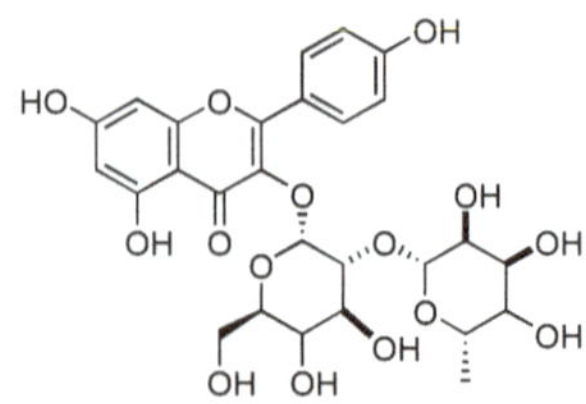

Kaempferol-3-neohesperidoside (**32**)

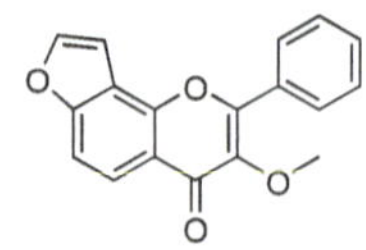

Karanjin (**33**)

FIGURE 4.5 Chemical structure of flavonoid molecules (**30–33**).

(Yang et al., 2013, pp. 374–379). However, (−)-aromadendrin (**28**), isomericitrin (**29**) (Fig. 4.4), myricetin 3-*O*-rhamnoside (**30**), and quercetin 3-*O*-glucopyranoside (**31**) (Fig. 4.5) did not show any GLUT4 translocation activity (Wang et al., 2014, p. 3096). A known compound, kaempferol 3-neohesperdoside (**32**) (Fig. 4.5) isolated from *Cyathea phalerata* Martius, was found to be effective for the glucose uptake with the concentration ranges 0.0001–2080 μM. Compound **32** found to increase the glucose uptake significantly by the activation of GLUT4 translocation to the plasma membrane through PKC isoform. The glucose uptake (35% and 21%) was found at 1 nM and 100 nM, respectively, in rat soleus muscle cells (Zanatta et al., 2008, p. 532).

Another well-known furanoflavonoid, karanjin (**33**) (Fig. 4.5) isolated from *Pongamia pinnata*, stimulated glucose uptake in dose-dependent manner. At 10 μM, it showed significant effect and reached maximal level at 25 μM. Karanjin treatment caused around 50% increase in surface level of GLUT4myc. Its effect on GLUT4myc translocation was found to be time dependent and significant stimulation was observed only after 12–16 h of incubation (Jaiswal et al., 2011, p. 22).

3.4 Isoflavonoids and Chalcones

Recent studies revealed that isoflavonoids and chalcones also possess significant GLUT4 translocation activity. In 2005, Anandharajan et al. reported an isoflavone glycoside (7-*O*-α-L-rhamnopyranosyloxy-4′-methoxy-5-hydroxyisoflavone) (**34**) (Fig. 4.6) from *Pterocarpus marsupium* methanol extract as a GLUT4 translocation activator. Elevation in expression of GLUT4 transcripts was observed by both isoflavone and methanolic extract of *P. marsupium,* which was found to be comparable with insulin and rosiglitazone. Densitometry scanning revealed that increase in GLUT4 transcripts

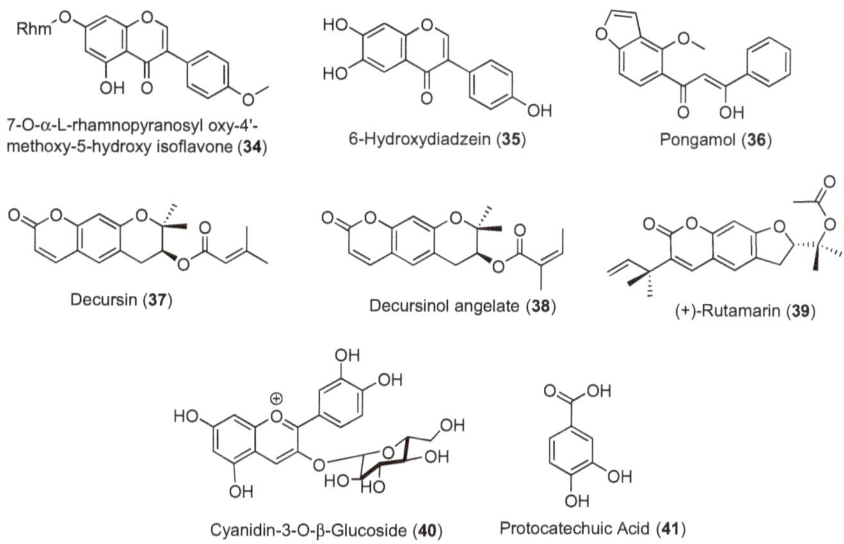

FIGURE 4.6 Chemical structure of isoflavonoids (**34–35**), chalcones (**36**), coumarins (**37–39**), cyanidin-3-*O*-*β*-glucoside (**40**), and protocatechuic acid (**41**).

by 2.9- and 2.4-fold for extract and pure compound, respectively, as compared to insulin control (3.4-fold) and rosiglitazone (3.2-fold). Western blot analysis revealed that elevation in protein expression of GLUT4 by *P. marsupium* methanol extract and the isoflavone **34** was comparable with insulin (Anandharajan et al., 2005, p. 253). Another interesting isoflavonoid, 6-hydroxydaidzein (**35**) (Fig. 4.6), enhanced adipocyte differentiation and improved insulin sensitivity through PPARγ activation. The downstream responses to PPARγ activation include increase in the C/EBPα mRNA expression and GLUT4 translocation, resulting in enhanced adipocyte differentiation, glucose uptake, and insulin sensitivity. It is expected that 6-hydroxydaidzein may also exert such effects on some other cell types because of its interaction with PPARγ signaling (Chen et al., 2013, p. 10714). A furanochalcone pongamol (**36**) (Fig. 4.6) was isolated from chloroform fraction of *P. pinnata* and found to enhance glucose uptake in a dose-dependent manner. It excreted maximum increase in the level of L6-GLUT4*myc* myotubes to the cell surface at 10 μM concentration. Compound **36** induced increase in glucose uptake in L6 myotubes due to increase in GLUT4 translocation to plasma membrane by PI3K/AKT-dependent pathway (Tamrakar et al., 2011, p. 98).

3.5 Coumarins

Two coumarins decursin (**37**) and decursinol angelate (**38**) (Fig. 4.6) were isolated from *Angelica gigas* Nakai and investigated for their blood glucose, glucose transporter, and adenosine monophosphate-activated protein kinase

(AMPK) expression levels in STZ-induced diabetic rats. Decursinol angelate significantly upregulated the expression of GLUT4 and phosphorylation of AMPK (*p*-AMPK) in skeletal muscle of rats at 2 mg/kg dose for 21 days, whereas the effect of decursin (2 mg/kg) was found to be very weak (Ok et al., 2011, p. 301). A coumarin, rutamarin (**39**) (Fig. 4.6), promotes insulin-induced GLUT4 translocation and also significantly stimulated the GLUT4 translocation in fully differentiated 3T3-L1 adipocytes as well as induced the GLUT4 expression, thus acting as a dual-functional inducer. Compound **39** induced the GLUT4 translocation by inhibiting PTP1B either with or without insulin stimulation. Subsequently, it was demonstrated that rutamarin acts as PTP1B inhibitor, inducer of basal GLUT4 translocation and insulin-induced GLUT4 translocation through PI3K–AKT/PKB (protein kinase B) pathway while it also acts as an agonist of retinoid X receptor a (RXRa) with potential increase in GLUT4 expression (Zhang et al., 2012, p. e31811).

Scazzocchio et al. found that an anthocyanin, cyanidin-3-*O*-β-glucoside (**40**) and a phenolic compound, protocatechuic acid (**41**) (Fig. 4.6) exerted insulin-like activity in human omental adipocytes. These compounds induce glucose uptake associated with enhancement of GLUT4 translocation and adiponectin secretion caused by the increased activity of PPARγ (Scazzocchio et al., 2011, p. 2234).

3.6 Anthraquinones

Anthraquinones are a class of naturally occurring quinone compounds based on the 9, 10-anthraquinone skeleton. Anand and his colleagues in 2010 reported GLUT4 translocation activity of aloe emodin-8-*O*-glycoside (**42**), which was found to be able to translocate GLUT4 from light microsomes to plasma membrane, and the majority of the GLUT4 translocation to plasma membrane upon treatment of compound **42** was observed similar to insulin (Anand et al., 2010, p. 3170). Two anthraquinones chrysophanol-8-*O*-β-D-glucopyranoside (**43**) and chrysophanol (**44**) (Fig. 4.7) were isolated from *Korean rhubarb* rhizome and enhanced the glucose transport in myotubes with moderate cytotoxicity. The compound **43** significantly enhanced insulin-stimulated glucose transport by IR activation but did not influence GLUT4 expression while the compound **44** influenced insulin-responsive glucose transport by increasing IR phosphorylation and GLUT4 mRNA expression (Lee and Sohn, 2008, p. 2154). Sennidin A (**45**) and B (**46**) (Fig. 4.7) are aglycones of sennoside A and B, respectively, which were evaluated for their antidiabetic potential. Sennidin A was found to induce the translocation of GLUT4 in rat adipocytes but this induction was comparatively lower than that of induced insulin. Compound **45** stimulated GLUT4 translocation through the induced phosphorylation of AKT. Compound **46** was found to be very less effective in the stimulation of translocation of GLUT4 (Abe et al., 2006, p. 1027).

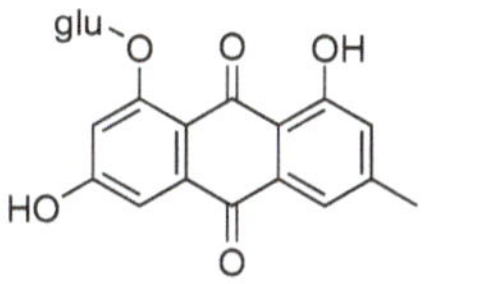

Aloe emodin-8-O-glycoside (**42**)

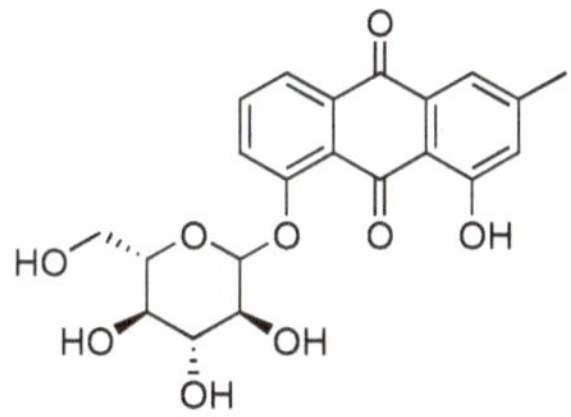

Chrysophanol-8-O-glycoside (**43**)

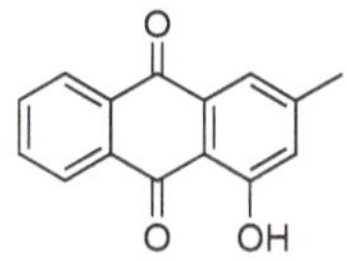

Chrysophanol (**44**)

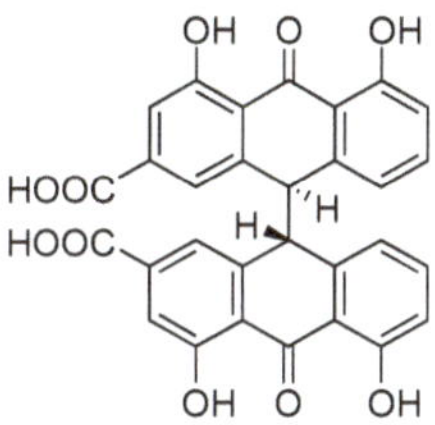

Sennidin A (**45**)

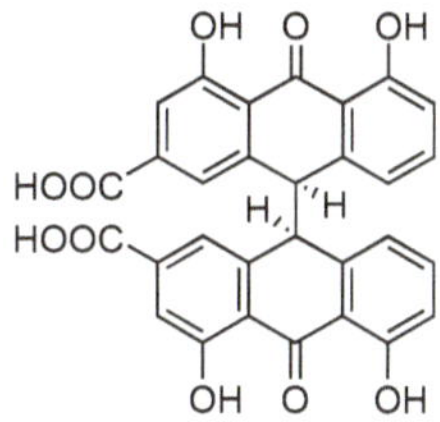

Sennidin B (**46**)

FIGURE 4.7 Chemical structure of anthraquinones (**42–46**).

3.7 Quinones and Naphthoquinones

An interesting quinone derivative embelin (**47**) (Fig. 4.8) was isolated from *Embelia ribes* Burm and found to reduce body weight gain, blood glucose, and plasma insulin in treated diabetic rats at 50 mg/kg dose. Embelin also controlled insulin-mediated glucose uptake in epididymal adipose tissue via translocation and activation of GLUT4 in PI3K/*p*-AKT signaling cascade. Embelin bound to PPARγ thereby exhibiting stable binding interactions to the active sites of PI3K, *p*-AKT, and GLUT4 (Gandhi et al., 2013, p. 2243). In 2011, Oeberg et al. reported that shikonin (**48**) (Fig. 4.8), a naphthoquinone analogue, stimulates the GLUT4 translocation from intracellular vesicles to the cell surface in L6 myoblasts via insulin-independent pathway, and the mechanism involves an increase in intracellular calcium levels and GLUT4 translocation. Shikonin significantly induced GLUT4 translocation to the same extent as insulin (Oeberg et al., 2011, p. e22510). Plumbagin (**49**) (Fig. 4.8), a naphthoquinone was isolated from *Plumbago zeylanica* L. root and showed its effect on GLUT4 translocation in STZ-induced diabetic rats. The compound **49** significantly restored the blood glucose and all other biochemical parameters to near normal level. The compound exerts its effect on diabetes restoration via an increase in

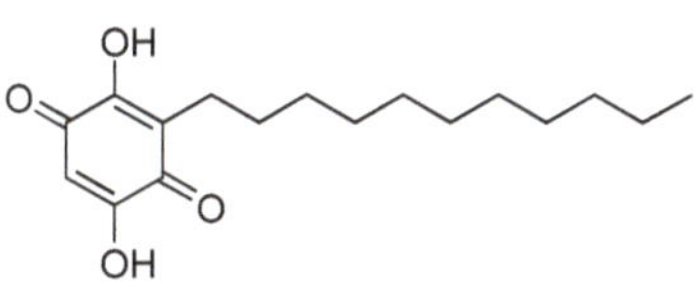

Embelin (**47**)

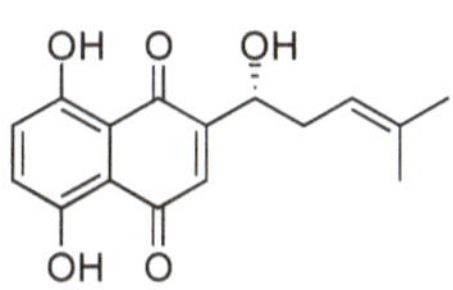

Shikonin (**48**)

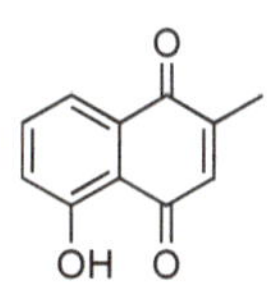

Plumbagin (**49**)

FIGURE 4.8 Chemical structure of quinone (**47**) and naphthoquinones (**48–49**).

GLUT4 mRNA and protein expression levels as demonstrated by treatment of compound **49** in diabetic rats (Sunil et al., 2012, p. 4356).

3.8 Biphenyls and Lignans

Choi et al. (2012) isolated two biphenylic compounds, honokiol (**50**) and magnolol (**51**) (Fig. 4.9), from *Magnolia officinalis*. Biological investigation demonstrated that compounds **50** and **51** enhance glucose uptake in L6 myotubes. The underlying mechanisms involved in the enhancement of glucose uptake possible through stimulated GLUT4 translocation to the cell surface. Moreover, compounds **50** and **51** stimulated glucose uptake blocked by the PI3K inhibitor, wortmannin (Choi et al., 2012, p. 372).

Fargesin (**52**) (Fig. 4.9), a lignan isolated from *Flos Magnoliae,* dose-dependently stimulated glucose consumption in L6 myotubes, which was accompanied by enhanced GLUT4 translocation to the cell surface. Wortmannin, a PI3K inhibitor blocked the glucose uptake stimulated by fargesin. In fargesin-treated cells, plasma membrane GLUT4 expression was increased compared with that in control cells without differences in total GLUT4 expression and it increased basal glucose uptake and GLUT4 translocation in muscle cells by activating the PI3K–AKT pathway (Choi et al., 2013, p. 320). Fragrin A (**53**) (Fig. 4.9), a natural lignan was tested for GAL-4/PPARγ *trans* activation assay. PPARγ ligand-binding activity of fragrin A increased 4.7-, 6.6-, and 7.3-fold at 3, 5, and 10 μM concentrations, respectively. Fragrin A also enhanced adipocytes differentiation and increased the expression of PPARγ target genes such as adipocytes fatty acid-binding protein (aP2), lipoprotein lipase (LPL), and phosphoenolpyruvate carboxykinase (PEPCK). Furthermore, it significantly increased the expression level of GLUT4 (Lee et al., 2008, p. 1146).

3.9 Steroids

Ginsenoside Rh2 (**54**) (Fig. 4.10), isolated from *Panax ginseng*, was evaluated to reduce plasma glucose in STZ diabetic rats administered intravenously at

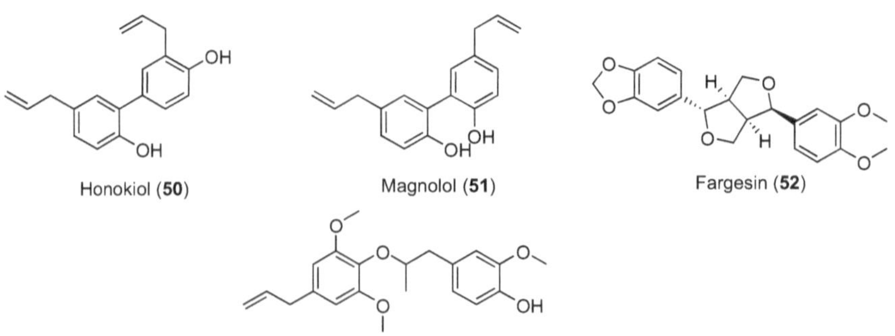

Honokiol (**50**) Magnolol (**51**) Fargesin (**52**) Fragrin A (**53**)

FIGURE 4.9 Chemical structure of biphenyls (**50–51**) and lignans (**52–53**).

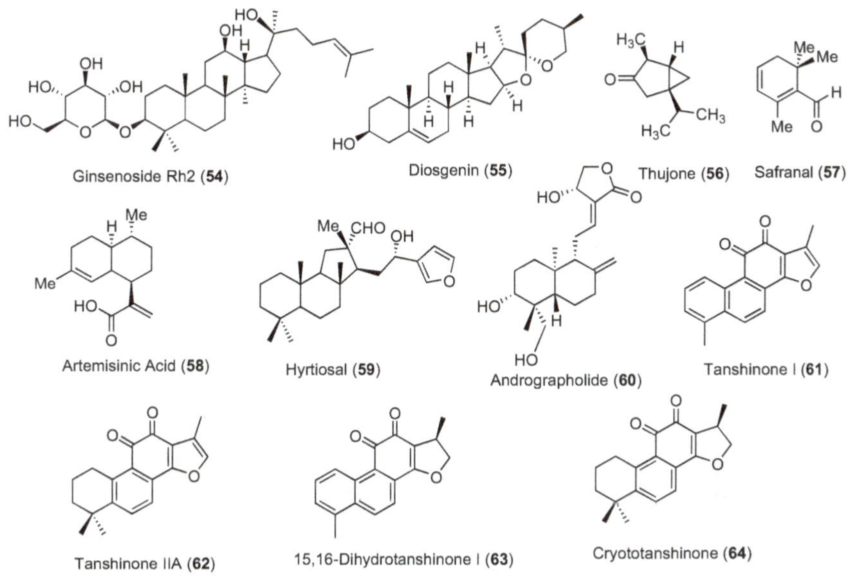

FIGURE 4.10 Chemical structure of steroids (**54–55**) and terpenoids (monoterpenes, sesquiterpenes, and diterpenes) (**56–64**).

1.0 mg/kg three times daily. It reduced plasma glucose level in STZ diabetic rats based on an increase in β-endorphin secretion, which activates opioid μ-receptors thereby resulting in increased protein expressions of GLUT4 in skeletal muscles of rats with dysfunctional pancreatic β-cells (Lai et al., 2006, p. 9).

Dehydroepiandrosterone improves hyperglycemia in rats with STZ-induced type 1 DM. Diosgenin (**55**) (Fig. 4.10), a major component of *Dioscorea* genus, is structurally similar to dehydroepiandrosterone. The administration of diosgenin in STZ-induced type 1 DM at the dose of 3 mg/kg increases phosphorylation of AKT, PKC as well as GLUT4 translocation significantly. This increased GLUT4 translocation level is due to blockage of 5α-reductase inhibitor by diosgenin in type 1 diabetic rats (Sato et al., 2014, p. 152).

3.10 Terpenes

3.10.1 Monoterpenes

Monoterpenes contain 10 carbons (two units of isoprene) attached to each other in head to tail manner generally. Thujone (**56**) (Fig. 4.10), a major component of several medicinal herbs, partly normalized insulin-stimulated GLUT4 translocation to the cell surface. In addition, it also enhanced GLUT4 intrinsic activity (Alkhateeb and Bonen, 2010, p. R804). Safranal (**57**) (Fig. 4.10), obtained from *Crocus sativus* L., promoted glucose uptake through GLUT4 translocation in C2C12 myotubes in an insulin-independent manner. Glucose uptake-enhancing efficiency of compound **57** was further determined by the cellular uptake of a fluorescent D-glucose derivative, 2-NBDG. The maximum increase in the

2-NBDG uptake was about 260% at 20 μM safranal, similar to those of insulin-treated cells in type 2 diabetic KK-Ay mice (Maeda et al., 2014, p. 1177).

3.10.2 Sesquiterpenes

Artemisinic acid (**58**) (Fig. 4.10) inhibited the generation of intracellular lipids during hormone-induced adipocyte differentiation and conformation was observed in the real-time PCR analysis of GLUT4, which is a marker gene of late-stage preadipocyte differentiation. The presence of artemisinic acid during hAMSC differentiation reduced GLUT4 and VEGF expression indicated that artemisinic acid showing an antiadipogenic effect (Lee et al., 2012a, p. 2488).

The possible effects of hyrtiosal (**59**) (Fig. 4.10), isolated from marine sponge *Hyrtios erectus*, on GLUT4 membrane translocation was explored by Sun et al. Hyrtiosal showed extensive cellular effects on PI3K/AKT activation, glucose transport, and TGFb/Smad2 signaling. Furthermore, hyrtiosal enhances AKT activation and GLUT4 membrane translocation in PTP1B-overexpressed CHO cells (Sun et al., 2007, p. 187).

3.10.3 Diterpenoids

Andrographolide (**60**) (Fig. 4.10) improved the glucose uptake in isolated soleus muscle of type 1 DM-infected rats by increasing the GLUT4 expression. Insulin resistance is responsible for the distraction of GLUT4 translocation. Andrographolide and ethanolic extract of *Andrographis paniculata* showed potent hypoglycemic effects in insulin-resistant rats. Besides these treatments, metformin could restore the diminished GLUT4 expression in insulin-resistant rats (Nugroho et al., 2011, p. 486; Brahmachari, 2014, p. 1).

Total extract of danshen, the dried root of *Salvia miltiorrhiza* Bunge (Labiatae) (1–10 μg/mL) and the four tanshinones (10 μM) did not show any activity for GLUT4 translocation in 3T3-L1 cells, but the total extract and tanshinone I (**61**), tanshinone IIA (**62**), and 15, 16-dihydrotanshinone (**63**) except cryptotanshinone (**64**) (Fig. 4.10) enhance the activity of insulin (1 nM) on the tyrosine phosphorylation of the IR and also the activation of the downstream kinases AKT, ERK1/2, and GSK3β. In the presence of insulin, the three tanshinones were demonstrated by the IR-downstream signaling and the translocation of GLUT4 in the adipocytes (Jung et al., 2009, p. 327).

3.10.4 Triterpenes

3-*β*-(2-Carboxybenzoyloxy) oleanolic acid (NPLC441) (**65**), an oleanolic acid ester, enhances GLUT4 expression in 3T3-L1adipocytes. NPLC441 increases GLUT4 gene transcription in dose-dependent manner and Western blotting indicates that the expression levels of GLUT4 protein considerably increased. The upregulation of GLUT4 gene transcription and protein expression in 3T3-L1 adipocytes may be due to NPLC441 (Lin et al., 2008, p. 1251). Damulin A (**66**) and damulin B (**67**) strongly activated AMPK in cultured L6 myotube cells. Dose-dependent increment of 2-deoxyglucose (2-DG) uptake and GLUT4 translocation to the cytoplasmic membrane in L6 myotube cells by damulin A, damulin B. The increased 2-DG uptake was observed at a concentration of 150 μM, 12 μM, respectively (Nguyen et al., 2011, p. 6254). In 2007, Dey et al. reported that insulin triggered GFP-GLUT4 translocation to the cell membrane from cytosol is suppressed by palmitate, and also administration of lupinoside PA4 (**68**) terminates the inhibition of insulin stimulation by free fatty acids (FFAs); this allowed GFP-GLUT4 translocation to the membrane (Dey et al., 2007, p. 149). 1*β*-Hydroxy-2-oxopomolic acid (**69**) was tested against the expression of adipocytes (3T3-L1), and among the tested adipocytes, GLUT4 mRNA expression was found to be increased after 4 days and it was markedly decreased after 8 days by treatment of compound **69** at 50, 100 μM (Ahn et al., 2012, p. 643). 3*β*-Taraxerol (**70**), a triterpenoid showed time-dependent insulin-stimulated glucose uptake through translocation and activation of the GLUT4 in an IRTK- and PI3K-dependent manner. This activation was comparable to that of rosiglitazonc (50 μM) and insulin (100 μM), which showed maximum GLUT4 translocation (Sangeetha et al., 2010, p. 359). Platyconic acid (**71**) most effectively increased insulin-stimulated glucose uptake in 3T3-L1 adipocytes and also improved insulin sensitivity. In addition, it stimulated the expression of adiponectin and PPARγ in adipose tissue, enhanced insulin signaling, and improved GLUT4 translocation into the membranes (Kwon et al., 2012, p. 529). Phytochemical investigation of the ethyl acetate fraction of the leaves of a Chinese medicinal plant, *Eucalyptus citriodora* Hook resulted in the isolation of seven triterpenoids *viz.* asiatic acid (**72**), betulinic acid (**73**), corosolic acid (**74**), euscaphic acid (**75**), madasiatic acid (**76**), oleanic acid (**77**), and ursolic acid (**78**) (Fig. 4.11). Among them, compound **72** showed GLUT4 translocation activity with 0.47-folds of enhancement within 10 min while euscaphic acid, madasiatic acid, oleanic acid, and ursolic acid exhibited moderate GLUT4 translocation activity with 0.70–1.06-folds improvement within 10 min. Betulinic and corosolic acids were the most active compounds, showing 2.38- and 1.78-folds GLUT4 translocation enhancement, respectively (Wang et al., 2014, p. 3096).

3.11 Phenols and Polyphenol

Rhodomyrtosone E (**79**), a xanthene-like derivative isolated for the first time from ethyl acetate fraction of Chinese medicinal herb, *E. citriodora* Hook was found to possess weak GLUT4 translocation effects with 0.49-folds of enhancement within 10 min at 10 μM concentration in L6 skeletal muscle cells (Wang et al., 2014, p. 3096). Brazilian green propolis material collected by bees from unique Brazilian plant *Baccharis dracunculifolia* (rosemary) is rich in a phenolic compound called artepillin C (**80**). Artepillin C showed adipocyte differentiation and glucose regulation in 3T3-L1 adipocytes. It stimulates adipocyte differentiation and enhances glucose transport in adipocytes by inducing the expression and translocation of GLUT1 and GLUT4 vesicles in 3T3-L1

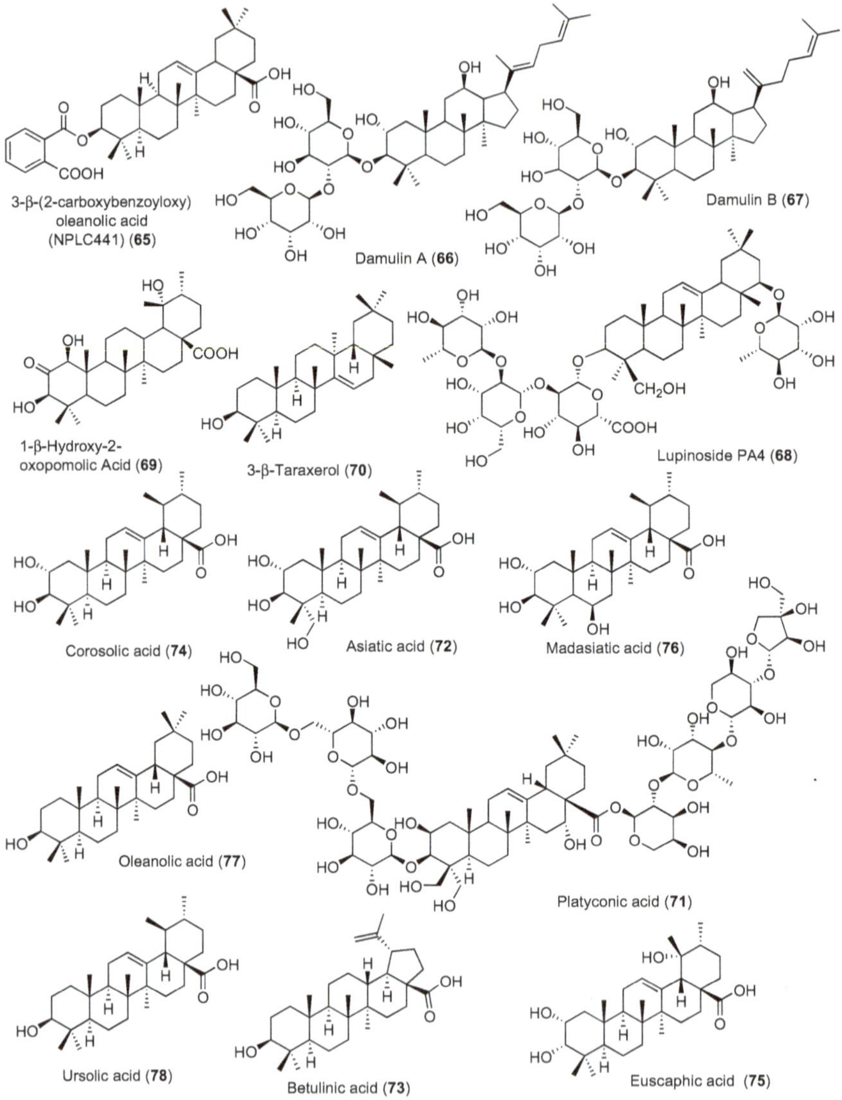

FIGURE 4.11 Chemical structure of triterpenes (**65–78**).

adipocytes. Artepillin C enhances the glucose consumption in 3T3-L1 adipocytes ($p < .05$). The basal glucose consumptions in cells treated with artepillin C at 10 and 30 μM were 1.7- and 1.9-fold higher than that in vehicle-treated cells, respectively. It also increases the mRNAs and protein expression of both GLUT1 and GLUT4 in fully differentiated adipocytes. In addition, artepillin C significantly increases the amounts of GLUT1 and GLUT4 proteins in the crude plasma membrane fraction (Choi et al., 2011, p. 925). Methyl caffeate (**81**), a methyl ester of caffeic acid, was evaluated for its effects on histological changes in the pancreas and GLUT4 expression in muscles of STZ-induced diabetic rats. Treatment with methyl caffeate in diabetic rats significantly increased GLUT4 mRNA expression compared to diabetic control rats. Methyl caffeate-treated diabetic rats showed significant ($p \leq .05$) redistribution of GLUT4 protein expression on the plasma membrane compared to diabetic control rats. Furthermore, four cinnamic acid derivatives viz., ferulic acid (**82**), eugenol (**83**), 4-hydroxycinnamic acid (**84**), and caffeic acid (**85**) (Fig. 4.12), have significantly increased expressions of GLUT4 and PI3K, whereas chlorogenic (**86**) and cinnamic acids (**87**) have significant effects on the expression of the PPARγ

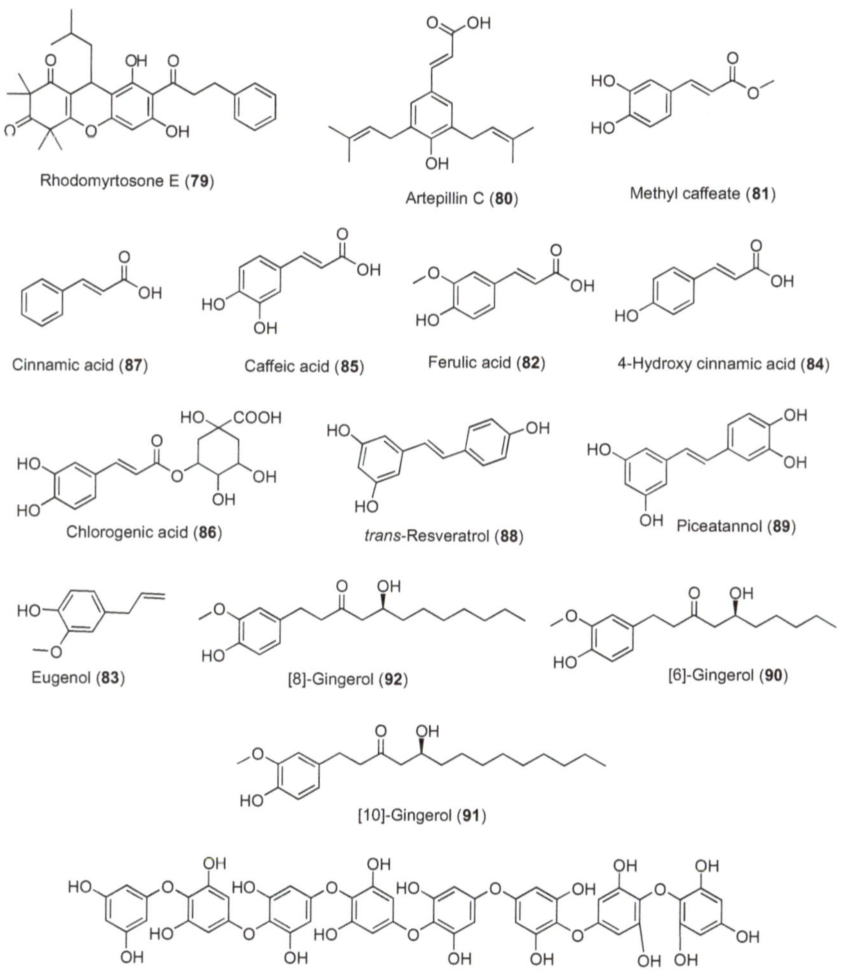

FIGURE 4.12 Chemical structure of phenols (**79–92**) and polyphenol (**93**).

gene. These hydroxycinnamic acid derivatives act differently on the expressions of the downregulators of the insulin cascade. Hence, it could be hypothesized that the 4-hydroxycinnamic acid derivatives enhance glucose uptake mediated by PI3K-dependent GLUT4 translocation, whereas chlorogenic and cinnamic acids increase glucose uptake via PPARγ-mediated GLUT4 translocation (Prabhakar and Doble, 2011, p. 9835). The synergic effect of chlorogenic acid, ferulic acid, and berberine was evaluated in the combination of two marked drugs metformin and 2,4-thiazolodinedione (THZ) on the uptake of 2-DG by L6 myotubes. Three different combinations of different concentrations of chlorogenic acid, ferulic acid, and berberine with metformin or THZ have shown synergistic effect in the uptake of 2-DG with a maximum of 5.0 and 5.3, 4.98 and 5.11, 4.1 and 4.7 times from the base value (without the drugs or the natural products), respectively. Chlorogenic acid and berberine increase the expressions of GLUT4 and PPARγ gene while ferulic acid enhanced the expression of GLUT4 and PI3K transcripts significantly. It indicates that chlorogenic acid enhances glucose uptake by increasing GLUT4 expression via PI3K-independent pathway whereas ferulic acid increases glucose uptake by PI3K-dependent pathway (Prabhakar and Doble, 2009, p. 1119). Another member of naturally occurring phenolic compounds, *trans*-resveratrol (**88**) has been evaluated to increase the glucose uptake in C2C12 myotubes by activating AMPK in a dose-dependent manner with a maximum of approximately twofold at 100 μM (Park et al., 2007, p. 222), and it can also activate AKT, thereby imparting significant role in GLUT4 translocation and activation of glucose uptake (Penumathsa et al., 2008, p. 2350). Piceatannol (**89**), a resveratrol derivative, was found to have the glucose uptake stimulatory effect, AMPK phosphorylation, and GLUT4 translocation by Western blotting analyses in L6 myotubes under the absence of insulin. Hence, resveratrol and its derivatives may have the preventive and remedial effects for type 2 diabetes (Minakawa et al., 2012, p. 469). Gingerols, the active constituents of the rhizomes of *Zingiber officinale,* Roscoe (ginger) were evaluated for their glucose uptake activity. At 40 μM, (*S*)-[6]-gingerol (**90**) and (*S*)-[10]-gingerol (**91**) showed approximately 60% increment in glucose uptake, while (*S*)-[8]-gingerol (**92**) displayed about a threefold increase in glucose uptake at 160 μM. At their increased concentrations, both (*S*)-[6]-gingerol and (*S*)-[10]-gingerol showed toxicity to the cells; however, its homologue (*S*)-[8]-gingerol showed highest potency and much lower toxicity toward cells and the test compound exhibited a concentration-dependent promotion of glucose uptake regardless of the presence or absence of insulin. This indicates that (*S*)-[8]-gingerol enhances glucose uptake via a distinct pathway independent of insulin. The activity of (*S*)-[8]-gingerol was found to be associated primarily with an increased expression of GLUT4 protein distributed on the surface of L6 myotube plasma membrane (Li et al., 2012, p. 1549).

Octaphlorethol A (**93**) (Fig. 4.12), a novel polyphenolic compound isolated from marine algae *Ishige foliacea*, stimulates glucose uptake (102%, 106%, 115%, and 140% at 6.25, 12.5, 25, and 50 μM, respectively, in L6 skeletal muscles). Glucose uptake induced by compound **93** peaked after 2h and then gradually decreased for the subsequent 10h. A treatment time of 2h resulted in a particularly significant increase in glucose uptake (187% of control uptake) relative to that induced by insulin (164% of control uptake). Subsequent treatments were therefore performed using time point and concentration. The translocation of GLUT4 was observed after L6 myotubes treated with octaphlorethol A for 2h. GLUT4 translocation to the plasma membrane of L6 myotubes was markedly increased by the treatment of compound **93** and it also induced increase in glucose uptake was dependent on both PI3K/AKT and AMPK activation while independent of insulin sensitivity (Lee et al., 2012b, p. 576).

3.12 Iridoids

Maurya and coworkers have discovered antidiabetic potential of *Kigelia pinnata* for the first time. The ethanolic extract of *K. pinnata* twigs was evaluated for GLUT4 translocation modulatory effect in L6-GLUT4myc myotubes. Cells were treated with ethanolic extract (10 µg/mL) for 16h under culture conditions before the measurement of surface level of GLUT4myc. Ethanolic extract of *K. pinnata* increases in the surface of GLUT4myc level as compared to control by 27% ($p < .05$).

On the basis of these preliminary findings, they carried out phytochemical investigation and purified 10 compounds, rehmaglutin C (**95**), 7-hydroxy viteoid II (**96**), leonuride (**97**) or ajugol (**97**), catalpol (**98**), specioside (**99**), verminoside (**100**), shoreaphenol, 4-hydroxy cinnamic acid, and caffeic acid along with one new iridoid named as (1*S*,2*R*,5*R*)-5-(2-hydroxyethyl)-3,4-bis(hydroxymethyl) cyclopent-3-ene-1,2-diol (**94**) (Fig. 4.13). All isolated compounds were evaluated for antidiabetic activity. Among them, compounds **94**, **98**, **99**, and **100** showed significant stimulation in the surface level of GLUT4myc with respective percentage of 36.3 ($p < .01$), 50.0 ($p < .01$), 41.8 ($p < .01$), and 45.6 ($p < .01$) at 10 µM

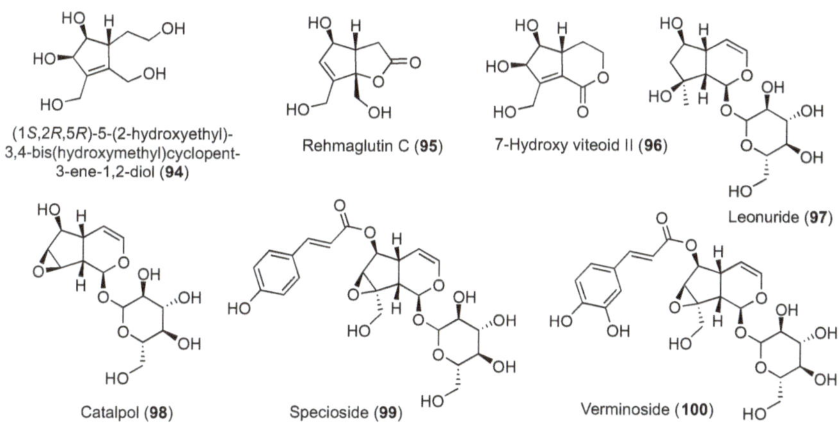

FIGURE 4.13 Chemical structure of iridoids (**94–100**).

concentration while rest of the compounds possess weak glucose uptake stimulation via the activation of translocation of GLUT4 in L6 skeletal muscles at same dose. The cell cytotoxicity of the compounds on L6-GLUT4myc cells was evaluated by MTT and found to exert no adverse effects (Khan et al., 2012, p. 125).

3.13 Polysaccharides

Wang et al. reported a polysaccharide MDG-1 (**101**) (Fig. 4.14), isolated from the water extract of *Ophiopogon japonicus,* a traditional Chinese medicine has been used to treat type 2 diabetes. To locate the hypoglycemic effect of MDG-1, KKAy mice were administered with 75 mg/kg MDG-1, 300 mg/kg MDG-1, 2 mg/kg rosiglitazone, or distilled water for 8 weeks and showed significant decrease in blood glucose levels in both MDG-1- and rosiglitazone-treated groups in a dose-dependent manner. The blood glucose levels of the mice that were fed and fasted decreased by 44.44% and 32.72%, respectively, after treatment of 300 mg/kg of MDG-1. The reduction in blood glucose levels of KKAy mice at 300 mg/kg MDG-1 was similar to 2 mg/kg of rosigilitazone. MDG-1 treatment significantly enhanced GLUT4 protein levels in KKAy mice. These results indicate that MDG-1 has a positive effect on skeletal muscle insulin transduction (Wang et al., 2012, p. 347).

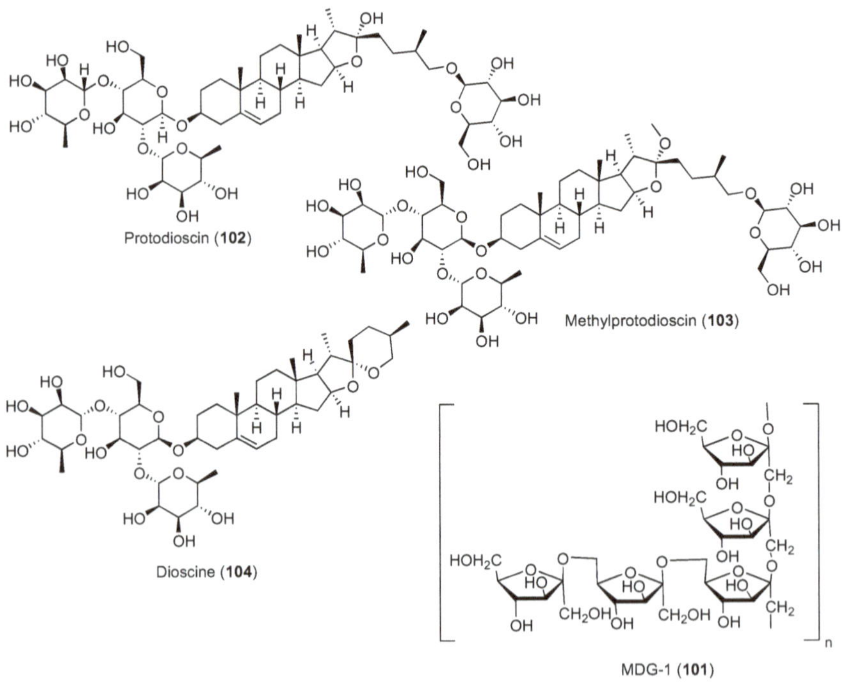

FIGURE 4.14 Chemical structure of saponins (**102–104**) and polysaccharide (**101**).

3.14 Saponins

A famous traditional Chinese herbal medicine named as *Dioscorea nipponica* Makino belongs to the family Dioscoreaceae. Traditionally, this plant has been used for relieving cough and asthma; eliminating rheumatic aches; alleviating pain; and improving blood circulation. Saponins [protodioscin (**102**), methyl-protodioscin (**103**), dioscine (**104**)] (Fig. 4.14) are the major constituents of this herb having antidiabetic activity. The increased GLUT4 protein expression and *p*-AKT ($p < .05$) indicate that the product played a direct role in peripheral glucose uptake. The potent antidiabetic activity of saponins on type 2 DM is through the improvement of insulin resistance, promoting insulin secretion, and adjusting insulin signaling pathway (Yu et al., 2015, p. 228).

3.15 Tannins

Gallotannin 1 (**105**) and gallotannin 2 (**106**) (Fig. 4.15), isolated from butanolic fraction of *Capparis moonii* fruit, were analyzed for the effect of glucose uptake in L6 myotubes with varying concentrations (1 pg/mL–1 μg/mL) for 24 h. A dose-dependent increase in glucose uptake was observed for both the

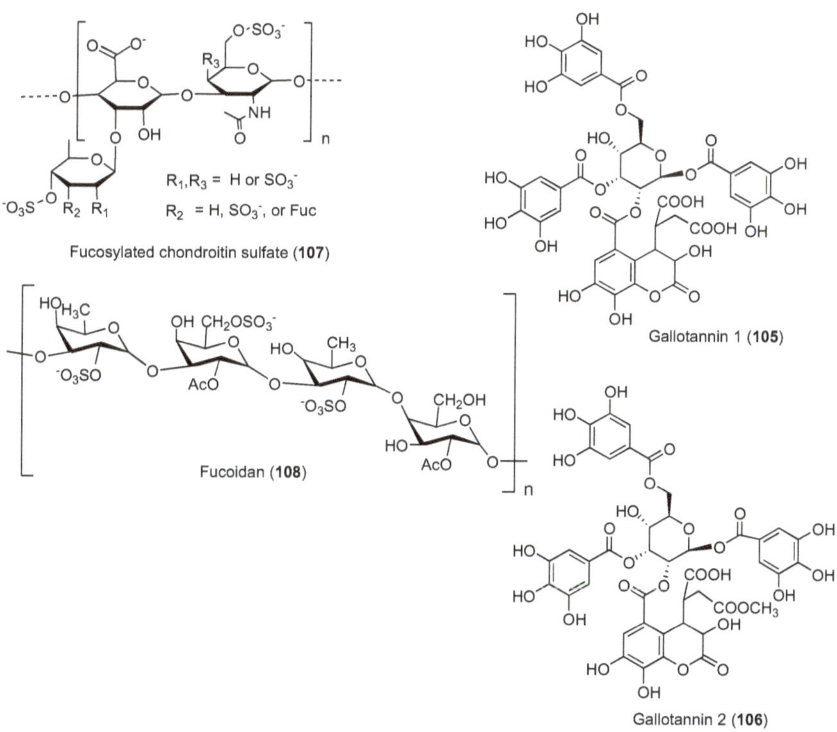

FIGURE 4.15 Chemical structure of tannins (**105–106**) and sulfonated sugars (**107–108**).

compounds. Gallotannin 1 at 10 ng/mL produces a maximal glucose uptake of 223% over the control cells, whereas under similar conditions positive control, rosiglitazone at 50 μM showed 240% glucose uptake. Gallotannin **2** was found to produce a maximal glucose uptake at 100 ng/mL with 219% over the control cells while under the similar conditions positive control, rosiglitazone at 50 μM showed 203% glucose uptake. These compounds **105** and **106** were tested on PI3K and GLUT4 mRNA expression in L6 cells. The expression of PI3K and GLUT4 mRNA was significantly increased upon treatment with gallotannin 1 (10 ng/mL) and gallotannin 2 (100 ng/mL), thereby indicates that these compounds exert glucose transportation activity through PI3K-mediated insulin signaling pathway (Kanaujia et al., 2010, p. 3940).

3.16 Sulfonated Sugars

It is very interesting to know that some sulfonated sugars from marine source possess antidiabetic potentials. Sea cucumber, a rich source of sulfonated sugars (Fig. 4.15) (approximately 27–32%) and used as a traditional marine nutrient food in China, and the output value that has reached 20 billion Chinese Yuan in 2012. Fucosylated chondroitin sulfate (CHS) (**107**) is one such component of sulfated polysaccharide having 2,4-*O*-disulfated fucose branches, which are responsible for their bioactivities. Wang et al. first time scientifically proved the importance of sea cucumber in disposal of blood glucose level to skeletal muscles and their antidiabetic activity. They also isolated the active component fucosylated chondroitin sulfate (CHS) from sea cucumber. The beneficial effects of CHS administration on fasting glucose level and glucose tolerance were observed with significant improvement in CHS-treated mice by 19.59% and 10.88%, respectively. Furthermore, the synergic effect was also studied in combination treatment with rosiglitazone and observed the significant decrease in blood glucose level. CHS reduces blood glucose level via promoting GLUT4 translocation by 1.7-fold rather than increasing total GLUT4 protein expression in skeletal muscles. Insulin-stimulated GLUT4 translocation was increased by CHS in high fat, high sucrose diet mice via promoting the phosphorylation of PKB at Ser473 by 1.5-fold and PI3K at p85 by 5.0-fold, and Thr308 by 1.6-fold in skeletal muscle (Hu et al., 2013, p. 1639; Hu et al., 2014a, p. H1424).

Fucoidan (fucan sulfate) (**108**) is a sulfated polysaccharide isolated from different brown algae such as *Fucus vesiculosus*, *Ecklonia kurome*, and *Cladosiphon okamuranus*. The expression of diabetes-related genes in mouse cell lines 3T3-L1 adipocytes was evaluated by fucan sulfate. The expression level of GLUT4, PPARγ, and CCAAT/enhancer binding protein α (C/EBPα) was increased after treatment of fucoidan at different concentration ranges from 10 to 500αα ppm in a dose-dependent manner. Fucoidan is seemed to enhance insulin sensitivity by increasing the expression level of diabetes-related genes in 3T3-L1 adipocytes and hence, might be a therapeutic agent for the treatment of hyperglycemia in type 2 diabetes (Kim et al., 2007, p. 212).

3.17 Miscellaneous

Pterosin A (**109**), isolated from a fern plant *Hypolepis punctata* (Thunbx.) Mett., showed effect on glucose metabolism. It increases the reduced level of GLUT4 translocation from cytosol to membrane in skeletal muscle and also elevates expression of PEPCK expression in the liver of diabetic mice. The GLUT4 translocation in muscle cells is activated by pterosin A via AMPK-regulated signaling pathway (Hsu et al., 2013, p. 628). A popular herbal tea, which is commonly known as rooibos in South Africa having major constituent phenylpyruvic acid-2-O-glucoside (**110**) (Fig. 4.16), an enolic phenylpyruvic acid glucoside derivative enhances glucose uptake in Chang cells at concentration ranges from 1.0 to 31.6 μM (EC_{50} 3.6 μM). In a gene expression analysis, treatment with compound **110** significantly increases the expressions of GLUT1 and GLUT2 mRNA in the liver. These are the major glucose transporters in the liver and hence it is suggested that treatment of compound **110** enhances the capacity of liver to absorb glucose from the portal circulation. In obese insulin-resistant rats, oral administration of compound **110** lowers fasting glucose concentrations and improves mRNA expression of glucokinase, glucose transporter 1 and 2, peroxisome proliferator-activated receptor alpha (PPARα), and suppressor of cytokine signaling 3 in the liver (Muller et al., 2013, p. 2216).

4. NATURAL PRODUCT DERIVATIVES AS GLUT4 TRANSLOCATION ACTIVATORS

Throughout the history of drug discovery and development, natural sources, especially plant-derived medicines, have been used for the treatment of a wide spectrum of diseases. In the early 1800s, advances in scientific knowledge and techniques led to the discovery of pure compounds, commonly referred to as natural products (NPs). Among several isolated natural compounds, morphine became the first commercial, pure natural product by E. Merck in 1826 and later in 1899, first semisynthetic drug aspirin was commercially sold by Bayer. Hence, a new era was started in medicine, where drugs could be acquired from plants in pure form as well as by chemical modifications to achieve more activity and good pharmacokinetic results (Newman et al., 2000, p. 215). Natural product-derived molecules contributed about 22% new chemical entities in search of various lead molecules (Newman and Cragg, 2012, p. 311). Due to

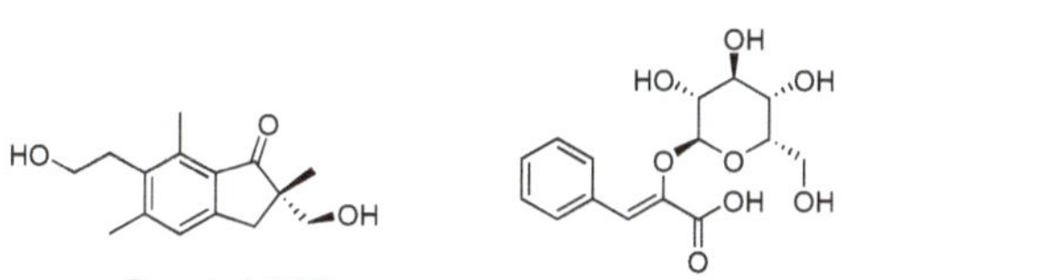

FIGURE 4.16 Chemical structure of the compounds pterosin A (**109**) and PPAG (phenylpyruvic acid-2-*O*-glucoside) (**110**).

such major contribution of natural product analogues in drug discovery, therefore, we have also included some natural product derivatives with their GLUT4 translocation activity.

4.1 Chemical Modification of Lupeol

Lupeol (**111**) is a lupane-type triterpene possesses several biological activities. It was isolated from stem bark of *Bombax ceiba* in a quantitative yield. To enhance the activity, chemical modifications were carried out by Maurya et al. at the isopropylene moiety of lupeol via allylic oxidation by using selenium dioxide to afford α, β-unsaturated aldehyde and esterification at C-3 carbon with different acids to get corresponding ester derivatives. Lupeol was treated under reflux condition in ethanol with SeO_2 for 24h yielded compound **112** (Scheme 4.1), which showed potent GLUT4 activity in L6 skeletal muscle cells. Compound **112** showed significant stimulation of glucose uptake with respective percentage stimulation of 173.1 ($p< .001$) via activating GLUT4 translocation compared to positive control whereas the acylated derivative **114** and cinnamoyl ester derivative (**115**) showed 1.12- and 0.98-fold glucose uptake at 10 μM concentration, respectively. Similarly, a pyrazole derivative **113** of compound **112** was prepared on reaction with hydrazine hydrate in KOH and again esterification-provided compound **116**. Compound **113** showed only 0.16-fold glucose uptake whereas its esterified derivative compound **116** displayed 1.07-fold glucose uptake activity at 10 μM concentration. It was achieved that the lupeol analogues showed notable increase in glucose uptake activity via GLUT4 translocation (Khan et al., 2014, p. 2674).

The same group of investigators also synthesized some other analogues of lupeol via Grubbs metathesis and ferric chloride and evaluated for their glucose

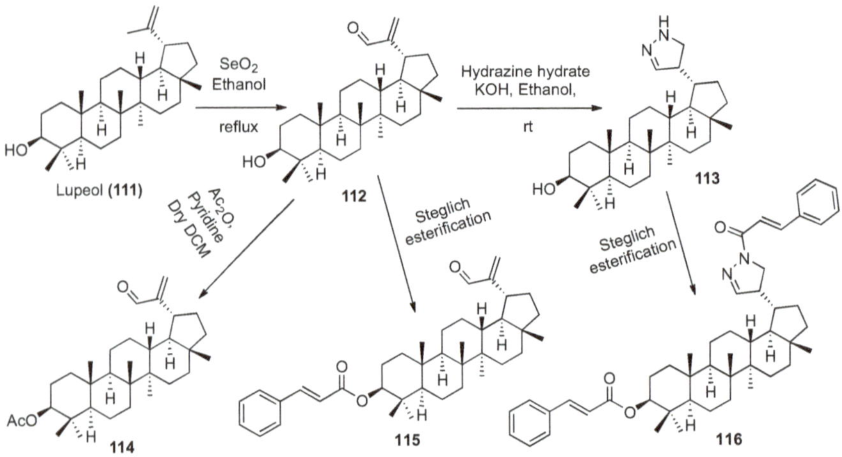

SCHEME 4.1 Typical procedure for the synthesis of lupeol analogues (**112–116**).

uptake activity but these molecules showed weak glucose uptake stimulatory activity in a dose-dependent study.

4.2 Chemical Modification of 4-Hydroxyisoleucine

4-Hydroxyisoleucine (**117**) is an amino acid and first isolated from fenugreek (*Trigonella foenum-graecum*) seeds in good amount. 4-Hydroxyisoleucin and its amide derivatives showed significant glucose uptake stimulatory activity in L6 skeletal muscle cells in dose-dependent manner. The amide derivatives of 4-hydroxyisoleucin were synthesized in two steps; first protection of amino group with Boc anhydride followed by lactonization of free acid with hydroxy group present at C-4 position.

Different amides were prepared via lactone ring opening under diisobutylaluminum hydride (DIBAL-H) and further reaction with different amines in THF at room temperature (Scheme 4.2). Synthesized compounds (**119**, **120**) showed maximal stimulation of glucose uptake with 109.5% (2.10-folds versus control) and 105.25% (2.05-folds versus control), respectively, at 10 μM concentration by activation of GLUT4 translocation (Korthikunta et al., 2015, p. 66).

4.3 Chemical Modification of Kojic Acid

The discovery of nonpeptidyl insulin mimetic small molecule demethylasterriquinone B1 (**121**) (Fig. 4.17) isolated from microbial sources such as *Aspergillus terreus* and *Pseudomassaria* fungi became a major breakthrough in drug discovery. This discovery of such small molecules from natural sources is capable of mimicking the biological function of a protein hormone by interacting with the IR and causing activation of IR tyrosine kinase (IR-TK) (Zhang et al., 1999, p. 974; Nankar and Doble, 2013, p. 748). Despite these promising attributes, this molecule has not yet entered in clinical trials. A major problem concerned with the safety of such pharmaceutical candidates is out of the presence of a quinone substructure. The replacement of quinone subunit of demethylasterriquinone B1 with kojic acid yielded indolylkojic acid (**122**) (Fig. 4.17), which showed potent glucose uptake stimulatory effect in skeletal muscle cells via the stimulation of translocation of GLUT4 (Henriksen et al., 1994, p. 1196; Leney and Tavare, 2009, p. 1).

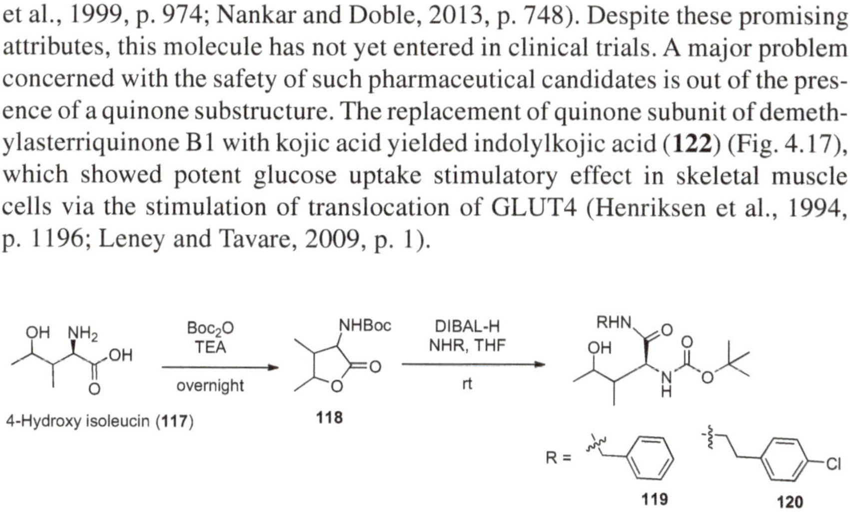

SCHEME 4.2 General procedure for the synthesis of 4-hydroxyisoleucin derivatives (**118–120**).

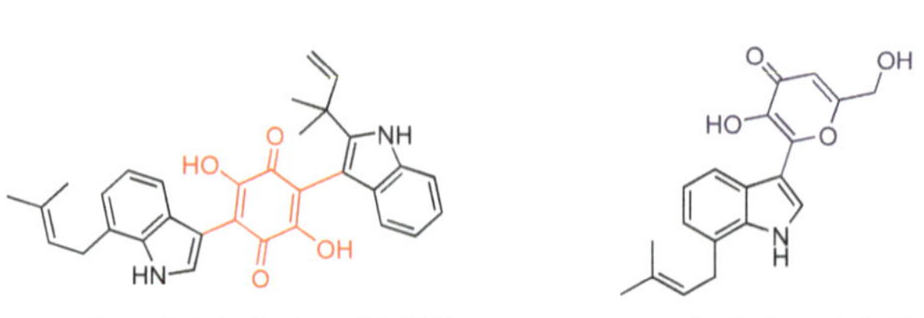

FIGURE 4.17 Chemical structure of insulin mimetic demethylasterriquinone B1 (**121**) and its indolylkojic acid-derived analogue (**122**).

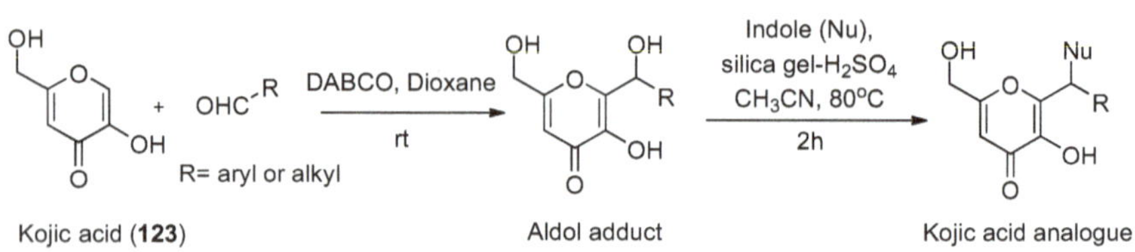

SCHEME 4.3 Typical procedure for the synthesis of kojic acid analogues (**124–130**).

In the search of potent activator of GLUT4, Mukherjee et al. also synthesized a library of kojic acid derivatives via nucleophilic substitution reaction of indole on various aldol adduct of kojic acid.

Kojic acid (**123**) was isolated from its natural source *Aspergillus flavus* in quantitative yield. Kojic acid on reaction with aldehydes provided their corresponding aldol adducts. These aldol adducts undergo nucleophilic substitution reaction with indole to furnish indole derivatives of kojic acid (Scheme 4.3).

Among all the synthesized compounds, compounds **124–130** (Fig. 4.18) showed significant glucose uptake activity. These kojic acid analogues exerted their biological activity by enhancing GLUT4 translocation to the cell surface via PI3K-dependent signaling pathway similar to the insulin mode of action. Compounds **124** and **125** synthesized by using electron-donating benzaldehyde show significant glucose uptake stimulation with respective 1.21- and 1.25-fold stimulation ($p < .05$) over control. Compound **126** having substituted indole showed better activity of 1.25-fold ($p < .05$) increase in glucose uptake than other substituted indoles. Compounds **127**, **128**, **129**, and **130** showed significant stimulation of glucose uptake with respective 1.33-, 1.38-, 1.34-, and 1.21-fold increase at 10 μM concentration (Sharma et al., 2014, p. 727).

5. CONCLUDING REMARKS

This chapter focuses on a range of natural products of different classes isolated from living organisms along with their glucose uptake activity either via the activation of GLUT4 translocation or protein expression. Antidiabetic activity has been observed from different class of compounds such as alkaloids, flavonoids, isoflavonoids, steroids, terpenoids, phenols, poylphenols, saponins, and

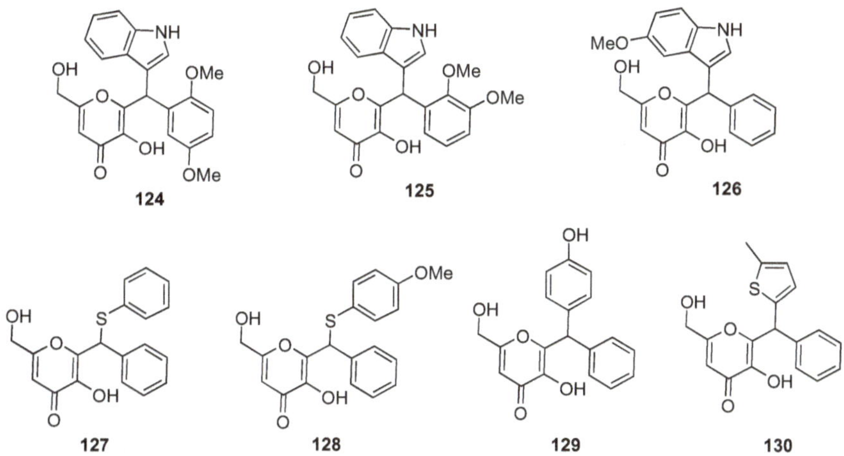

FIGURE 4.18 Chemical structure of kojic acid analogues (**124–130**), which showed significant GLUT4 translocation activity.

tannins. Natural products have been found to be effective in the activation of translocation of GLUT4 -mediated treatment of diabetes. The chemical transformations of bioactive molecules have provided a wider window of discovering potent GLUT4 translocation activator with improved pharmacological profile. Natural products and their analogues, having a good history of supplying leads for numerous drugs in use, are thus strongly anticipated to provide novel and promising candidates in the ongoing drug discovery and development process for having potent and efficacious antidiabetic drugs.

LIST OF ABBREVIATIONS

2-DG 2-DG uptake
2-NBDG 2-(*N*-(7-Nitrobenz-2-oxa-1,3-diazol-4-yl)amino)-2-deoxyglucose
AMPK Adenosine monophosphate-activated protein kinase
C/EBPα CCAAT enhancer-binding proteins
CaMKII Ca^{2+}/cal modulin-dependent protein kinase II
DIBAL-H Diisobutylaluminum hydride
DIO Diet-induced obese
GLUT4 Glucose transporter type 4
HPLC High-pressure liquid chromatography
IGF-1 Insulin growth factor 1
IR Insulin receptor
IRS-1 Insulin receptor substrate 1
IRTK Insulin receptor tyrosine kinase
LDM Low density microsome
LPL Lipoprotein lipase
PEPCK Phosphoenol pyruvate carboxykinase

PI3K Phosphatidylinositol-3 kinase
PKB Protein kinase B
PPARγ Peroxisome proliferator-activated receptor γ
PTP1B Protein tyrosine phosphatase 1B
RXRa Retinoid X receptor a
STZ Streptozotocin
THZ 2,4-Thiazolodinedione
TZQ-F Tangzhiqing formula
WAT White adipose tissue
VEGF Vascular endothelial growth factor

ACKNOWLEDGMENTS

Kapil Dev is thankful to Council of Scientific and Industrial Research (CSIR), New Delhi and E. Ramakrishna is grateful to University Grant Commission (UGC), New Delhi, for providing funds as senior research fellowships. CSIR-CDRI communication number for this manuscript is 9269.

REFERENCES

Abe, D., Saito, T., Sekiya, K., 2006. Sennidin stimulates glucose incorporation in rat adipocytes. Life Science 79 (11), 1027–1033.

Ahn, E.-K., Lee, J.A., Seo, D.-W., Hong, S.S., Oh, J.S., 2012. 1β-hydroxy-2-oxopomolic acid isolated from *Agrimonia pilosa* extract inhibits adipogenesis in 3T3-L1 cells. Biological & Pharmaceutical Bulletin 35 (5), 643–649.

Alkhateeb, H., Bonen, A., 2010. Thujone, a component of medicinal herbs, rescues palmitate-induced insulin resistance in skeletal muscle. American Journal of Physiology 299 (3, Pt. 2), R804–R812.

Anand, S., Muthusamy, V.S., Sujatha, S., Sangeetha, K.N., Bharathi Raja, R., Sudhagar, S., Poornima Devi, N., Lakshmi, B.S., 2010. Aloe emodin glycosides stimulates glucose transport and glycogen storage through PI3K dependent mechanism in L6 myotubes and inhibits adipocyte differentiation in 3T3L1 adipocytes. FEBS Letters 584 (14), 3170–3178.

Anandharajan, R., Pathmanathan, K., Shankernarayanan, N.P., Vishwakarma, R.A., Balakrishnan, A., 2005. Upregulation of Glut-4 and PPARγ by an isoflavone from *Pterocarpus marsupium* on L6 myotubes: a possible mechanism of action. Journal of Ethnopharmacology 97 (2), 253–260.

Anhe, G.F., Okamoto, M.M., Kinote, A., Sollon, C., Lellis-Santos, C., Anhe, F.F., Lima, G.A., Hirabara, S.M., Velloso, L.A., Bordin, S., Machado, U.F., 2012. Quercetin decreases inflammatory response and increases insulin action in skeletal muscle of ob/ob mice and in L6 myotubes. European Journal of Pharmacology 689 (1–3), 285–293.

Annapurna, H.V., Apoorva, B., Ravichandran, N., Arun, K.P., Brindha, P., Swaminathan, S., Vijayalakshmi, M., Nagarajan, A., 2013. Isolation and in silico evaluation of antidiabetic molecules of *Cynodon dactylon* (L.). Journal of Molecular Graphics & Modelling 39, 87–97.

Bailey, C.J., Day, C., 1989. Traditional plant medicines as treatments for diabetes. Diabetes Care 12 (8), 553–564.

Brahmachari, G., 2011. Natural products in drug discovery: impacts and opportunities – an assessment. In: Brahmachari, G. (Ed.), Bioactive Natural Products: Opportunities & Challenges in Medicinal Chemistry. World Scientific Publishing Co., Singapore, pp. 1–199.

Brahmachari, G., 2014. Natural bioactive flavonoids – recent developments in research: a look through. In: Brahmachari, G. (Ed.), Natural Bioactive Molecules: Impacts and Prospects. Narosa Publishing House, New Delhi, pp. 1–118.

Butler, M.S., 2005. Natural products to drugs: natural product derived compounds in clinical trials. Natural Product Reports 22 (2), 162–195.

Chen, X., Lei, K., Dong, M., Fang, Z., Jin, L., 2010. Effect of puerarin on myocardial damage in STZ-induced diabetic rats. Zhongguo Bingli Shengli Zazhi 26 (4), 650–655.

Chen, L., Li, Q.-Y., Shi, X.-J., Mao, S.-L., Du, Y.-L., 2013. 6-Hydroxydaidzein enhances adipocyte differentiation and glucose uptake in 3T3-L1 cells. Journal of Agricultural and Food Chemistry 61 (45), 10714–10719.

Choi, S.-S., Cha, B.-Y., Iida, K., Lee, Y.-S., Yonezawa, T., Teruya, T., Nagai, K., Woo, J.-T., 2011. Artepillin C, as a PPARγ ligand, enhances adipocyte differentiation and glucose uptake in 3T3-L1 cells. Biochemical Pharmacology 81 (7), 925–933.

Choi, S.-S., Cha, B.-Y., Lee, Y.-S., Yonezawa, T., Teruya, T., Nagai, K., Woo, J.-T., 2012. Honokiol and magnolol stimulate glucose uptake by activating PI3K-dependent Akt in L6 myotubes. BioFactors 38 (5), 372–377.

Choi, S.-S., Cha, B.-Y., Choi, B.-K., Lee, Y.-S., Yonezawa, T., Teruya, T., Nagai, K., Woo, J.-T., 2013. Fargesin, a component of Flos Magnoliae, stimulates glucose uptake in L6 myotubes. Journal of Natural Medicines 67 (2), 320–326.

Dey, D., Pal, B.C., Biswas, T., Roy, S.S., Bandyopadhyay, A., Mandal, S.K., Giri, B.B., Bhattacharya, S., 2007. A Lupinoside prevented fatty acid induced inhibition of insulin sensitivity in 3T3 L1 adipocytes. Molecular and Cellular Biochemistry 300 (1&2), 149–157.

Farsi, E., Ahmad, M., Hor, S.Y., Ahamed, M.B.K., Yam, M.F., Asmawi, M.Z., 2014. Standardized extract of *Ficus deltoidea* stimulates insulin secretion and blocks hepatic glucose production by regulating the expression of glucose-metabolic genes in streptozitocin-induced diabetic rats. BMC Complementary and Alternative Medicine 14, 1–25.

Fujimori, K., Shibano, M., 2013. Avicularin, a plant flavonoid, suppresses lipid accumulation through repression of C/EBPα-Activated GLUT4-mediated glucose uptake in 3T3-L1 cells. Journal of Agricultural and Food Chemistry 61 (21), 5139–5147.

Gandhi, G.R., Stalin, A., Balakrishna, K., Ignacimuthu, S., Paulraj, M.G., Vishal, R., 2013. Insulin sensitization via partial agonism of PPARγ and glucose uptake through translocation and activation of GLUT4 in PI3K/*p*-Akt signaling pathway by embelin in type 2 diabetic rats. Biochimica et Biophysica Acta 1830 (1)), 2243–2255.

Gao, J., Li, J., An, Y., Liu, X., Qian, Q., Wu, Y., Zhang, Y., Wang, T., 2014. Increasing effect of Tangzhiqing formula on IRS-1-dependent PI3K/AKT signaling in muscle. BMC Complementary and Alternative Medicine 14, 198/191–198/119 119 pp.

Guo, C., Zhang, C., Li, L., Wang, Z., Xiao, W., Yang, Z., 2014. Hypoglycemic and hypolipidemic effects of oxymatrine in high-fat diet and streptozotocin-induced diabetic rats. Phytomedicine 21 (6), 807–814.

Henriksen, J.E., Alford, F., Handberg, A., Vaag, A., Ward, G.M., Kalfas, A., Beck-Nielsen, H., 1994. Increased glucose effectiveness in normoglycemic but insulin-resistant relatives of patients with non-insulin-dependent diabetes mellitus. A novel compensatory mechanism. Journal of Clinical Investigation 94 (3), 1196–1204.

Hsu, F.-L., Huang, C.-F., Chen, Y.-W., Yen, Y.-P., Wu, C.-T., Uang, B.-J., Yang, R.-S., Liu, S.-H., 2013. Antidiabetic effects of pterosin A, a small-molecular-weight natural product, on diabetic mouse models. Diabetes 62 (2), 628–638.

Hu, S., Chang, Y., Wang, J., Xue, C., Shi, D., Xu, H., Wang, Y., 2013. Fucosylated chondroitin sulfate from *Acaudina molpadioides* improves hyperglycemia via activation of PKB/GLUT4 signaling in skeletal muscle of insulin resistant mice. Food & Function 4 (11), 1639–1646.

Hu, S., Chang, Y., He, M., Wang, J., Wang, Y., Xue, C., 2014a. Fucosylated chondroitin sulfate from sea cucumber improves insulin sensitivity via activation of PI3K/PKB pathway. Journal of Food Science 79 (7), H1424–H1429.

Hu, S., Xu, L., Shi, D., Wang, J., Wang, Y., Lou, Q., Xue, C., 2014b. Eicosapentaenoic acid-enriched phosphatidylcholine isolated from *Cucumaria frondosa* exhibits anti-hyperglycemic effects via activating phosphoinositide 3-kinase/protein kinase B signal pathway. Journal of Bioscience and Bioengineering 117 (4), 457–463.

Hung, H.Y., Qian, K., Morris-Natschke, S.L., Hsu, C.S., Lee, K.H., 2012. Recent discovery of plant-derived anti-diabetic natural products. Natural Product Reports 29 (5), 580–606.

Iaconelli, A., Gastaldelli, A., Chiellini, C., Gniuli, D., Favuzzi, A., Binnert, C., Mace, K., Mingrone, G., 2010. Effect of oral sebacic acid on postprandial glycemia, insulinemia, and glucose rate of appearance in type 2 diabetes. Diabetes Care 33 (11), 2327–2332.

IDF, 2013. International Diabetes Federation. IDF Diabetes Atlas, sixth ed. International Diabetes Federation, Brussels, Belgium. [Online] Available from: http://www.diabetesatlas.org.

Jaiswal, N., Yadav, P.P., Maurya, R., Srivastava, A.K., Tamrakar, A.K., 2011. Karanjin from *Pongamia pinnata* induces GLUT4 translocation in skeletal muscle cells in a phosphatidylinositol-3-kinase-independent manner. European Journal of Pharmacology 670 (1), 22–28.

Jung, S.H., Seol, H.J., Jeon, S.J., Son, K.H., Lee, J.R., 2009. Insulin-sensitizing activities of tanshinones, diterpene compounds of the root of *Salvia miltiorrhiza* Bunge. Phytomedicine 16 (4)), 327–335.

Kanaujia, A., Duggar, R., Pannakal, S.T., Yadav, S.S., Katiyar, C.K., Bansal, V., Anand, S., Sujatha, S., Lakshmi, B.S., 2010. Insulinomimetic activity of two new gallotannins from the fruits of *Capparis moonii*. Bioorganic & Medicinal Chemistry 18 (11), 3940–3945.

Kappel, V.D., Zanatta, L., Postal, B.G., Silva, F.R.M.B., 2013. Rutin potentiates calcium uptake via voltage-dependent calcium channel associated with stimulation of glucose uptake in skeletal muscle. Archives of Biochemistry and Biophysics 532 (2), 55–60.

Khan, M.F., Dixit, P., Jaiswal, N., Tamrakar, A.K., Srivastava, A.K., Maurya, R., 2012. Chemical constituents of *Kigelia pinnata* twigs and their GLUT4 translocation modulatory effect in skeletal muscle cells. Fitoterapia 83 (1), 125–129.

Khan, M.F., Maurya, C.K., Dev, K., Arha, D., Rai, A.K., Tamrakar, A.K., Maurya, R., 2014. Design and synthesis of lupeol analogues and their glucose uptake stimulatory effect in L6 skeletal muscle cells. Bioorganic & Medicinal Chemistry Letters 24 (12), 2674–2679.

Kim, K.-J., Lee, O.-H., Lee, H.-C., Kim, Y.-C., Lee, B.-Y., 2007. Effect of fucoidan on expression of diabetes mellitus related genes in mouse adipocytes. Food Science and Biotechnology 16 (2), 212–217.

Korthikunta, V., Pandey, J., Singh, R., Srivastava, R., Srivastava, A.K., Tamrakar, A.K., Tadigoppula, N., 2015. In vitro anti-hyperglycemic activity of 4-hydroxyisoleucine derivatives. Phytomedicine 22 (1), 66–70.

Ku, H.-C., Tsuei, Y.-W., Kao, C.-C., Weng, J.-T., Shih, L.-J., Chang, H.-H., Liu, C.-W., Tsai, S.-W., Kuo, Y.-C., Kao, Y.-H., 2014. Green tea (−)-epigallocatechin gallate suppresses IGF-I and IGF-II stimulation of 3T3–L1 adipocyte glucose uptake via the glucose transporter 4, but not glucose transporter 1 pathway. General and Comparative Endocrinology 199, 46–55.

Kwon, D.Y., Kim, Y.S., Ryu, S.Y., Choi, Y.H., Cha, M.-R., Yang, H.J., Park, S., 2012. Platyconic acid, a saponin from *Platycodi radix*, improves glucose homeostasis by enhancing insulin sensitivity *in vitro* and *in vivo*. European Journal of Nutrition 51 (5), 529–540.

Lai, D.-M., Tu, Y.-K., Liu, I.M., Chen, P.-F., Cheng, J.-T., 2006. Mediation of β-endorphin by ginsenoside Rh2 to lower plasma glucose in streptozotocin-induced diabetic rats. Planta Medica 72 (1), 9–13.

Lee, M.S., Sohn, C.B., 2008. Anti-diabetic properties of chrysophanol and its glucoside from rhubarb rhizome. Biological & Pharmaceutical Bulletin 31 (11), 2154–2157.

Lee, J.Y., Kim, B.-R., Oh, H.I., Shen, L., Kim, N.B., Hwang, J.-K., 2008. PPARγ ligand-binding activity of fragrin a isolated from mace (the aril of *Myristica fragrans* Houtt.). Food Science and Biotechnology 17 (6)), 1146–1150.

Lee, M.-J., Rao, Y.K., Chen, K., Lee, Y.-C., Tzeng, Y.-M., 2009. Effect of flavonol glycosides from *Cinnamomum osmophloeum* leaves on adiponectin secretion and phosphorylation of insulin receptor-β in 3T3-L1 adipocytes. Journal of Ethnopharmacology 126 (1), 79–85.

Lee, H.-H., Kim, K.-J., Lee, O.-H., Lee, B.-Y., 2010a. Effect of Pycnogenol on glucose transport in mature 3T3-L1 adipocytes. Phytotherapy Research 24 (8), 1242–1249.

Lee, Y.-S., Cha, B.-Y., Saito, K., Yamakawa, H., Choi, S.-S., Yamaguchi, K., Yonezawa, T., Teruya, T., Nagai, K., Woo, J.-T., 2010b. Nobiletin improves hyperglycemia and insulin resistance in obese diabetic ob/ob mice. Biochemical Pharmacology 79 (11), 1674–1683.

Lee, J., Kim, M.-H., Lee, J.-H., Jung, E., Yoo, E.-S., Park, D., 2012a. Artemisinic acid is a regulator of adipocyte differentiation and C/EBP δ expression. Journal of Cellular Biochemistry 113 (7), 2488–2499.

Lee, S.-H., Kang, S.-M., Ko, S.-C., Lee, D.-H., Jeon, Y.-J., 2012b. Octaphlorethol A, a novel phenolic compound isolated from a brown alga, *Ishige foliacea*, increases glucose transporter 4-mediated glucose uptake in skeletal muscle cells. Biochemical and Biophysical Research Communications 420 (3), 576–581.

Leney, S.E., Tavare, J.M., 2009. The molecular basis of insulin-stimulated glucose uptake: signalling, trafficking and potential drug targets. Journal of Endocrinology 203 (1), 1–18.

Li, Y., Tran, V.H., Duke, C.C., Roufogalis, B.D., 2012. Gingerols of *Zingiber officinale* enhance glucose uptake by increasing cell surface GLUT4 in cultured L6 myotubes. Planta Medica 78 (14), 1549–1555.

Liao, Z., Wu, Z., Wu, M., 2012. *Cirsium japonicum* flavones enhance adipocyte differentiation and glucose uptake in 3T3-L1 cells. Biological & Pharmaceutical Bulletin 35 (6), 855–860.

Lin, Z., Zhang, Y., Zhang, Y., Shen, H., Hu, L., Jiang, H., Shen, X., 2008. Oleanolic acid derivative NPLC441 potently stimulates glucose transport in 3T3-L1 adipocytes via a multi-target mechanism. Biochemical Pharmacology 76 (10), 1251–1262.

Maeda, A., Kai, K., Ishii, M., Ishii, T., Akagawa, M., 2014. Safranal, a novel protein tyrosine phosphatase 1B inhibitor, activates insulin signaling in C2C12 myotubes and improves glucose tolerance in diabetic KK-Ay mice. Molecular Nutrition & Food Research 58 (6), 1177–1189.

Minakawa, M., Miura, Y., Yagasaki, K., 2012. Piceatannol, a resveratrol derivative, promotes glucose uptake through glucose transporter 4 translocation to plasma membrane in L6 myocytes and suppresses blood glucose levels in type 2 diabetic model db/db mice. Biochemical and Biophysical Research Communications 422 (3), 469–475.

Morral, N., 2003. Novel targets and therapeutic strategies for type 2 diabetes. Trends in Endocrinology and Metabolism 14 (4), 169–175.

Muller, C.J.F., Joubert, E., Pheiffer, C., Ghoor, S., Sanderson, M., Chellan, N., Fey, S.J., Louw, J., 2013. Z-2-(β-D-glucopyranosyloxy)-3-phenylpropenoic acid, an α-hydroxy acid from rooibos (*Aspalathus linearis*) with hypoglycemic activity. Molecular Nutrition & Food Research 57 (12), 2216–2222.

Nankar, R.P., Doble, M., 2013. Non-peptidyl insulin mimetics as a potential antidiabetic agent. Drug Discovery Today 18 (15–16), 748–755.

Naresh, G., Jaiswal, N., Sukanya, P., Srivastava, A.K., Tamrakar, A.K., Narender, T., 2012. Glucose uptake stimulatory effect of 4-hydroxypipecolic acid by increased GLUT 4 translocation in skeletal muscle cells. Bioorganic & Medicinal Chemistry Letters 22 (17), 5648–5651.

Newman, D.J., Cragg, G.M., 2012. Natural products as sources of new drugs over the 30 years from 1981 to 2010. Journal of Natural Products 75 (3), 311–335.

Newman, D.J., Cragg, G.M., Snader, K.M., 2000. The influence of natural products upon drug discovery. Natural Product Reports 17 (3), 215–234.

Nguyen, P.H., Gauhar, R., Hwang, S.L., Dao, T.T., Park, D.C., Kim, J.E., Song, H., Huh, T.L., Oh, W.K., 2011. New dammarane-type glucosides as potential activators of AMP-activated protein kinase (AMPK) from *Gynostemma pentaphyllum*. Bioorganic & Medicinal Chemistry 19 (21), 6254–6260.

Nugroho, A.E., Andrie, M., Susilowati, R., Nurrochmad, A., Lukitaningsih, E., Pramono, S., 2011. Ethanolic extracts of *A. paniculata* (Burm. F.) Nees and its active compound, andrographolide, decrease the expression of glucose transporters (GLUT 4) in high fructose-fat fed rats. International Journal of Phytomedicine 3, 486–497.

Oeberg, A.I., Yassin, K., Csikasz, R.I., Dehvari, N., Shabalina, I.G., Hutchinson, D.S., Wilcke, M., Oestenson, C.-G., Bengtsson, T., 2011. Shikonin increases glucose uptake in skeletal muscle cells and improves plasma glucose levels in diabetic Goto-Kakizaki rats. PloS One 6 (7), e22510.

Ok, S., Lee, J.H., Kim, I.H., Kang, J.S., 2011. Glucose transporters and AMP-activated protein kinase modulation effects of decursin and decursinol angelate on diabetic rats. Yakhak Hoechi 55 (4)), 301–308.

Over, B., Wetzel, S., Grutter, C., Nakai, Y., Renner, S., Rauh, D., Waldmann, H., 2013. Natural-product-derived fragments for fragment-based ligand discovery. Nature Chemistry 5 (1), 21–28.

Park, C.E., Kim, M.J., Lee, J.H., Min, B.I., Bae, H., Choe, W., Kim, S.S., Ha, J., 2007. Resveratrol stimulates glucose transport in C2C12 myotubes by activating AMP-activated protein kinase. Experimental & Molecular Medicine 39 (2), 222–229.

Penumathsa, S.V., Thirunavukkarasu, M., Zhan, L., Maulik, G., Menon, V.P., Bagchi, D., Maulik, N., 2008. Resveratrol enhances GLUT-4 translocation to the caveolar lipid raft fractions through AMPK/Akt/eNOS signalling pathway in diabetic myocardium. Journal of Cellular and Molecular Medicine 12 (6A), 2350–2361.

Prabhakar, P.K., Doble, M., 2009. Synergistic effect of phytochemicals in combination with hypoglycemic drugs on glucose uptake in myotubes. Phytomedicine 16 (12), 1119–1126.

Prabhakar, P.K., Doble, M., 2011. Interaction of cinnamic acid derivatives with commercial hypoglycemic drugs on 2-deoxyglucose uptake in 3T3-L1 adipocytes. Journal of Agricultural and Food Chemistry 59 (18), 9835–9844.

Qi, L.W., Liu, E.H., Chu, C., Peng, Y.B., Cai, H.X., Li, P., 2010. Anti-diabetic agents from natural products-an update from 2004 to 2009. Current Topics in Medicinal Chemistry 10 (4), 434–457.

Ross, P.L., Huang, Y.N., Marchese, J.N., Williamson, B., Parker, K., Hattan, S., Khainovski, N., Pillai, S., Dey, S., Daniels, S., Purkayastha, S., Juhasz, P., Martin, S., Bartlet-Jones, M., He, F., Jacobson, A., Pappin, D.J., 2004. Multiplexed protein quantitation in *Saccharomyces cerevisiae* using amine-reactive isobaric tagging reagents. Molecular & Cellular Proteomics 3 (12), 1154–1169.

Salimifar, M., Fatehi-Hassanabad, Z., Fatehi, M., 2013. A review on natural products for controlling type 2 diabetes with an emphasis on their mechanisms of actions. Current Diabetes Reviews 9 (5), 402–411.

Sangeetha, K.N., Sujatha, S., Muthusamy, V.S., Anand, S., Nithya, N., Velmurugan, D., Balakrishnan, A., Lakshmi, B.S., 2010. 3β-taraxerol of *Mangifera indica*, a PI3K dependent dual activator of glucose transport and glycogen synthesis in 3T3-L1 adipocytes. Biochimica et Biophysica Acta 1800 (3), 359–366.

Sato, K., Fujita, S., Iemitsu, M., 2014. Acute administration of diosgenin or dioscorea improves hyperglycemia with increases muscular steroidogenesis in STZ-induced type 1 diabetic rats. Journal of Steroid Biochemistry and Molecular Biology 143, 152–159.

Scazzocchio, B., Vari, R., Filesi, C., D'archivio, M., Santangelo, C., Giovannini, C., Iacovelli, A., Silecchia, G., Li Volti, G., Galvano, F., Masella, R., 2011. Cyanidin-3-O-β-glucoside and protocatechuic acid exert insulin-like effects by upregulating PPARγ activity in human omental adipocytes. Diabetes 60 (9), 2234–2244.

Schluter, A., Barbera, M.J., Iglesias, R., Giralt, M., Villarroya, F., 2002. Phytanic acid, a novel activator of uncoupling protein-1 gene transcription and brown adipocyte differentiation. Biochemical Journal 362 (1), 61–69.

Shapiro, K., Gong, W.C., 2002. Natural products used for diabetes. Journal of the American Pharmaceutical Association (Washington) 42 (2), 217–226.

Sharma, D.K., Pandey, J., Tamrakar, A.K., Mukherjee, D., 2014. Synthesis of heteroaryl/aryl kojic acid conjugates as stimulators of glucose uptake by GLUT4 translocation. European Journal of Medicinal Chemistry 85, 727–736.

Soares de Moura, R., da Costa, G.F., Moreira, A.S.B., Queiroz, E.F., Dal Col Moreira, D., Garcia-Souza, E.P., Resende, A.C., Moura, A.S., Teixeira, M.T., 2012. *Vitis vinifera* L. grape skin extract activates the insulin-signalling cascade and reduces hyperglycaemia in alloxan-induced diabetic mice. Journal of Pharmacy and Pharmacology 64 (2), 268–276.

Sun, T., Wang, Q., Yu, Z., Zhang, Y., Guo, Y., Chen, K., Shen, X., Jiang, H., 2007. Hyrtiosal, a PTP1B inhibitor from the marine sponge *Hyrtios erectus*, shows extensive cellular effects on PI3K/AKT activation, glucose transport, and TGFβ/Smad2 signaling. ChemBioChem 8 (2), 187–193.

Sunil, C., Duraipandiyan, V., Agastian, P., Ignacimuthu, S., 2012. Antidiabetic effect of plumbagin isolated from *Plumbago zeylanica* L. root and its effect on GLUT4 translocation in streptozotocin-induced diabetic rats. Food and Chemical Toxicology 50 (12), 4356–4363.

Tamrakar, A.K., Jaiswal, N., Yadav, P.P., Maurya, R., Srivastava, A.K., 2011. Pongamol from *Pongamia pinnata* stimulates glucose uptake by increasing surface GLUT4 level in skeletal muscle cells. Molecular and Cellular Endocrinology 339 (1–2), 98–104.

Tzeng, Y.-M., Chen, K., Rao, Y.K., Lee, M.-J., 2009. Kaempferitrin activates the insulin signaling pathway and stimulates secretion of adiponectin in 3T3-L1 adipocytes. European Journal of Pharmacology 607 (1–3), 27–34.

Wang, L.-Y., Wang, Y., Xu, D.-S., Ruan, K.-F., Feng, Y., Wang, S., 2012. MDG-1, a polysaccharide from *Ophiopogon japonicus* exerts hypoglycemic effects through the PI3K/Akt pathway in a diabetic KKAy mouse model. Journal of Ethnopharmacology 143 (1), 347–354.

Wang, C., Yang, J., Zhao, P., Zhou, Q., Mei, Z., Yang, G., Yang, X., Feng, Y., 2014. Chemical constituents from *Eucalyptus citriodora* Hook leaves and their glucose transporter 4 translocation activities. Bioorganic & Medicinal Chemistry Letters 24 (14), 3096–3099.

Yang, Y., Wolfram, J., Boom, K., Fang, X., Shen, H., Ferrari, M., 2013. Hesperetin impairs glucose uptake and inhibits proliferation of breast cancer cells. Cell Biochemistry and Function 31 (5), 374–379.

Yu, H., Zheng, L., Xu, L., Yin, L., Lin, Y., Li, H., Liu, K., Peng, J., 2015. Potent effects of the total saponins from *Dioscorea nipponica* Makino against streptozotocin-induced type 2 diabetes mellitus in rats. Phytotherapy Research 29 (2), 228–240.

Zanatta, L., Rosso, A., Folador, P., Figueiredo, M.S.R.B., Pizzolatti, M.G., Leite, L.D., Silva, F.R.M.B., 2008. Insulinomimetic effect of kaempferol 3-neohesperidoside on the rat soleus muscle. Journal of Natural Products 71 (4), 532–535.

Zhang, B., Salituro, G., Szalkowski, D., Li, Z., Zhang, Y., Royo, I., Vilella, D., Diez, M.T., Pelaez, F., Ruby, C., Kendall, R.L., Mao, X., Griffin, P., Calaycay, J., Zierath, J.R., Heck, J.V., Smith, R.G., Moller, D.E., 1999. Discovery of a small molecule insulin mimetic with antidiabetic activity in mice. Science 284 (5416), 974–977.

Zhang, Y., Zhang, H., Yao, X.-G., Shen, H., Chen, J., Li, C., Chen, L., Zheng, M., Ye, J., Hu, L., Shen, X., Jiang, H., 2012. (+)-Rutamarin as a dual inducer of both GLUT4 translocation and expression efficiently ameliorates glucose homeostasis in insulin-resistant mice. PloS One 7 (2), e31811.

Chapter 5

Carbohydrate-Based Antidiabetic Agents From Nature*

S. Mishra[1], A.S. Singh[1], N. Mishra[1], H. Pandey[2], V.K. Tiwari[1]
[1]Banaras Hindu University, Varanasi, Uttar Pradesh, India; [2]Sam Higginbottom Institute of Agriculture, Technology & Sciences, Allahabad, India

1. INTRODUCTION

Diabetes is worldwide dilemma that has escalated exponentially in the past few decades and it is expected to affect over 439 million people by 2030 (Huizinga and Rothman, 2006, p. 481; Shaw et al., 2010, p. 4). Diabetes contributes highly to the ever-rising mortality and morbidity rate over a wide range of age group across the globe. This complex metabolically controlled endocrinopathy is originated from both absolute and relative deficiency of a peptide hormone known as insulin, which governs the glucose uptake in cells, which in turn controls the glucose levels in blood. As the insulin is the key hormone that maintains blood glucose level, its absence creates a condition of increased blood glucose level known as hyperglycemic state. The curse associated with this state triggers other related long-term problems like impaired lipid and protein metabolism along with delayed healing of injuries and increased risk of affecting vital organs like brain, heart, kidney, and eye (Arya et al., 2008, p. 221).

On account of severity of diabetes epidemic, the scientific community is highly motivated to design and synthesize molecules that can target the machinery involved in metabolism of carbohydrate and glucose level in blood. As we know the key to cure diabetes is to check levels of glucose, a simple polysaccharide which is vital to all living beings, this immediately draws the attention of carbohydrate chemist toward developing glycoconjugates that acts on specific target to maintain glucose levels. The significance of carbohydrate-based molecules is well recognized in scientific community as a number of antidiabetic drugs are

* This chapter is dedicated to Dr R P Tripathi, Senior Scientist at Central Drug Research Institute, Lucknow on the occasion of his 60th birthday celebration.

Discovery and Development of Antidiabetic Agents From Natural Products.
http://dx.doi.org/10.1016/B978-0-12-809450-1.00005-3

fabricated with carbohydrate moiety that is fundamentally important for pharmacological activities of drug. For example, acarbose (**1**) and voglibose (**2**) are well known α-glucosidase inhibitors used to treat diabetes, and both are associated with saccharide moieties (Tarleton, 2005, p. 1379; Hewitt et al., 2002, p. 13434). So it can be inferred that imbalance of a carbohydrate (glucose) can cause not only diabetes but carbohydrate-based entities can also treat diabetes. On the basis of mechanism of action, carbohydrate-based antidiabetic agents can be broadly visualized in terms of antidiabetic glucosidase inhibitors and novel sodium glucose cotransporter 2 (SGLT2) inhibitors along with other antidiabetic agents with different mode of action. Another interesting class of sugar, where ring oxygen is replaced by *N*-atom, commonly known as iminosugar, for example miglitol (**3**), emiglitate (**4**), etc., are well known α-glucosidase inhibitor and considered as promising molecules to treat diabetes (Fig. 5.1). A large number of glycoconjugates are entering frequently in the plethora of antidiabetic agents on account of their lower side effects and better efficiency. As diabetes is intensifying and altering its nature with changing human genotype, the medication strategy should also be modified with the same pace. As the previously available drugs like metformin, sulfonylureas, and glitazone are prone to inflict side effects, on the other hand, carbohydrate-derived drugs can serve as potential tool because of its inherent properties like substantial solubility, minimized toxicity, and enhanced pharmacokinetics that make them suitable for biological system and impose lesser side effects. The additional advantage associated with carbohydrate-based drugs can be realized in terms of their molecular diversity on account of number of rings, linkages, and functional groups along with desired conformability in biological system (Nogueira et al., 2009, p. 149; Cao et al., 2011, p. 411; Cipolla et al., 2010, p. 587; Cipolla et al., 2011, p. 1002; Cipolla and Peri, 2011, p. 39; Lonngren, 1989, p. 1313; Etayo et al., 2010, p. 1775; Tiwari et al., 2012, p. 1497). The carbohydrate-derived antidiabetic agents are developing by both natural and synthetic strategies. This chapter briefly highlights the recent progress in designing, synthesizing, or isolating carbohydrate-based antidiabetic agents emphasizing on their mechanism of action, structure, and synthetic approach to some important leading carbohydrate-based antidiabetic agents.

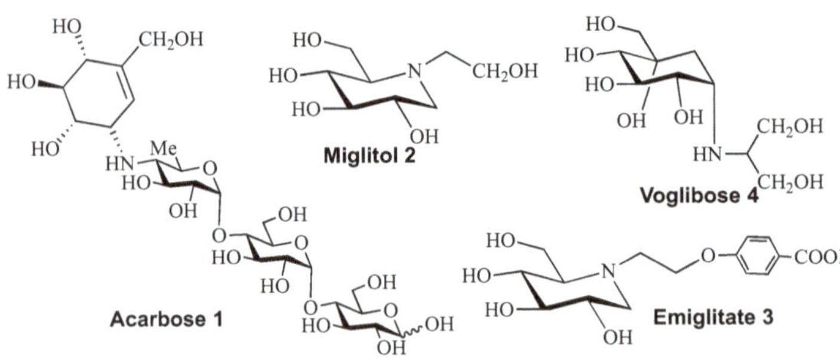

FIGURE 5.1 Antidiabetic glucosidase inhibitor drugs.

2. DIABETES: DEFINITION AND CLASSIFICATION

Today's world is posed at the risk of many fatal diseases, among which diabetes is expanding its horizon at a tremendous rate. As per reports released by International Diabetes Federation (IDF) in 2013, 382 million people across the globe are affected by diabetes (IDF Diabetes Atlas, 2014). On account of its susceptibility with emerging era, diabetes is one of the most common metabolic diseases characterized by raised level of glucose in blood. This hyperglycemic state is resulted by several pathogenic processes, which induce autoimmune response to destroy pancreatic β-cells that are solely responsible for insulin secretion. Due to insufficient levels of insulin, glucose metabolism is affected in target cells resulting in abnormalities, which mark the resistance to action of insulin, commonly referred as insulin resistance (IR). Insulin hormone governs the regulation of glucose transport in vital insulin-dependent organs or tissues, i.e., liver, skeletal muscle adipocytes, and brain. IR is a complicated metabolism-related disorder where abnormal levels (endogenous/exogenous) of insulin generate an intensified biological response, which is visualized in terms of weakened or disturbed sensitivity to insulin-related glucose metabolism and transport. As a result of impaired insulin action, insulin-resistant organs do not use glucose adequately that, in turn, triggers hyperglycemic state. Although the scientific community is focusing on its research to reveal the precise molecular mechanism of IR, the primary reason is yet not deciphered completely. The evidence as achieved so far, however, suggests that in insulin-resistant condition, disorder arises from the pathway of insulin-signaling system not from the insulin hormone or insulin receptor. Extensive research to explore the root behind the cause of IR, several targets, and their related mechanism has been anticipated to contribute in the IR process. These mechanisms and pathways involve elevated levels of free fatty acids in plasma, severe inflammations, abnormal gene expression, mitochondrial function, defective oxidation, and nitration processes (Fig 5.2) (Kokil et al., 2015, p. 4719).

Prolonged hyperglycemic state leads to chronic stages of diabetes that can affect vital organs of body like blood vessels, heart, kidney, and nerves. Hyperglycemia is associated with various physiological dysfunctions including polyphagia, polyuria, polydypsia, and considerable weight loss (American Diabetes Association, 2008, p. S55). The probable conditions that can arise as a result of long and complicated diabetic state include retinopathy, nephropathy, neuropathy, and cardiovascular and cerebrovascular diseases (Aguilar, 2011, p. 408).

Diabetes is broadly classified in two types—insulin-dependent type 1 diabetes mellitus (T1DM) and non-insulin-dependent type 2 diabetes mellitus (T2DM). Type 2 constitutes around 90% of all cases whereas type 1 is less common, as a result T2DM is widely focused by various researchers to develop its effective medication. T1DM generally includes autoimmune-mediated diabetes, which is caused by absolute deficient insulin secretion out of damaged β-cells of pancreas and less common idiopathic diabetes with no known origin

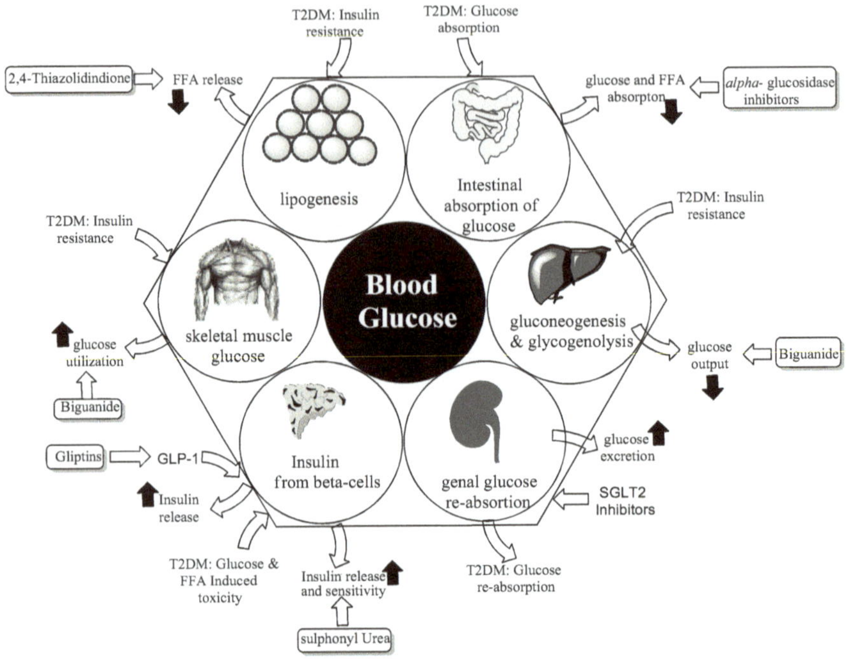

FIGURE 5.2 Interpretation of underlying causes of insulin resistance (Kokil et al., 2015, p. 4719).

but characterized by insulinopenia and ketoacidosis. On the other hand, T2DM is caused by IR marked by insulin secretion-related defects where the insulin deficiency is relative and many other factors are responsible for developed IR, although exact origin is yet not specified. As diabetes is a condition of impaired glycemic state, so it can be triggered by many factors, viz. gestational diabetes mellitus (GDM) is developed as a result of glucose intolerance in pregnant women that may or may not continue in postgestation period. Maturity onset diabetes of young (MODY) caused by genetic defects in action of β-cells can develop hyperglycemia in early stages of life (around 25 years). In fact, insulin action is also altered because of genetic defects in insulin receptor that results in hyperglycemic state. Apart from these, endocrinopathy, drugs, and several infections can also trigger diabetes in some conditions.

3. IMPORTANCE OF CARBOHYDRATE IN ANTIDIABETIC DRUG DEVELOPMENT

The commonly available drugs to treat diabetes are of several groups including insulin secretagogues (like sulfonylureas), Glucagon-like Peptide-1(GLP-1) agonists, Dipeptidyl Peptidase-4 (DPP-4) inhibitors, meglitinides, insulin sensitizers (like thiazolidinediones, biguanides), glucosidase inhibitors, and SGLT2 inhibitors. Most of the marketed hypoglycemic agents along with diet control

TABLE 5.1 Different Classes of Antidiabetic Drugs

S.N.	Medication Class	Drugs Used	Adverse Effects
1.	Sulfonylureas	Acetohexamide, chlorpropamide, glipizide, gliclazide, tolazamide	Possible increased incidence of cancer, weight gain, hypoglycemia
2.	Biguanides	Buformin, proguanil, metformin, phenform	Gastrointestinal disturbance, lactic acidosis, hepatotoxicity
3.	DPP-IV inhibitors	Saxagliptin, sitagliptin, vildagliptin	Possible association with exfoliative dermatitis, infection related to nasopharyngitis and upper respiratory tract
4.	2,4-thiazalidinedione	Rosiglitazone, troglitazone, ciglitazone	Bone fractures, edema, bladder cancer, congestive heart failure
5.	Meglitinides	Nateglinide and repaglinide	Weight gain, hypoglycemia
6.	GLP-1 receptor agonists	Liraglutide and exenatide	Nausea and vomiting, possible association with acute pancreatitis, thyroid cancer

measures are often not sufficient to suppress the metabolic abnormalities associated with diabetes. In addition to this, risks of hypoglycemic state and resulting side effects limit their use, for example, insulin, sulfonylureas, and glitazones increase IR by facilitating weight gain and pioglitazone causes edema, cardiovascular difficulties (Clar et al., 2012, p. 1). So, there is urgent requirement for more potent and safer hypoglycemic agents (Table 5.1) (Aguilar, 2011, p. 408; Complications Trial Research Group, 1993, p. 977). The emerging need of efficacious and safe antidiabetic drugs has opened the door for carbohydrate-based molecules or glycoconjugates to combat diabetes. Glucosidase inhibitors came into picture in the form of carbohydrate-based drugs like acarbose, miglitol, and voglibose. Recently, carbohydrate-based SGLT2 inhibitors like dapagliflozin and canagliflozin have also established the milestone in lowering the blood glucose levels. These novel antidiabetic chemotherapeutics are designed with incorporation of carbohydrate moieties to have better efficacy and less adverse side effects.

On account of pervasiveness and abundance of carbohydrates among various biomolecules, they serve as vital source of energy and effective structural building blocks. Their significant roles in many cellular and intracellular events as cell surface receptors, bacterial adhesives, and signal molecules, make them desired candidates for drug development. Carbohydrate-based compounds can serve as potential pharmaceutical and diagnostic agents because of undermined capabilities including vast structural diversity in terms of functional groups, rings, and linkages. The optimum pharmacokinetics associated with glycoconjugates is attributed to the fact that they are advantageous by properties like hydrophilicity and less toxicity that makes them available in biological systems effortlessly. As discussed earlier, it can be inferred that carbohydrate moieties are not only pliable to being attached with other scaffolds but also found to be fundamentally significant in exhibiting desired biological activity. The emerging era of drug development needs chemotherapeutics with high efficacy, low toxicity, and fewer side effects, and hence, carbohydrates offer the crucial tools for developing biologically important molecules, diverse glycoconjugates, and molecular scaffolds by scrutinizing their functional and structural aspects (Mishra et al., 2016).

The importance of carbohydrates in the treatment of diabetes can be understood with recent ongoing research, where natural products have shown enormous potential from the very beginning. Bharti et al. (2015) reported the importance of fructooligosaccharides and isomaltooligosaccharides as dietary components to reduce blood glucose levels (Bharti et al., 2015, p. 317). In fact, these compounds were found to be antihyperglycemic in nature by acting on several targets that govern the onset of diabetes, among the docked oligosaccharides, particularly nystose and pannose were remarkably efficacious (Narender et al., 2011, p. 155). Further, very recently, Wang et al. (2015) suggested that a type II arabinogalactan found in leaves of green tea (*Camellia sinensis*) named as 7WA facilitated insulin secretion when glucose levels were high. The structural analysis revealed that 7WA consists of galactopyranosyl moieties with 1,3- and 1,6-linkage with branching at third oxygen of 1,6-linked galactose moiety along with fourth and sixth oxygen of 1,3-linked galactose moiety (Wang et al., 2015, p. 98).

One of the key targets to reduce secondary complications associated with diabetes includes inhibition of advanced glycation end product (AGE). The available AGE inhibitors include pyridoxamine that inhibited retinopathy and neuropathy in diabetic rats. Using the same target as inspiration, a group of scientist isolated astragalosides from crude extract obtained from roots of *Astragalus radix* that served as potential AGE inhibitors (Motomura et al., 2009, p. 7666).

Apart from cure of diabetes and its complication, the precautionary aspect focuses on life style interventions including proper dietary habits. Recent ongoing research is also edging toward this aspect also; a report suggests that carbohydrate sweeteners obtained from sugarcane, corn syrups, and palm may inhibit the vital enzymes related to diabetes. Brown sugar (dark muscovado) found in Peru and Mauritius has radical scavenger characteristics that can be correlated with a modest inhibition of yeast α-glucosidase. On the other hand, date sugar displayed high α-glucosidase activity in addition to α-amylase and angiotensin-converting enzyme (ACE) inhibitory activities. This study can provide potential tool for extracting information to design improved dietary sweeteners for diabetic patients (Ranilla et al., 2008, p. 337).

4. GLYCOSIDASE INHIBITORS AS ANTIDIABETIC AGENTS

Glycosidases are imperative group of enzymes that plays key role in oligosaccharide biosynthesis in cell organelles (endoplasmic reticulum and Golgi apparatus) by acting on glycosidic bond in glycoproteins and facilitates carbohydrate digestion. Various glycosidases target catalytic hydrolysis of glycosidic linkages during digestion of carbohydrate. Glycosidase inhibitors can be used as a potential tool to control blood glucose levels as they prevent the hydrolysis of polysaccharides and thereby delay or check carbohydrate metabolism. This strategy is widely applied to develop antidiabetic agents to treat type 2 non-insulin-independent diabetes mellitus (NIDDM). The α-glucosidase inhibitors with antidiabetic characteristics are intended with some common structural aspects with a class of biologically active entities that act as mimic of sugar moieties. Usually, they bear ring-containing structure with amine group and some hydroxyl substituents arranged in three-dimensional manners similar to basic carbohydrate framework. These structural similarities are used as a tool to modify functional aspects of mimicking entities; this concept is very well visualized in case of carbasugars with carbocyclic framework and iminosugars with nitrogen-containing heterocycle with different linked functionalities. On account of their remarkable enzymatic actions and bioactivity, they can create a class of glucosidase inhibitors targeting diabetes (Fig 5.3) (Trapero and Llebaria, 2012, p. 10345).

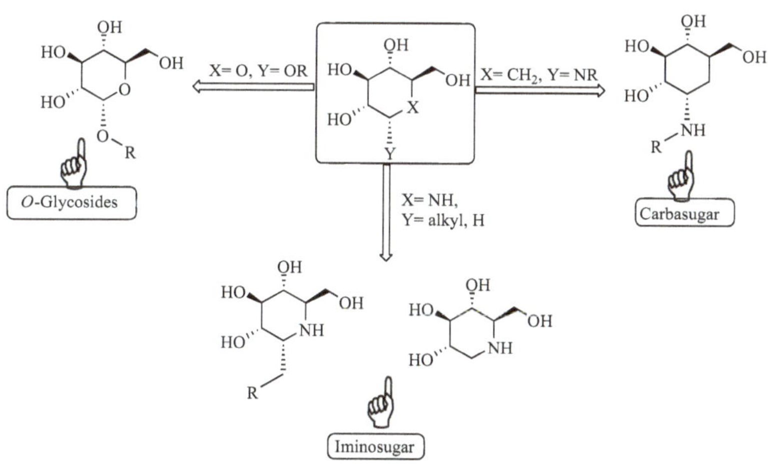

FIGURE 5.3 Strategic modification in carbohydrate framework to develop α-glucosidase inhibitors.

Glucosidase inhibitors are broadly classified into three categories namely, reversible α-glucosidase inhibitors, irreversible glucosidase inhibitors, and powerful sucrose inhibitors (Sundaram et al., 1998, p. 24). Reversible α-glucosidase inhibitors include acarbose (**1**), which is a pseudotetrasaccharide obtained from *Actinoplanes* strain SE50 by making their cultural broths, it consists of a cyclitol unit attached with α-(1-4) linkage with D-glucopyranose unit, which in turn linked to α-(1-4) maltose. It binds in a reversible manner to the α-glucosidase enzyme's binding site for saccharide units in brush border of small intestine resulting in the prevention of hydrolytic cleavage of oligosaccharides, thereby decreasing the glucose levels in blood. Another reversible α-glucosidase inhibitor, miglitol (**3**) exhibit glycemic control in the same manner as acarbose but with improved activity. Later the developed analog of miglitol, known as emiglitate (**4**), exhibited more lipophilicity than miglitol, however, displayed lower inhibitory activity. So, we can conclude that these drugs act on membrane-bound α-glucosidases present over brush border of intestinal cells and thereby prevent the breakdown of oligosaccharides into simpler forms and maintain the desired blood glucose level. Irreversible glucosidase inhibitors include gasternospermin. Powerful sucrose inhibitors include voglibose (**2**), which is administered to treat diabetes mellitus in postprandial condition. They maintain lower glycemic state by facilitating delayed absorption of glucose. These drugs are efficacious with few side effects like abdominal disturbances including flatulence, diarrhea, and stomach aching that can be visualized in terms of constrained digestion of complex carbohydrates in intestinal region; consequently, when they pass through colon, they eventually get digested by bacteria residing in this region and leads to these side effects.

4.1 Novel Carbohydrate-Based α-Glucosidase Inhibitors With Antidiabetic Activity

As discussed earlier, phenylureas have been used as antidiabetic agents but they encounter some severe side effects so taking the structural aspect in mind, a new approach was implemented to alter the pharmacokinetics positively, which results in the synthesis of a class of glycosylureas. Compounds (**5–7**) were accessed for their α-glucosidase inhibitory activity and were found to display significant activity (Tewari et al., 2003, p. 2911). Further, a series of α-mannosylated phenols (**8,9**) have been synthesized with potential inhibitory activity over intestinal α-glycosidase of rats (Khan et al., 2004, p. 107) (Fig. 5.4).

In recent years, Cu-catalyzed click chemistry inspired 1,4-disubstituted 1,2,3-triazolyl moiety coupled with carbohydrate scaffold has found to be the potential inhibitors of wide range of glycosidases (Tiwari et al., 2016, p. 3086). Using this approach, library of 1,2,3-triazolyl glycoconjugates were synthesized (Fig. 5.5) and scrutinized for α-glucosidase inhibition at 500 μM, where β-D-ribosyl derivatives (**10–12**), α-D-galactosyl derivative (**13**), and α-D-xylosyl derivative (**14**) were found to be promising glucosidase inhibitors (Ferreira et al., 2010, p. 2364).

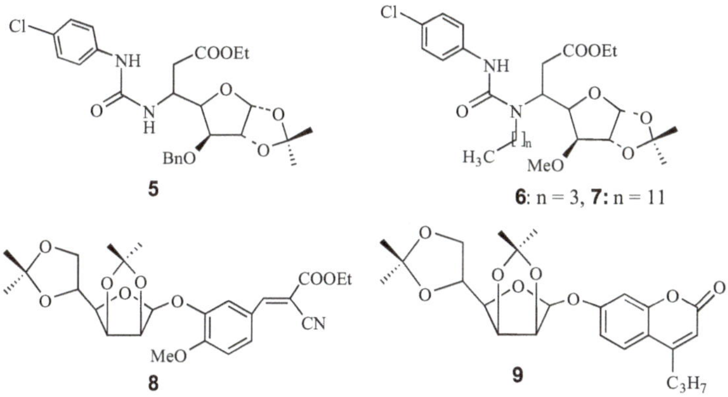

FIGURE 5.4 Potential α-glucosidase inhibitors.

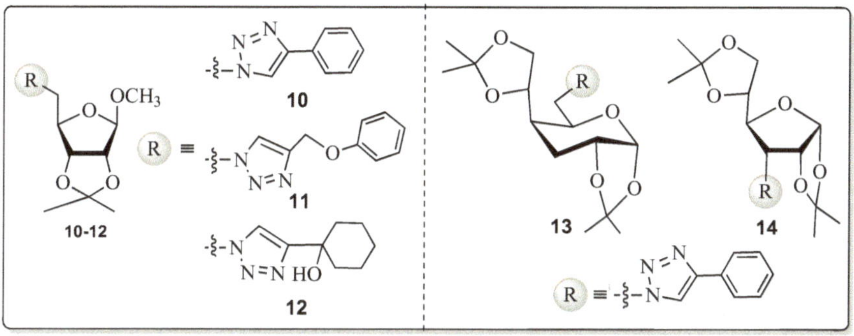

FIGURE 5.5 1,2,3-triazoles displaying α-glucosidase inhibitory activity.

A series of triazolyl glycoconjugates with several linked sugars or chromenone substituent were developed using click strategy and scrutinized for their α-glucosidase activity in addition to some other enzymes. At 100 μM concentration, triazolyl glycoconjugates (**15–18**) displayed significant α-glucosidase inhibitory activity referenced with acarbose and sodium *ortho*vandate. Further, glycohybrids (**19–20**) were synthesized and scrutinized for α-glucosidase inhibitory activity and they were found to be noticeably active (Anand et al., 2011, p. 16; Singh et al., 2010, p. 1641) (Fig. 5.6).

Recently, a class of iminosugars with branched carbon structures derived from natural products, 2,5-Dideoxy-2,5-imino-D-mannitol (DMDP) (**21**), 2,5-Dideoxy-2,5-imino-D-glucitol (DGDP) (**22**), and 1,4-Dideoxy-1,4-imino-D-arabinitol (DAB) (**23**), namely, isoDMDP (**24**), isoDGDP (**28**), and isoDAB (**26**) and their enantiomeric structures (**27**), (**28**), (**29**), (**30**), and (**31**) were developed using Aldol reaction and all these compounds have been comparatively studied to their linear chain counterparts for α-glucosidase inhibitory activity in rat intestine. Among synthesized compounds, L-isoDMDP (IC_{50} = 0.19 μM) was found to be specifically most potential α-glucosidase inhibitor in comparison to

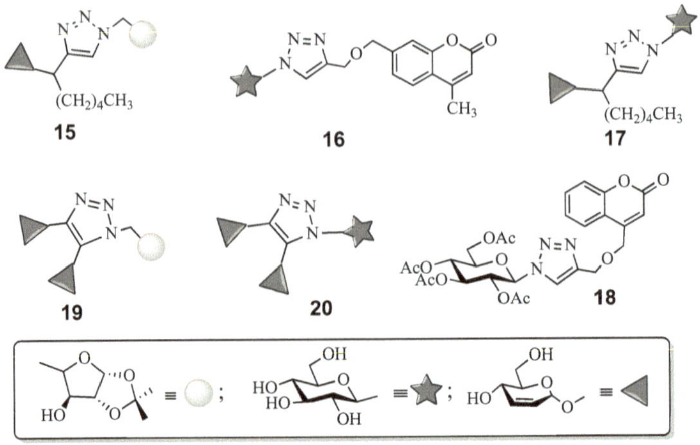

FIGURE 5.6 Triazolyl glycohybrid-based α-glucosidase inhibitor.

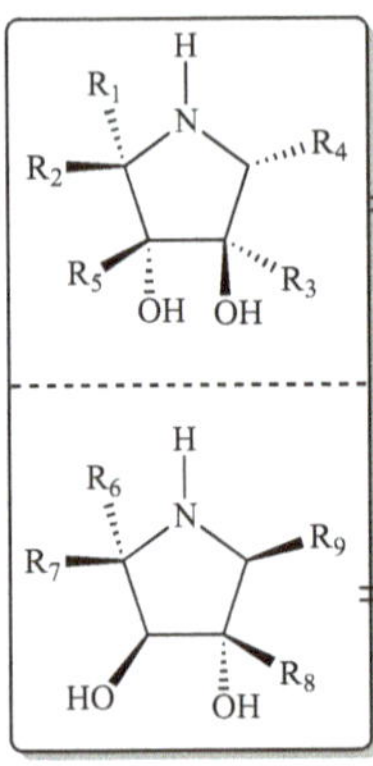

(**21**) R_2,R_4= CH_2OH; R_1,R_3,R_5=H DMDP (IC_{50} 250 μM)
(**22**) R_1,R_4 = CH_2OH; R_2,R_3,R_5=H DGDP (IC_{50} 138 μM)
(**23**) R_2 = CH_2OH; R_1,R_3,R_4,R_5=H DAB (IC_{50} 55 μM)
(**24**) R_2,R_3 = CH_2OH; R_1,R_4,R_5=H isoDMDP (No inhibition)
(**25**) R_1,R_3 = CH_2OH; R_2,R_4,R_5=H isoDgDP (No inhibition)
(**26**) R_5 = CH_2OH; R_1,R_2,R_3,R_4=H DAB (IC_{50} 24 μM)

(**27**) R_6, R_8 = CH_2OH; R_7,R_9 = H L- isoDMDP (IC_{50} 0.19 μM)
(**28**) R_7, R_8 = CH_2OH; R_6,R_9 = H L- isoDGDP (IC_{50} 175 μM)
(**29**) R_6, R_9 = CH_2OH; R_7,R_8 = H L- DMDP (IC_{50} 1.2 μM)
(**30**) R_7, R_9 = CH_2OH; R_6,R_8 = H L- DGDP (IC_{50} 48 μM)
(**31**) R_6 = CH_2OH; R_7,R_8,R_9 = H LAB (IC_{50} 0.93 μM)

FIGURE 5.7 Iminosugars and derived potential α-glucosidase inhibitors.

miglitol for balancing the blood glucose level in mice (Jenkinson et al., 2013, p. 7380) (Fig. 5.7).

An aqueous extract from the leaves and stems of *Cyclopia genistoides* afforded a *C*- and *O*-glucosides with benzophenone moiety (**32**) along with *C*-glucoside (**33**) (Fig. 5.8). These compounds were tested for α-glucosidase inhibitory activity against α-glucosidase extracted from rat intestinal acetone powder and acarbose was used as control at 65 μM. Compounds **32** and **33** displayed 43% and 54% inhibitory activity, respectively, at 200 μM. Compound **33** was considerably active with *p* less than 0.05 value ($p<.05$) (Beelders et al., 2014, p. 2694).

Recently, several phenolics were obtained from flowers of *Edgeworthia gardneri* used as herbal tea in Tibet. These phenolics were isolated and accessed for α-glucosidase activity to test their antihyperglycemic activity (Fig. 5.9). Among these compounds, **34** with linked glycosyl unit was found to have remarkable α-glucosidase inhibitory activity; it potentially decreased the postprandial blood glucose levels in both streptozotocin-induced and normal

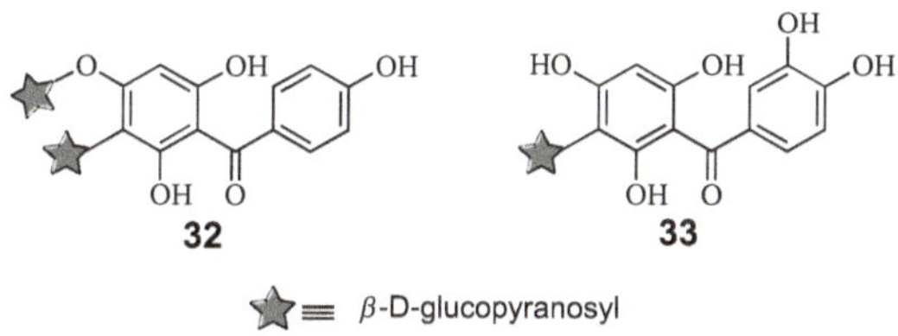

FIGURE 5.8 Glycoconjugate glucosidase inhibitors obtained from *Cyclopia genistoides*.

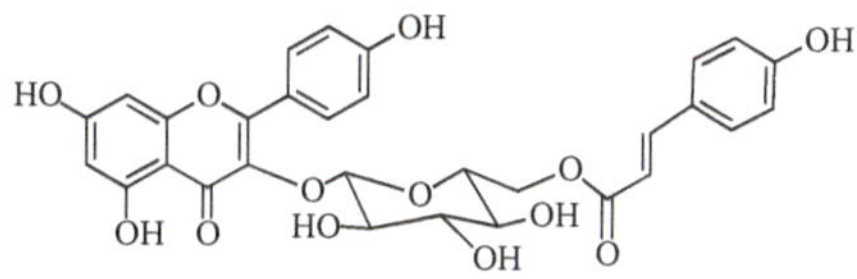

FIGURE 5.9 Glycosylated phenolic as α-glucosidase inhibitor.

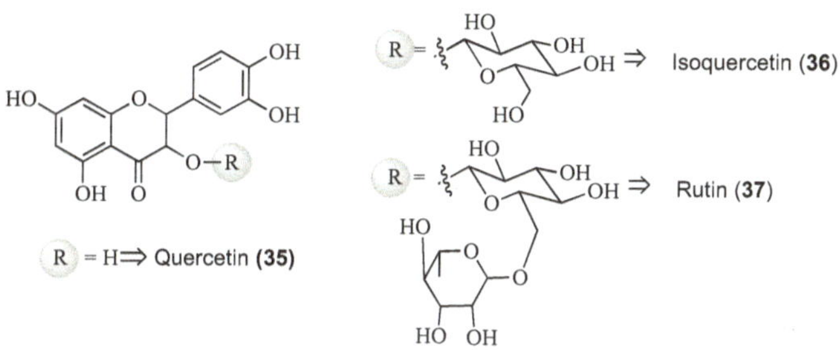

FIGURE 5.10 Quercetin-derived glycoconjugates with α-glucosidase inhibitory activity.

mice at a dose of 300 mg/kg; in addition to this, it also decreased blood glucose level in fasting streptozotocin-induced diabetic mice (Ma et al., 2015, p. 8162).

Flavonoids are abundant plant-derived, naturally occurring compounds with remarkable pharmacological properties. They are found to be potential hyperglycemic agents for treatment of diabetes (Odetti et al., 1990, p. 796; Srinivasan et al., 2005, p. 327). Many reports are available in literature that underline the effectiveness of flavonoids as α-glucosidase inhibitors (Tadera et al., 2006, p. 149; Kim et al., 2006, p. 302). Quercetin (**35**) and its glycoconjugates including monoglycosidic derivative isoquercetin (**36**) and diglycosidic derivative rutin (**37**) are flavonoids that are found to exhibit α-glucosidase inhibitory activity based on enzymatic kinetic measurement studies (Fig. 5.10). IC_{50} was obtained at 0.017, 0.185, and 0.196 μM in comparison to acarbose (IC_{50} = 0.09 μM) (Li et al., 2009, p. 11463).

Other flavanol-based glycoconjugates isolated from *Spiraea cantoniensis* plant, found in temperate regions of Japan, were found to be useful for their α-glucosidase inhibitory activity (Fig. 5.11), which is in accordance to the fact that the extract prepared from the flowers of this plant exhibited potential

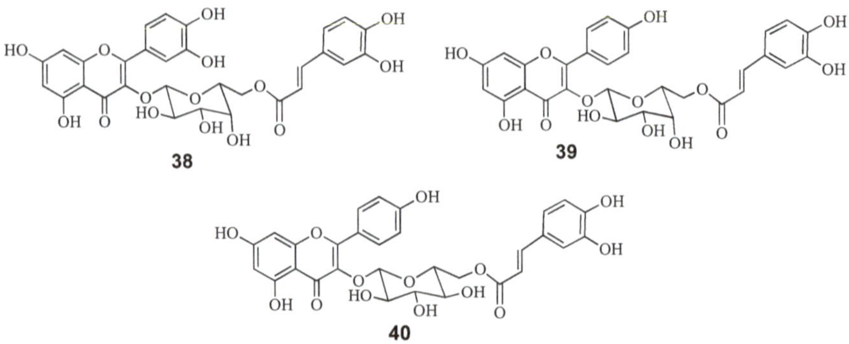

FIGURE 5.11 Flavanol glycoconjugates as potential glucosidase inhibitor.

maltase-inhibiting activity. Three carbohydrate-based caffeoylglycosides were isolated by enzyme assay-motivated fractionation technique, namely quercetin 3-*O*-(6-*O*-caffeoyl)-α-galactoside (**38**), kaempferol 3-*O*-(6-*O*-caffeoyl)-α-galactoside (**39**), and kaempferol 3-*O*-(6-*O*-caffeoyl)-α-glucoside (**40**), with potential inhibitory activity with respective IC_{50} values of 0.085, 0.35, and 0.47 μM against rat intestinal glucosidases (Yoshida et al., 2008, p. 4367).

A series of 10 xanthone glycoconjugates (**41–47**) isolated from *Swertia kouitchensis*, exhibited potential α-glucosidase inhibitory activity with IC_{50} values (μM) of 383 ± 18, 360 ± 39, 371 ± 22, 184 ± 23, 126 ± 23, 296 ± 52, and 451 ± 41 μM, respectively, and the efficacy was found to be better than the positive control, acarbose (IC_{50} = 627 ± 28 μM) (Fig 5.12) (Wan et al., 2013, p. 1248). It was found that *O*-Glc(6–1)-Xyl moiety present in the compounds displayed more inhibition than the other diglycoside substituents; moreover, hydroxyl functionality at C-1 or C-8 carbon exhibited greater inhibitory activities as in the case of compounds **44** and **45** (Wan et al., 2013, p. 1248).

Saponins include triterpenoids and steroidal glycosides, which are considered to be biologically active agents widely distributed in several herbs and known and found to have potential antihypoglycemic properties. Betavulgaroside (**49**), a known glucuronide saponin isolated from *Beta vulgaris* (from roots and leaves), displayed significant antihyperglycemic activity in rats (Rao and Gurfinkel, 2000, p. 211; Yoshikawa et al. 1995a, p. 1621). Elatoside E (**48**), another triterpenoid isolated from the root cortex of *Aralia elata* also displayed antihyperglycemic activity in rats by affecting plasma glucose levels as per the sugar tolerance tests (Yoshikawa et al., 1994, p. 1354) (Fig. 5.13).

5. CARBOHYDRATE-BASED SODIUM GLUCOSE COTRANSPORTER 2 INHIBITORS

According to the National Institute for Clinical Excellence, the typical medication prescribed for treatment involves polytherapy approach, which requires administration of first line drug metformin, then sulfonylurea or glitazone, and

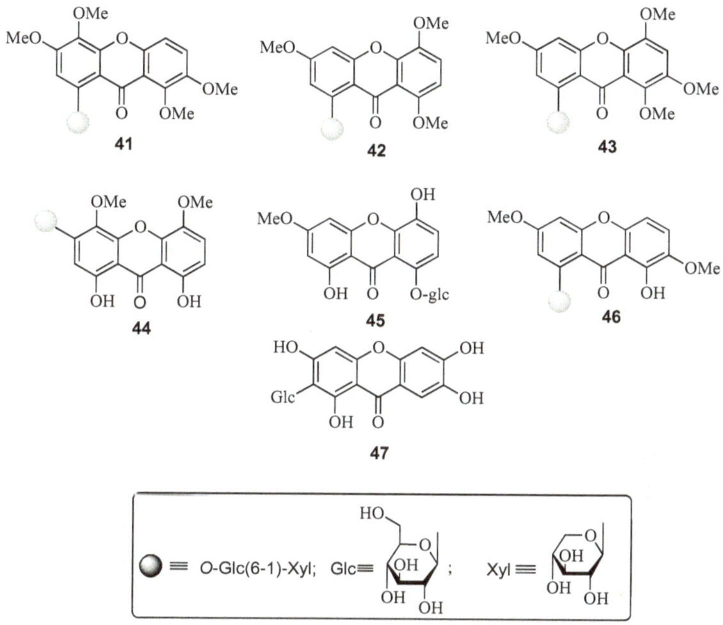

FIGURE 5.12 Xanthone glycoconjugates as α-glucosidase inhibitors.

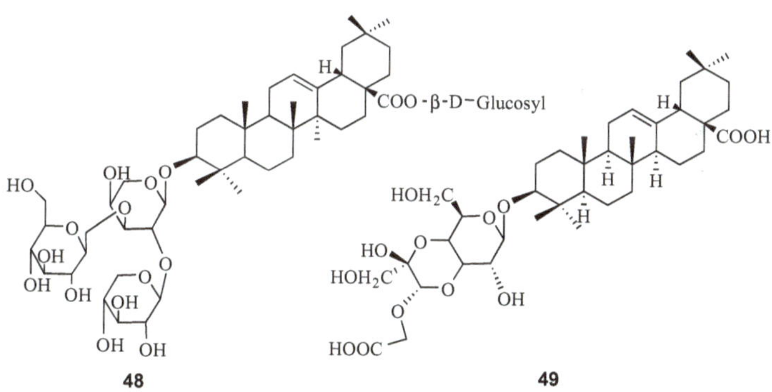

FIGURE 5.13 Antihypergylcemic agents obtained from *Beta vulgaris* and *Aralia elata*.

insulin. The adverse effects associated with them include weight gain that triggers IR, in addition to this they often lead hypoglycemia and even glitazones trigger cardiovascular diseases. So, we require antidiabetic agents that reduce blood glucose level with minimal side effects and cardiovascular disease-driven mortality (Mokdad et al., 2003, p. 76; Stone et al., 1989, p. 49). Recently, because of persistent nature of diabetes, a novel strategy has been evolved that involves targeting the low affinity, high capacity SGLT2 that mediates renal reabsorption of glucose via the proximal convoluted tubule. SLC5A2 gene governs the action of SGLT2 in which a natural mutation can lead to renal glycosuria as a result of defective SGLT2 protein, although this excretory removal of glucose does not significantly affect the excretory system. This mechanism can be used to develop therapeutic agents that can mimic the action of mutated SLC5A2 leading to excretory removal of glucose and thereby, maintaining the blood glucose levels. As this strategy of targeting SGLT2 does not involve interaction or alteration in glucose metabolism, so it can also be used along with other therapies for type 2 diabetes mellitus. Using the aforementioned mechanism, a novel family of drugs known as SGLT2 inhibitors have been synthesized, which are found to be clinically effective and comparatively safer for treatment of diabetes (Hanefeld and Forst, 2010, p. 2196; Santer et al., 2003, p. 2873). SGLT2 is found to be promisingly effective in treating hyperglycemia in individuals with poorly controlled blood glucose level by managing reduced level of HbA1c, weight loss (in addition, proper diet plan suggested), lowered blood pressure, and decreased fasting plasma glucose (FPG) levels (Clar et al., 2012, p. 1). Besides, several clinical studies revealed desired decrease in HbAc1 levels when SGLT2 inhibitors were administered in combination with drugs like metformin and glimepiride (Bailey et al., 2010, p. 2223; Nauck et al., 2011, p. 2015; Taskinen et al., 2011, p. 65; Strojek et al., 2011, p. 928).

Carbohydrate-based SGLT2 inhibitors form a massive class, in fact phlorizin (**50**), an *O*-aryl glycoside was first synthesized SGLT2 inhibitor but on account of its unstable metabolism and vulnerability to β-glycosidase, it is disapproved as an antidiabetic drug (Fig 5.14). Therefore, scientific community has been motivated toward synthesis of several *O*-glycosides in a systematic way to develop glycoconjugates with metabolic stability and greater efficacy in comparison to phlorizin.

Recently, a number of carbohydrate-based SGLT2 inhibitors have been designed and synthesized that can be potentially used as antidiabetic agents. Phlorizin was structurally modified by the group of Tanabe Seiyaku to develop another SGLT2 inhibitor, an *O*-aryl glucopyranoside, T-1095A (**51**), which when administered in diabetic rodents as its prodrug T-1095 (**52**) significantly decreased blood glucose levels but it was susceptible to β-glycosidases which restricted its utility (Fig 5.15) (Asano et al., 2004, p. 2717). Another *O*-glucopyranosides, sergliflozin (**54**) and its active form sergliflozin A (**53**), have also been developed with 296-fold high selectivity for SGLT2, although they were unable to make its way beyond phase II clinical trials (Katsuno et al., 2007, p. 323). In 2015, remogliflozin etabonate (**56**), a prodrug of remogliflozin (**55**),

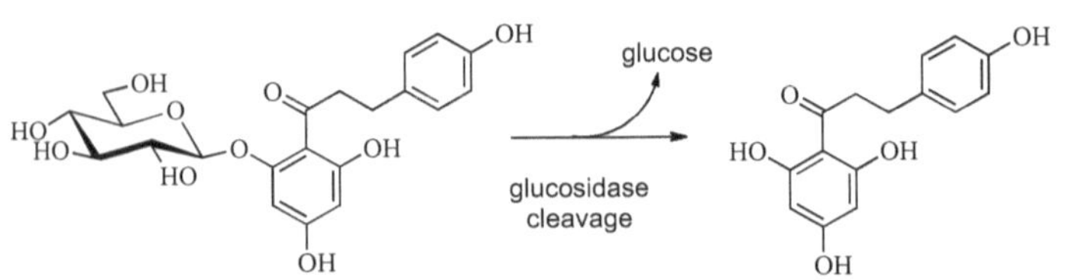

Phlorizin (**50**) Phloretin (de-glycosylated phorizin)

FIGURE 5.14 Phlorizin and its glycosidic cleavage leads to phloretin.

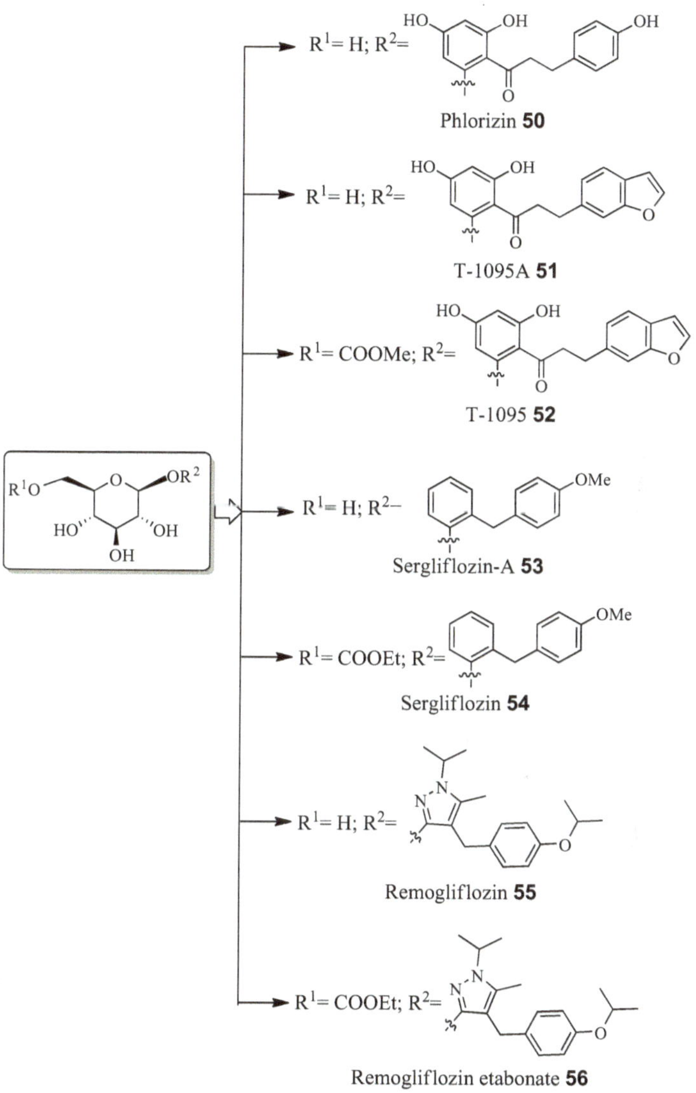

FIGURE 5.15 *O*-aryl glucopyranosides as potential SGLT2 inhibitors.

is also an *O*-glucopyranoside with attached pyrazolyl moiety; it was found to maintain glycemic levels in both rodents and humans and it is under phase II clinical trials.

The fact that *C*-glycosides are metabolically stable in comparison to *O*-glycosides, the interest of scientific community shifted toward its synthesis to get stable SGLT2 inhibitors. In an attempt, a *C*-aryl glucopyranoside, dapagliflozin (**57**), has been synthesized by companies Bristol-Myers Squibb and AstraZeneca and it was found to be potential via SGLT2 inhibitor. Dapagliflozin

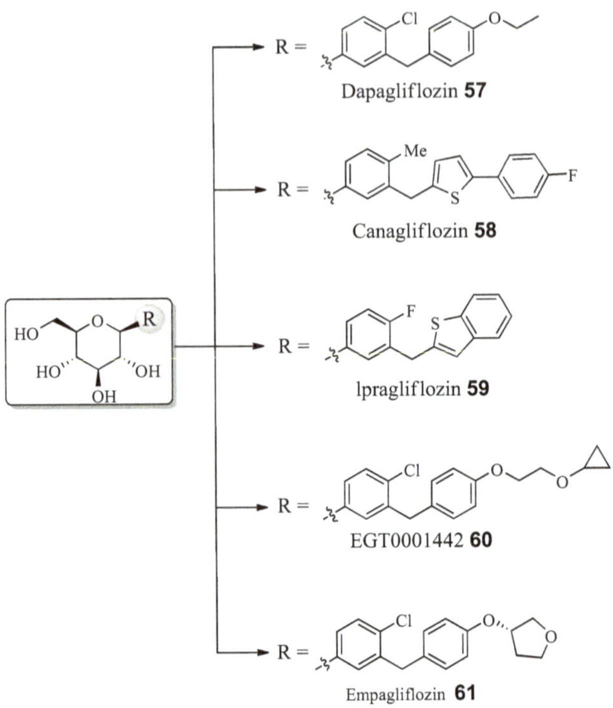

FIGURE 5.16 *C*-Aryl glucopyranosides as potential SGLT2 inhibitors.

was approved by FDA in 2014 for maintaining glycemic levels as its effect was profound in both postprandial and hypoglycemic state. The synthetic approach to dapagliflozin is discussed later in this chapter (Fig 5.16).

In March 2013, another *C*-aryl glucopyranoside named canagliflozin (**58**) (trade name, Invokana, Sulisent) was developed by Mitsubishi Tanabe and Johnson & Johnson with SGLT2 inhibition properties; presently it is under phase III clinical trials for treatment of diabetes and found to reduce HbA1c levels (0.77–1.16), although it is associated with risks of urinary tract infections (UTIs) and fungal infections (Nomura et al., 2010, p. 6355). Other modifications at C-aryl moiety of glucopyranoside lead to synthesis of other SGLT2 inhibitors like ipragliflozin (**59**) (trade name, Suglat), a D-glucitol, and empagliflozin (**61**) (trade name, Jardiance) were synthesized that were found to be potent SGLT2 inhibitors. Empagliflozin was developed by Boehringer Ingelheim and Eli Lilly Company (Grempler et al., 2012, p. 83; Sheridan, 2012, p. 899; Hardman and Dubrey, 2011, p. 133; Bays, 2009, p. 671; Washburn, 2009a, p. 1485; Washburn, 2009b, p. 1785). Another SGLT2 inhibitor EGT0001442 (**60**) developed by Theracos was found to significantly lower glycosylated HbA1c levels and hence blood glucose levels, presently it is under phase II clinical trials (Xu et al., 2014, p. 1236; Zhang et al., 2011, p. 284; Cai et al., 2011 patent).

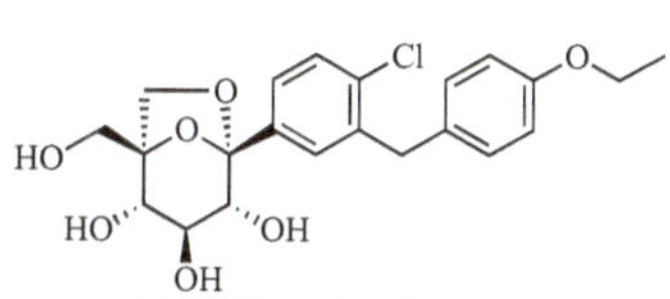

Ertuglifiozin **62**

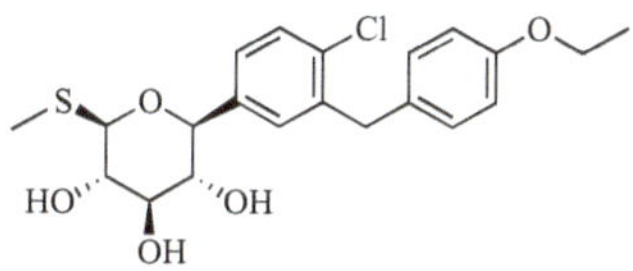

Sotagliflozin **63**

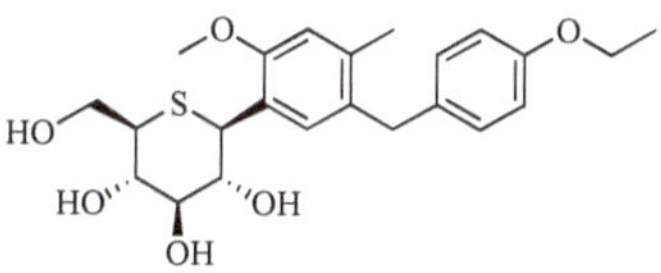

lusogliflozin **64**

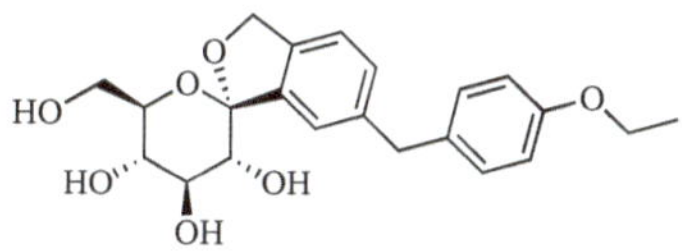

Tofogliflozin **65**

FIGURE 5.17 Some other carbohydrate-derived potential SGLT2 inhibitors.

In 2013, another carbohydrate-based diabetic agent ertugliflozin (**62**) (PF-04971729) was developed by Pfizer and was evaluated for treatment of type 2 diabetes and presently it is under phase III clinical trials; it belongs to new class of SGLT2 inhibitor having a dioxa-bicyclo[3.2.1]octane-bridged ketal motif (Mascitti et al., 2011, p. 2952) (Fig. 5.17). Sotagliflozin (**63**) (LX-4211), an xylopyranoside developed by Lexicon pharmaceuticals was found to be inhibitors of both SGLT1 and SGLT2 and exhibited potential antihyperglycemic activity under phase II clinical trials (Goodwin et al., 2009, p. 6201). Luseogliflozin (**64**) (TS-071), a thioglycoside, was recognized as both SGLT2 and SGLT1 inhibitors, it is approved by Pharmaceuticals and Medical Devices Agency of Japan (PMDA) for treatment of type 2 diabetes. It is under phase III clinical trials at Taisho pharmaceuticals. It represents a different class where a sulfur atom replaces oxygen in glucose moiety, which establishes the synthetic approach of metabolic stability of thio-*O*-glycosides (Kakinuma et al., 2010, p. 3247). Tofogliflozin (**65**) (CSG452), representative of novel class SGLT2 inhibitors featured by O-spiroketal C-aryl glucoside, developed by Kowa and Sanofi in collaboration with Chugai Pharma; it is under phase III clinical trials for treatment of diabetes (Ohtake et al., 2012, p. 7828).

So we can conclude that carbohydrate-based SGLT2 inhibitors are growing exponentially with cardinal modifications at aryl moiety and basic sugar unit to synthesize effective, metabolically stable, and safe antidiabetic agents. This strategy is encompassing various fields of synthetic research like gigantic branches of growing tree and as a result, a huge number of SGLT2 inhibitors are making their way to develop into potential antidiabetic agents. In further efforts to synthesize potential SGLT2 inhibitors (Fig. 5.18), Lee et al. (2011) developed novel *C*-aryl glucosides with cyclic diaryl polynoid moiety, using similar Pd-catalyzed cross-coupling reaction and the synthesized compounds (**66–70**) exhibited reasonable in vitro SGLT2 inhibition with IC_{50} value ranging between 59.5 and 103 nM (Kang et al., 2011, p. 3759).

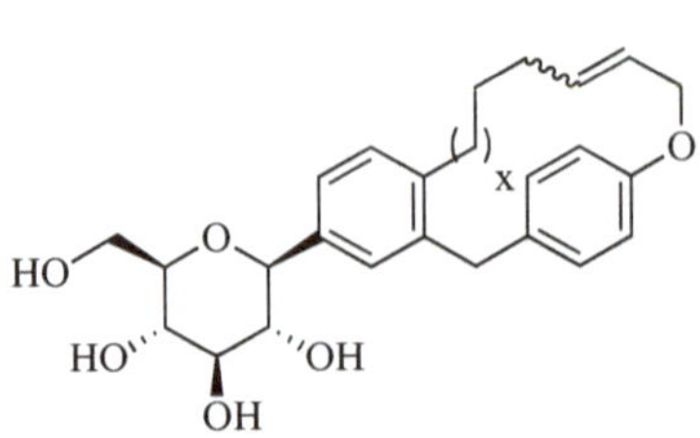

66; x= 0, IC_{50} 59.5 nM
67; x= 1, IC_{50} 89.6 nM
68; x= 2, IC_{50} 103.6 nM

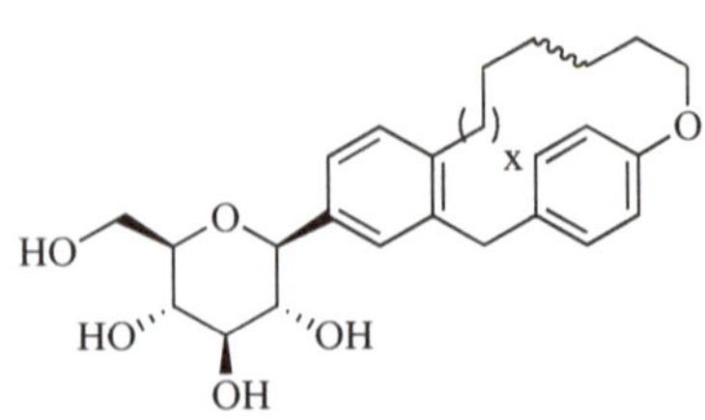

69; x= 1, IC_{50} 81.9 nM
70; x= 2, IC_{50} 99.7 nM

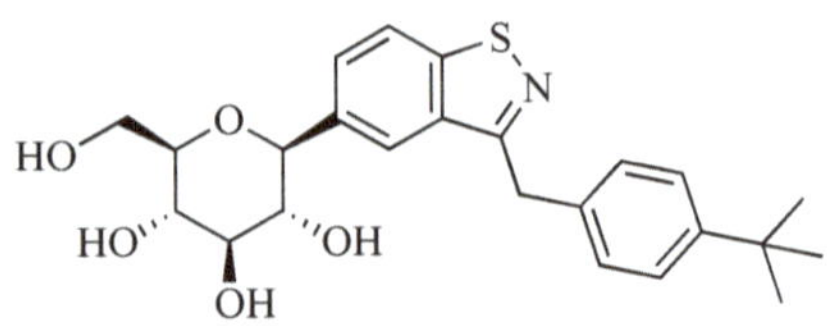

71

FIGURE 5.18 Benzisothiazole *C*-glycoside and cyclic diaryl polynoids as potential SGLT2 inhibitor.

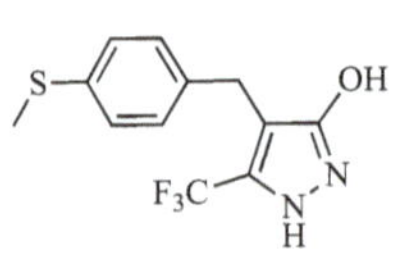

Way- 123783 (**72**)

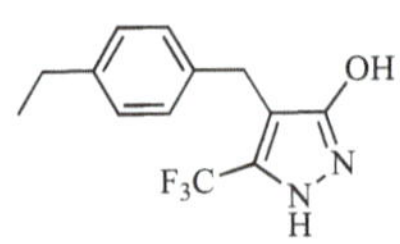

Way-2-12 (**73**)

Way-2-12 (**73**)

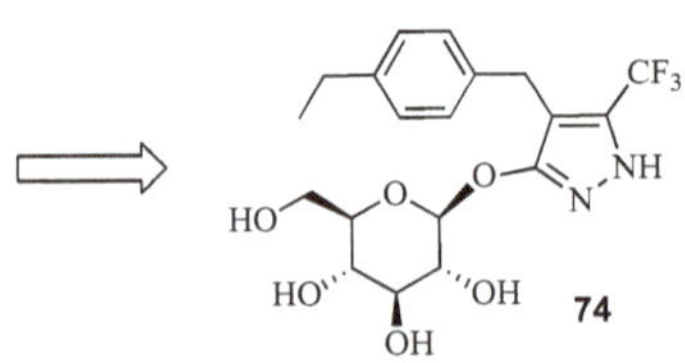

74

FIGURE 5.19 Pyrazole *O*-glucosides as potential SGLT2 inhibitor.

A series of glycoconjugates were developed by linking pyrazole moieties over *O*-glucuronides and *O*-glucosides by Ohsumi et al. (2003) to synthesize potential SGLT2 inhibitors (Fig. 5.19). The group synthesized O-glucoside and O-glucuronide derivatives of SGLT2 inhibitor WAY-123,783 (**72**), which enhanced glycosuria and eventually decreased glycemic levels in db/db mice as reported (Kees et al., 1995, p. 617). Among the synthesized compounds, compound with no carbohydrate moiety (**73**) exhibited weaker activity (30% inhibitory action, 100 μM); on the other hand, its glucoside (**74**) showed potential SGLT2 inhibitory activity (84% inhibitory action, 10 μM) and against rat kidney brush-border membrane vesicles (BBMVs) (Ohsumi et al., 2003, p. 2269).

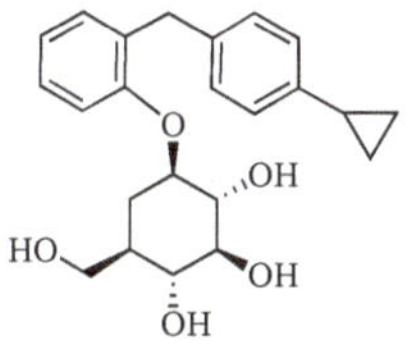

75

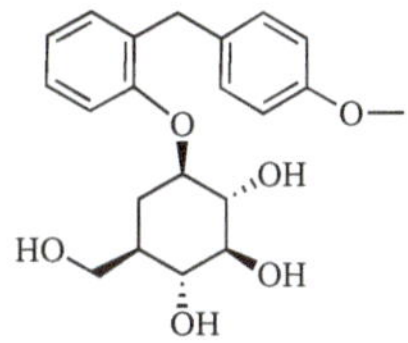

76

FIGURE 5.20 5a-Carba-β-D-glucopyranose conjugates as potential SGLT2 inhibitors.

Later, Ohtake et al. (2011) synthesized 5a-carba-β-D-glucopyranose conjugates by using coupling reaction between carbasugar and corresponding aglycone with inversion of first carbon of cyclohexyl ring (Fig. 5.20). Among the synthesized compounds, cyclopropyl containing compound (**75**) exhibited remarkable hSGLT2 inhibitory activity with IC_{50} value of 11Nm and 510-fold higher selectivity against hSGLT1 with reference phlorizin and sergliflozin (IC_{50} values 16, 260 nM, respectively). Another methoxy-substituted compound (**76**) with similar aglycone moiety as that of active form of sergliflozin exhibited threefold higher half life time than sergliflozin and lowered glycemic levels comparable to sergliflozin in db/db mice (Ohtake et al., 2011, p. 5334).

In another attempt to synthesize potential SGLT2 inhibitors, Lee et al. (2010) synthesized novel *C*-aryl glucosides with the goal of synthesizing antidiabetic agent that can be administered orally without a prodrug; their approach involved incorporation of heteroaromatic ring at distal ring position of known SGLT2 inhibitor dapagliflozin (**57**) to improve partition coefficient. They selected thidiazole moiety as the heterocylclic structure and synthesized 1,3,4-thiadiazolylmethylphenyl glucoside congeners (Fig. 5.21). Among the

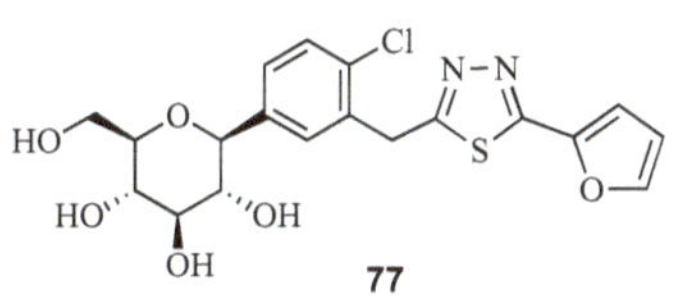

77

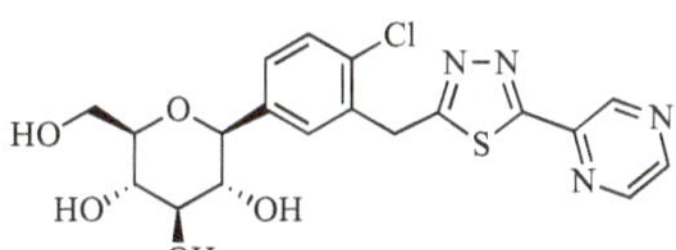

78

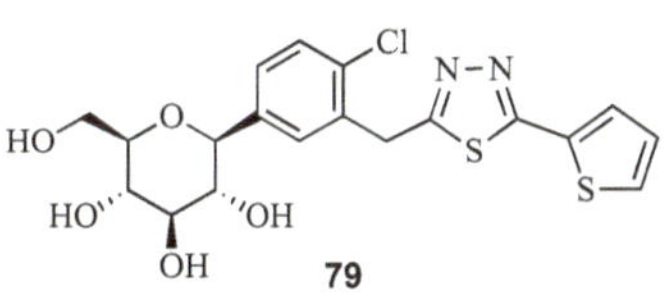

79

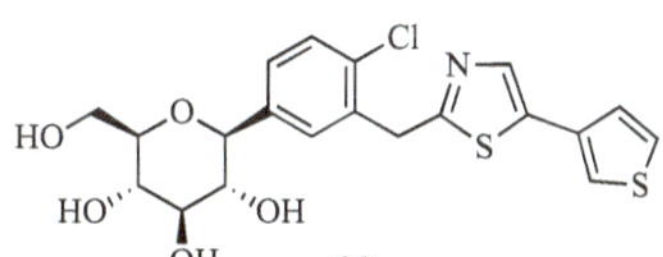

80

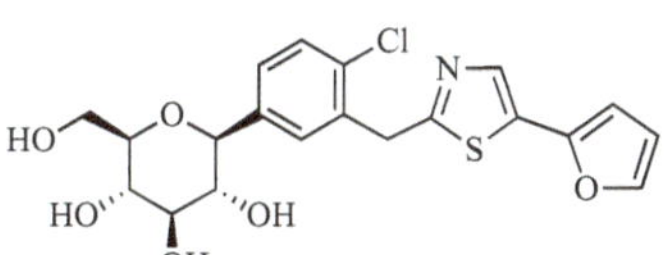

81

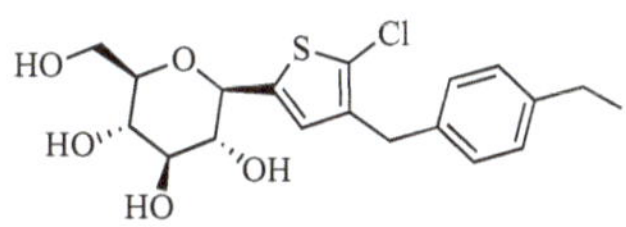

82

FIGURE 5.21 1,3,4-Thiadiazolylmethylphenyl glucoside congeners as SGLT2 inhibitors.

synthesized compounds, pyrazine-containing diaryl compound (**78**), furan-containing compound (**77**), and thiophene-containing compound (**79**) displayed significant in vitro SGLT2 inhibitory activity; in addition to this, the furan-containing compound (**77**) exhibited satisfactory antihyperglycemic effects as predicted by SAR studies. The research group further extended their work and designed two novel *C*-aryl glucoside-bearing thiazole moieties and the synthesized molecules; they were assayed for their in vitro SGLT2 inhibitory activity. Among the evaluated compounds, furanyl-containing thiazole glucoside (**81**) and thiophenyl-containing glucoside (**80**) exhibited significant SGLT2 inhibitory activity with IC_{50} values 0.720 and 0.772 nM, respectively (Lee et al., 2010, p. 2178). Further, these SGLT2 inhibitors were subjected under glucose excretion test in normal SD rats, the test revealed that thiazole (**80**) elevated 730-fold (10 mg/kg) urinary glucose excretion against control, on the other hand, thiazole (**81**) elevated 700-fold (10 mg/kg) urinary glucose excretion against the control (Song et al., 2011, p. 182). The research group further applied similar approach to design and synthesize novel *C*-aryl glucosides with thiophene moiety. Among the synthesized benzylthiophenyl compounds, 4-ethylbenzylthiophene (**82**) exhibited promising SGLT2 inhibitory activity IC_{50} value 4.47 nM (Lee et al., 2011, p. 5813).

Again, Zhang et al. (2005) synthesized a class of benzo-fused heteroaryl *O*-glucosides (Fig. 5.22) using a approach where the core entity of potent SGLT2 inhibitor TA-1095 A was modified by replacing its phenolic moiety with heteroaryl ring. Among the synthesized compounds, indole-*O*-glucoside (**83**) and benzimadazole-*O*-glucoside (**84**) displayed potential in vitro SGLT2 inhibitory activity with SGLT2 IC50 ± SEM (μM) = 0.024 ± 0.004 and IC_{50} = 0.039 ± 0.001, respectively (Zhang et al., 2005, p. 5202). Further, the research group extended their work to synthesize a series of indole-*O*-glucosides

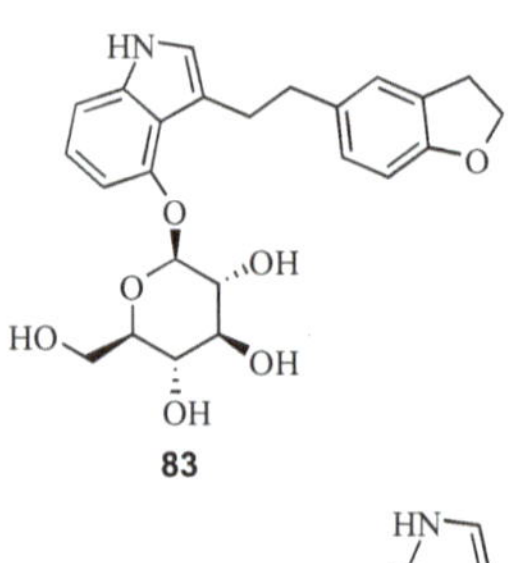

83

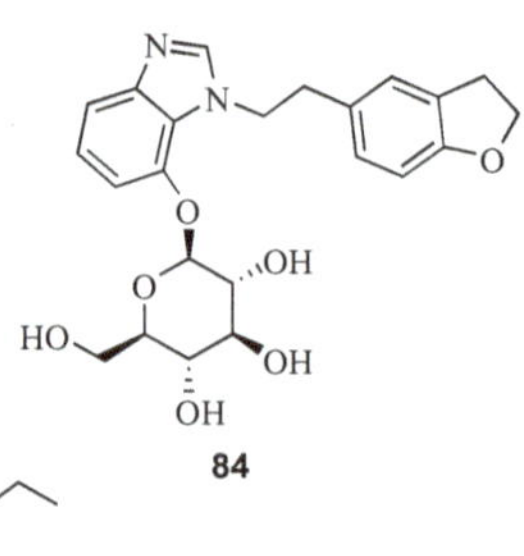

84

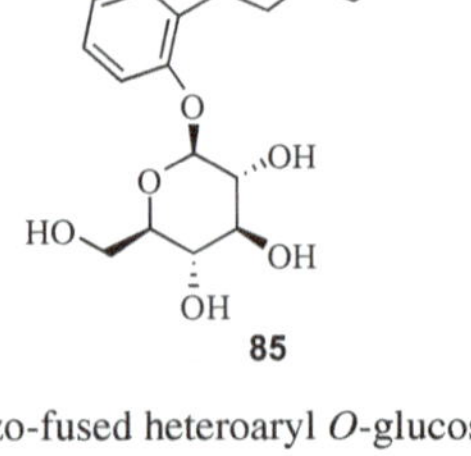

85

FIGURE 5.22 Benzo-fused heteroaryl *O*-glucosides as potent SGLT2 inhibitors.

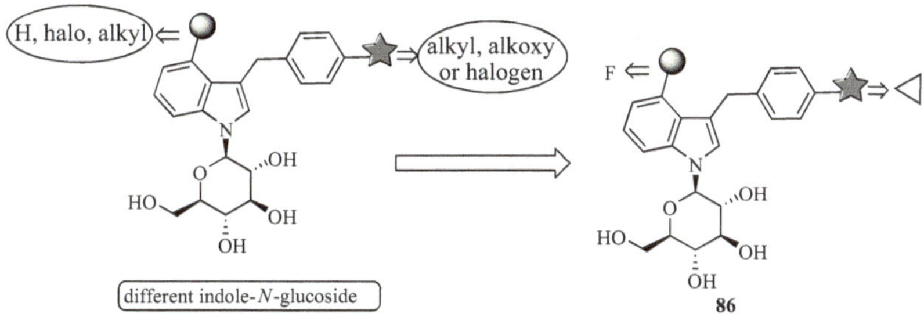

FIGURE 5.23 Novel indole-*N*-glucoside as potential SGLT2 inhibitor.

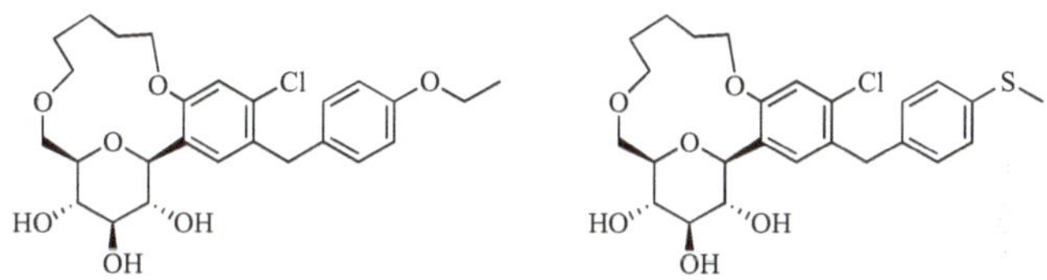

FIGURE 5.24 Macrocyclic C-aryl glycosides as potential SGLT2 inhibitors.

in which compound (**85**) exhibited selective SGLT2 inhibitory activity [IC_{50} (in μM) = 0.028 ± 0.004 as per evaluation results performed via administration in male Zucker diabetic fatty rats] (Zhang et al., 2006, p. 1696).

Recently, a series of novel 3-benzylindole-*N*-glucosides were designed and synthesized by Nomura et al. (2014), which were screened for their SGLT2 inhibitory activity (Fig. 5.23). Among the synthesized compounds, the fluoroindole-*N*-glucoside (**86**) displayed profound antidiabetic effect in high-fat diet-fed KK mice (Nomura et al., 2014, p. 51).

In further attempts, Kim et al. (2011) synthesized a novel class of macrocyclic *C*-aryl glycosides (Fig. 5.24) by using two separate ways of macrocyclization by implementing approach to link carbohydrate moiety with proximal cyclic structure and leading to development of potential SGLT2 inhibitors. Greatest in vitro activity was displayed by macrocycle with methylthiophenyl substituent at distal ring (**88**) and ethoxyphenyl at distal ring (**87**) with IC_{50} value 0.778 and 0.899 nM against hSGLT2 (Kim et al., 2011, p. 5468).

As the importance of *O*-spiroketal *C*-aryl glucosides has been established by the SGLT2 inhibitor tofogliflozin (**65**, Fig 5.17) discussed earlier in this chapter, synthetic chemists directed their interest toward synthesis of more such structures that can act as potential SGLT2 inhibitor. Lv et al. (2009) synthesized a series of novel *O*-spiroketal *C*-aryl glucosides by diversifying core spiro[isobenzofuran-1,20-pyran] moiety and further accessed their activity cell-based functional assays against hSGLT2 and hSGLT1. Among the synthesized compounds, fluoro-substituted compound **89** (Fig 5.25) facilitated significant glucose excretion (1220 mg of glucose per 200 g body weight)

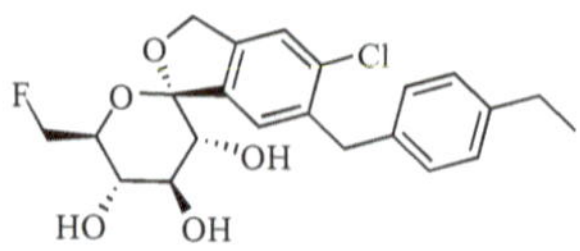

89

FIGURE 5.25 Novel *O*-spiroketal C-aryl glucoside as potential SGLT2 inhibitor.

in normal Sprague Dawley rats and displayed remarkable SGLT2 inhibitory activity (IC_{50} = 3.8 nM) (Lv et al., 2009, p. 6877). In attempts to synthesize potential SGLT2 inhibitors, Zhou et al. (2010) synthesized benzisothiazole and indolizine C-glucopyranosides using Pd-catalyzed cross-coupling reaction using bromo heterocyclic structure and glucal boronate. Cell-derived functional transporter assay technique was used to access the SGLT2 inhibitory activity, among the synthesized compound 4-t-butyl analogue exhibited remarkable SGLT2 inhibitory activity with IC_{50} of 10 nM (Zhou et al., 2010, p. 1923).

5.1 Synthetic Approach to Dapagliflozin (57): A Leading SGLT2 Inhibitor

Recently, Lemaire et al. (2012) designed a highly stereoselective and transition metal-free strategy to synthesize recognized potential SGLT2 inhibitor dapagliflozin (Lemaire et al., 2012, p. 1480). The synthesis is two-step process commences with direct cross-coupling reaction between with aryl bromide (**89**) and glycosyl bromide (**90**) that yielded tetrapivaloyl intermediate (**91**), which eventually undergoes smooth deprotection at room temperature, using MeONa/MeOH to give dapagliflozin (**57**) in excellent yield (Fig 5.26). The course of cross-coupling proceeded via metalating agent magnesate *n*-Hex $(n\text{-Bu})_2MgLi$ to accomplish selective Br/Mg exchange along with Mg/Zn exchange by means of $ZnBr_2$ 3LiCl that finally driven coupling. The group synthesized another *C*-glycosides SGLT2 inhibitor including canagliflozin using the same strategy

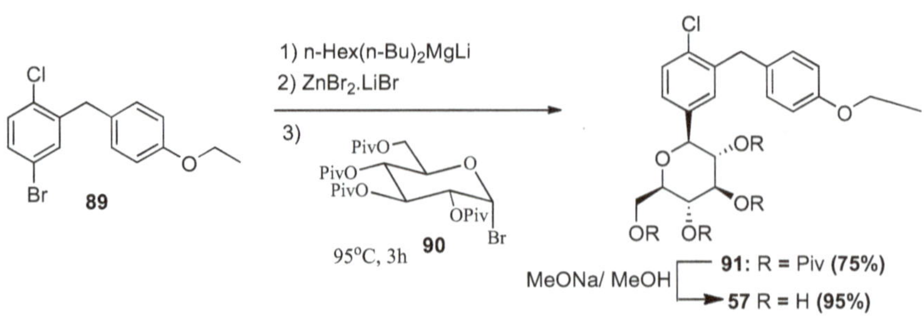

FIGURE 5.26 Synthetic strategy for synthesis of Dapagliflozin.

in fairly good yields, in addition to this, steric and electronic effects of aryl part were studied that evidenced ionic mechanism that involves anchimeric assistance (Lemaire et al., 2012, p. 1480).

6. OTHER IMPORTANT CARBOHYDRATE-BASED ANTIDIABETIC AGENTS

The search of potent antidiabetic agents motivates the chemists to follow the bidirectional pathway, one involves going deep in nature to discover and isolate the naturally occurring antihyperglycemic agents and other involves development of novel synthetic strategies to synthesize potential candidates that can serve the purpose of managing glycemic control. The rigorous attempts of scientific community to get carbohydrate-based natural products of therapeutic concern unlocked the door for biologically active antihyperglycemic compounds residing in the lap of nature. In this section, we will discuss some important antidiabetic glycoconjugates of natural origin. Plants served as vital resource in terms of drug development and they added a considerable number of marketed drugs that are either directly obtained or modified from them. It is estimated that 800 plants can possess antidiabetic activity on the basis of information reported in ethnobotanical studies. Present experimental evidence indicates that many herbs display antidiabetic potential. An extensive range of natural plant-derived chemical agents has confirmed activity in terms of their potential use in treating diabetes (Alarcon-Aguilara et al., 1998, p. 101; Saifi et al., 1971, p. 205; Mukherjee et al., 1972, p. 347; Coimbra et al., 1992, p. 320; Kar et al., 1999, p. 179; Jafri et al., 2000, p. 309; Bailey and Day, 1989, p. 553; Ivorra et al., 1988, p. 282; Marles and Farnsworth, 1995, p. 133).

The inherent property of carbohydrates to enhance the pharmacokinetic profile facilitates its administration with other available antidiabetic drugs. Waisundara et al. (2008) established the antioxidant property of carbohydrate-based baicalin (**92**, Fig 5.27), obtained from ethanolic extract of *Scutellaria baicalensis*, a herb of potential antidiabetic value (Waisundara et al., 2008, p. 517). The study scrutinized the antidiabetic effects of baicalin when administered in combination with metformin. The rats administered with

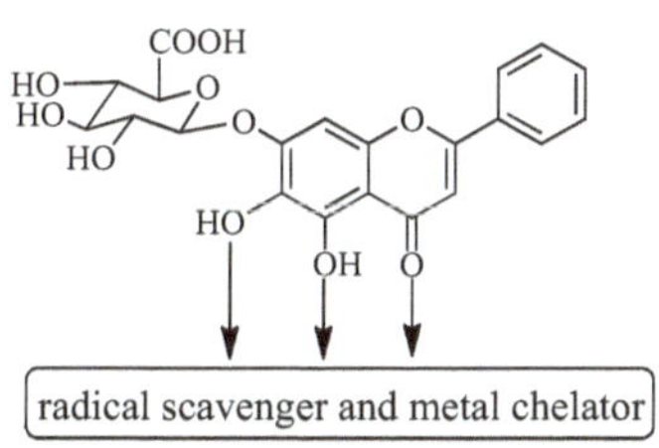

92

FIGURE 5.27 Baicalin, a potent anti diabetic agent.

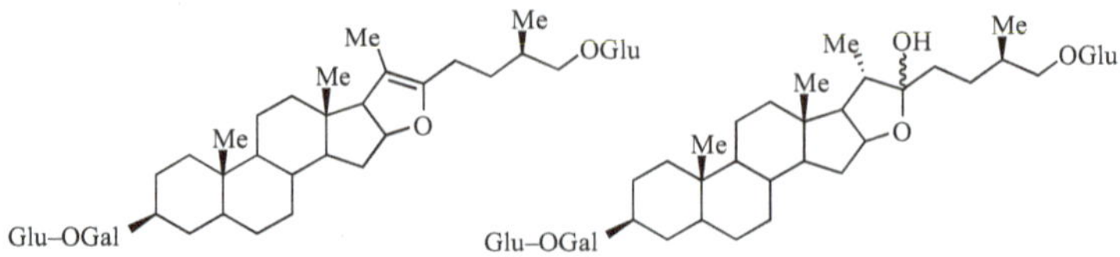

93 **94**

FIGURE 5.28 Glycosides obtained from *Anemarrhena asphodeloides* with potential antihyperglycemic activity.

metformin and baicalin had remarkably enhanced hepatic actions of catalase, superoxide dismutase, and glutathione peroxidase. The evidence suggests that baicalin alleviated oxidative activities and improved the antidiabetic activity of metformin by enhancing the antioxidant condition (Banerjee et al., 2009, p. 4068).

Two steroidal glycoconjugates were obtained from the rhizomes of *Anemarrhena asphodeloides* using fractionation techniques, namely, pseudoprotoimosaponin AIII, (**93**) and prototimosaponin AIII (**94**) (Fig 5.28). They potentially reduced blood glucose levels in streptozotocin-induced diabetic mice when administered in dose-dependent manner without altering glucose uptake by cells and insulin secretion, which, in turn, suggests that either hepatic gluconeogenesis or glycogenolysis can probably lowered the blood glucose level (Nakashima et al., 1993, p. 345).

Diabetes is associated with secondary complications which can be visualized in terms of first step involved in polyol pathway that is governed by aldose reductase (the enzyme that converts glucose into sorbitol). This step can be potentially targeted by agents that inhibit action of aldose reductase, which can result into delayed commencement of these complications (Oka and Kato, 2001, p. 465). Several rat lens aldose reductase (RLAR) has been isolated from various plants. For example, from *Momordica charantia*, a steroidal saponin charantin (**95**) was isolated known that mimicked insulin activity by checking the insulin secretion and hence, decreased blood glucose level (Ng et al., 1986, p. 107) (Fig. 5.29). Another steroidal saponin (**96**) was isolated from *Ganoderma applanatum* that was found to be potential RLAR inhibitor (Lee et al., 2005, p. 1103). Another saponin (**98**) was obtained from this herb that displayed RLAR inhibitory activity. Further, aerial part of *Chlorophytum nimonii* (Grah) Dalz A isolates characterized another steroidal saponin, chloragin (**97**) displaying significant antihyperglycemic property in streptozotocin diabetic rats (Lakshmi et al., 2009, p. 963).

Three elastosides G (**99**), H (**100**), and I (**101**) were isolated by Yoshikawa et al. from the edible material known as *Taranome* identified as immature shoot of *A. elata*; they served as remarkable antihyperglycemic agent in rats (Fig 5.30) (Yoshikawa et al. 1995b, p. 1878). Further, a gypenoside saponin obtained from *Gynostemma pentaphyllum* Makino of Cucurbitaceae family (**102**). This compound triggered secretion of insulin from isolates of islets of langerhans in rats, which, in turn, enhanced glucose tolerance when administered

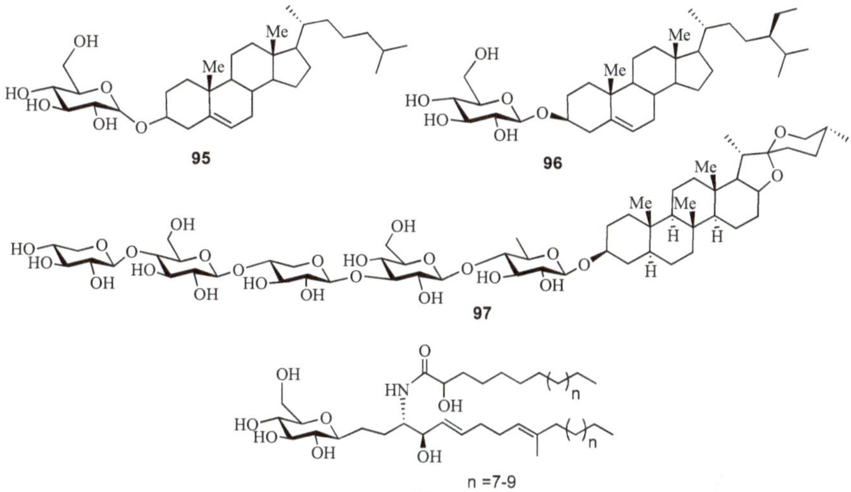
95
96
97
98

FIGURE 5.29 Steroidal saponins as rat lens aldose reductase (RLAR) inhibitors.

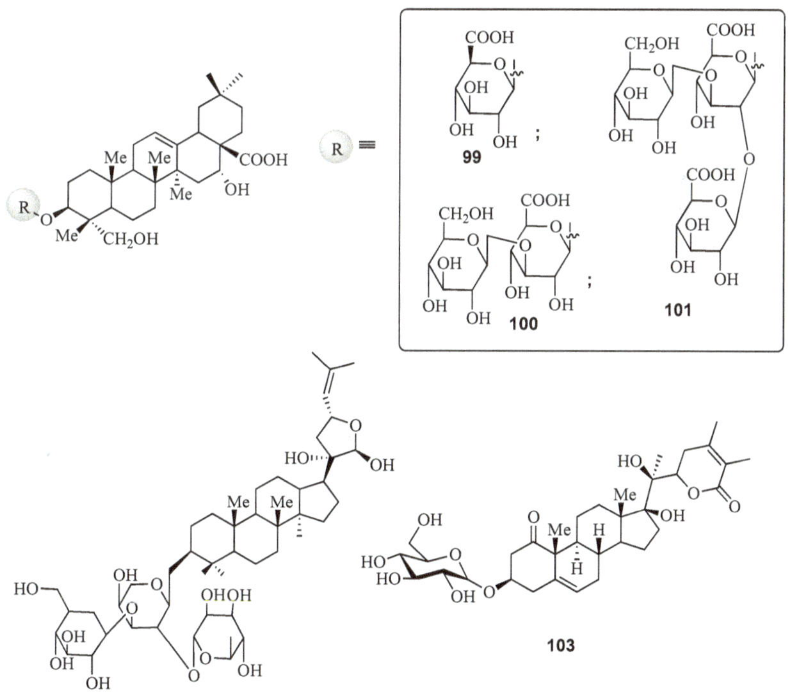
99
100
101
102
103

FIGURE 5.30 Carbohydrate-based antihyperglycemic agents obtained from *Aralia elata*, *Gynostemma pentaphyllum*, and *Withania somnifera*.

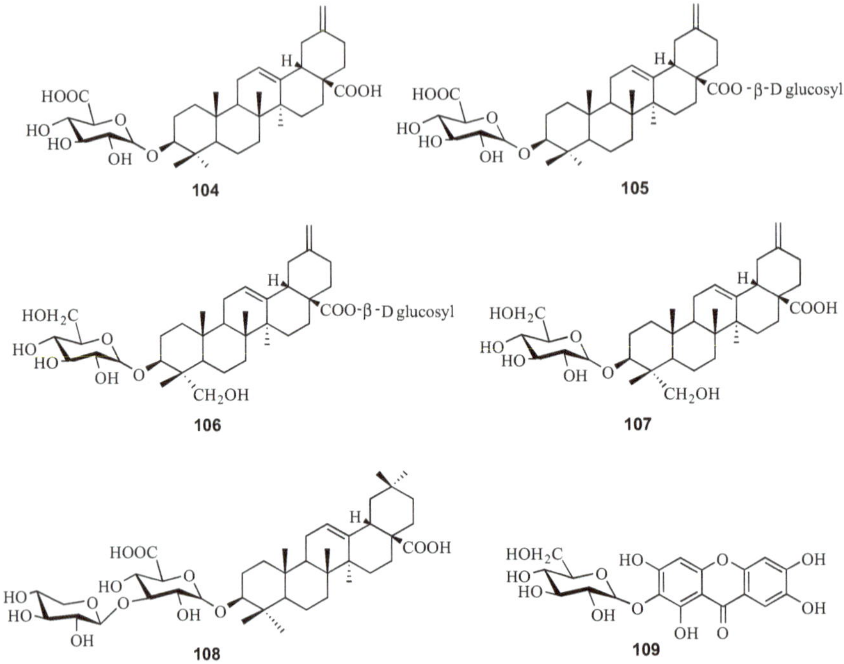

FIGURE 5.31 Antihyperglycemic agents obtained from *Boussingaultia baselloides*.

orally to rats (Norberg et al., 2004, p. 41361). The edible part of *Withania somnifera* yielded coagulin L (**103**) that displayed remarkable antihyperglycemic activity by inhibiting postprandial increase in glucose level in normoglycemic rats along with streptozotocin diabetic rats. The peripheral glucose levels were significantly reduced and glucose tolerance capacity was also enhanced in db/db mice by this compound (Maurya et al., 2008, p. 6534).

In an attempt to isolate more potent antihyperglycemic agents, a saponin along with four nor saponins have been isolated from the leaves of methanolic extract of the leaves of *Boussingaultia baselloides* (**104–108**) that exhibited significant antihyperglycemic activity (Fig. 5.31), among the isolated compounds boussingoside A1 (**108**) was found to be most potent antihyperglycemic agent in rats (Espada et al., 1990, p. 2039). The leaves of *Mangifera indica* yielded a xanthone glycoconjugate named mangiferin (**109**), a xanthone exhibited antihyperglycemic activity in streptozotocin diabetic rats. The magniferin was administered daily in doses of 10 and 20 mg/kg for a period of 28 days in diabetic rats and it was observed that extensive administration prominently enhanced glucose tolerance in rats loaded with glucose (Muruganandan et al., 2005, p. 492).

6.1 Synthetic Approach for Magniferin

Magniferin (**109**) along with other biologically active xanthone glycosides, isomagniferin and homomagniferin were synthesized by *C*-glycosylation of xanthene derivative (**111**), which was obtained by reduction and methylation of 1,3,6,7-tetrahydroxyxanthone derivative (**110**) with perbenzylglucopyranosyl trifluoroacetimidate (**112**) under the treatment of trimethyl silyl triflate (TMSOTf) that gives rise to 2-C-β-glycoside (**113**) whose further oxidation with 2,3-dichloro-5,6-dicyano-1,4-benzoquinone (DDQ), followed by deprotection in the presence of $(Ph_3P)_3RhCl$ and 1,4-diazabicyclo[2.2.2]octane (DABCO) in mixed solvent containing EtOH/PhMe/H_2O under refluxing condition leading to synthesis of two diols, one of them (**115**) was treated with 2N aq NaOH in 1,4-dioxane at 60°C to remove 7-O-mesyl moiety that yielded two triols; one of the triols (**116**) was treated with 1N BBr_3 in CH_2Cl_2 at room temperature facilitating cleavage of four *O*-benzyl positions on the carbohydrate moiety and the xanthone 6-O-methyl group leading to synthesis of Magniferin (**109**) in 94% yield (Fig 5.32) (Zhongtao et al., 2010, p. 5725).

The class carbohydrate-based antidiabetic agents obtained by plants are further enriched by flavonoid icariin (**117**) obtained from Epimedium (Berberidaceae), which was found to be helpful in weight loss and myocardial collagen with respect to diabetes-related complications (Bao and Chen, 2011, p. 1503). Another flavonoid glycoside, kaempferitrin (**118**) obtained from *Bauhinia forficata* was found to have stimulatory effect on uptake of glucose, which, in turn, significantly lowered blood glucose (Cazarolli et al., 2013, p. 1). Recently, Mahmoud et al. (2012) scrutinized beneficial effects of two flavonoid glycosides, namely, hesperidin (**119**) and naringin (**120**) obtained from citrus peel and citrus fruit, on STZ-induced diabetic rats were found to have antidiabetic and antioxidative activities (Mahmoud et al., 2012, p. 483) (Fig. 5.33).

7. CONCLUDING REMARKS

The prevailing threat of metabolic disease diabetes is rising at unexpected rate encircling wide range of age groups and exposing people at high risk of mortality and morbidity. Non-insulin-dependent type 2 diabetes is more common type of diabetes that obligates scientific community to synthesize and isolate efficacious and safe drugs. The prime cause of T2DM is relative deficiency of insulin facilitated by IR that eventually decreases blood glucose levels and hence, triggering a hyperglycemic state which when persists, affects vital organs leading to physiological dysfunctions. This noxious disease needs immediate attention of chemists because the available drugs in market are somehow prone to side effects and its nature is changing day by day, which necessitates the renewal of available antidiabetic agents. Recently, carbohydrate-based compounds emerged as biomolecules inherited with better solubility and pharmacokinetics along with less toxic effects, can serve as improved antidiabetic agents. The glycoconjugates with antidiabetic property are added to present collection of active agents by using both natural and synthetic strategies. The lap of nature enfolds a large number of carbohydrate-based antidiabetic agents, which are isolated by rigorously scrutinizing the isolation strategies used by carbohydrate chemist. Similarly, the synthetic chemists are focusing on designing and synthesizing target-based compounds with antidiabetic

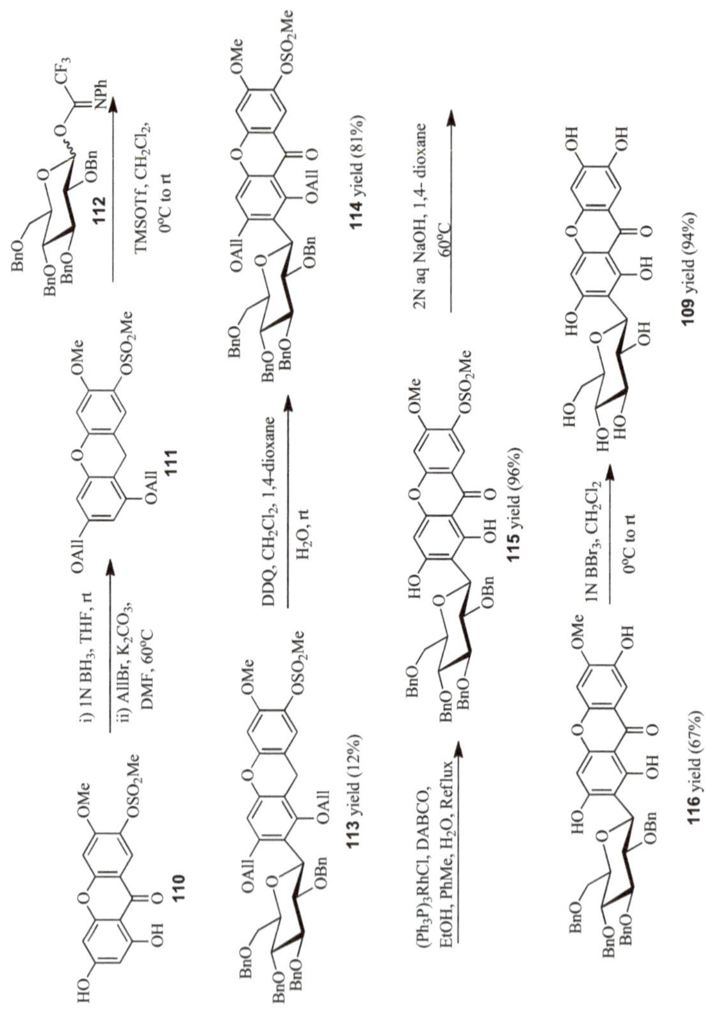

FIGURE 5.32 Synthetic route to synthesis of Magniferin.

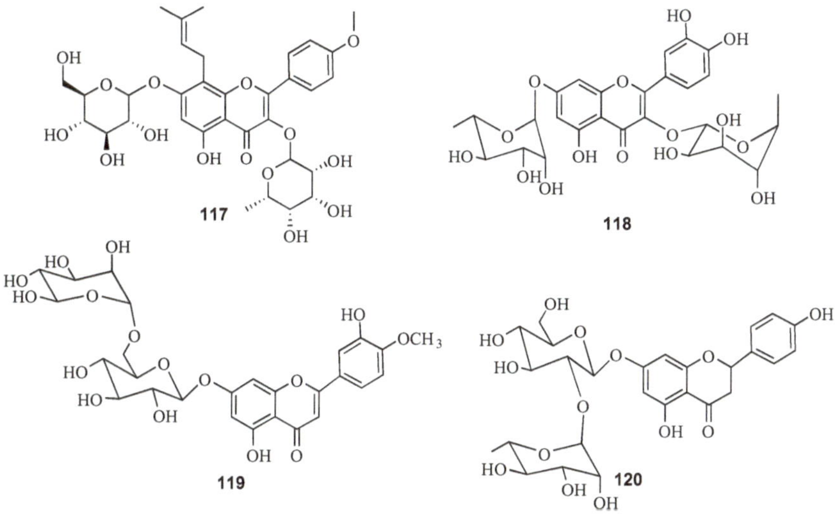

FIGURE 5.33 Antidiabetic flavonoid glycosides.

activity. In broad terms, carbohydrate-based glucosidase inhibitors and SGLT2 inhibitors are the major active classes for the treatment of diabetes along with other agents that directly affects insulin secretion. The literature reports that some glycoconjugates are even beneficial in minimizing diabetes-related secondary complications. This chapter offers a comprehensive account on the importance of recent carbohydrate-based antidiabetic agents with emphasis given on their structures and mechanism of action; in addition, the cause and classification of diabetes are also discussed in brief along with synthetic approach to some significant and recent antidiabetic drugs.

LIST OF ABBREVIATIONS

ACE Angiotensin- converting enzyme
AGE Advanced glycation eEnd product
BBMV Brush-border membrane vesicle
DABCO 1,4-diazabicyclo[2.2.2]octane
DDQ 2,3-Dichloro-5,6-dicyano-1,4-benzoquinone
FDA Food and drug administration
FPG Fasting plasma glucose
GDM Gestational diabetes mellitus
IDF International Diabetes Federation
IR Insulin rResistance
MODY Maturity onset diabetes of young
NIDDM Non- insulin- dependent diabetes mellitus
PMDA Pharmaceuticals and Medical Devices Agency of Japan
RLAR Rat lens aldose reductase

SGLT2 Sodium glucose cotransporter two2
SLC5A2 Solute cCarrier fFamily 5 (sodium/glucose cCotransporter), mMember two2
T1DM Type 1 diabetes mellitus
T2DM Type 2 diabetes mellitus
TMSOTf Trimethyl silyl triflate
UTI Urinary tract infection

ACKNOWLEDGMENTS

VKT thanks Banaras Hindu University (BHU) for the support and Council of Scientific & Industrial Research (CSIR), New Delhi for the funding.

REFERENCES

Aguilar, R.B., 2011. Evaluating treatment algorithms for the management of patients with type 2 diabetes mellitus: a perspective on the definition of treatment success. Clinical Therapeutics 33 (4), 408–824.

Aguirre, F., Brown, A., Cho, N.H., Dahlquist, G., The Diabetes Education Consultative Section, Dodd, S., Dunning, T., Hirst, M., Hwang, C., Magliano, D., Patterson, C., Scott, C., Shaw, J., Soltesz, G., Usher-Smith, J., Whiting, D., 2014. IDF diabetes atlas. In: Guariguata, L., Nolan, T., Beagley, J., Linnenkamp, U., Jacqmain, O. (Eds.), Int. Diabetes Federation., sixth ed. Brussels, Belgium. http://www.idf.org/diabetesatlas/introduction.

Alarcon-Aguilara, F.J., Roman-Ramos, R., Perez-Gutierrez, S., Aguilar-Contreras, A., Contreras-Weber, C.C., Flores-Saenz, J.L., 1998. Study of the anti-hyperglycemic effect of plants used as antidiabetics. Journal of Ethnopharmacology 61 (2), 101–110.

Anand, N., Jaiswal, N., Pandey, S.K., Srivastava, A.K., Tripathi, R.P., 2011. Application of click chemistry towards an efficient synthesis of 1,2,3-1H-triazolyl glycohybrids as enzyme inhibitors. Carbohydrate Research 346 (1), 16–25.

Arya, A.K., Kumar, L., Pokharia, D., Tripathi, K., 2008. Applications of nanotechnology in diabetes. Digest Journal of Nanomaterials and Biostructures 3 (4), 221–225.

Asano, T., Ogihara, T., Katagiri, H., Sakoda, H., Ono, H., Fujishiro, M., Anai, M., Kurihara, H., Uchijima, Y., 2004. Glucose transporter and Na^+/glucose cotransporter as molecular targets of anti-diabetic drugs. Current Medicinal Chemistry 11 (20), 2717–2724.

Bailey, C.J., Day, C., 1989. Traditional plant medicines as treatments for diabetes. Diabetes Care 12 (8), 553–564.

Bailey, C.J., Gross, J.L., Pieters, A., Bastien, A., List, J.F., 2010. Effect of dapagliflozin in patients with type 2 diabetes who have inadequate glycaemic control with metformin: a randomised, double-blind, placebo-controlled trial. Lancet 375, 2223–2233.

Banerjee, K., Oulkar, D.P., Patil, S.B., Jadhav, M.R., Dasgupta, S., Patil, S.H., Bal, S., Adsule, P.G., 2009. Multiresidue determination and uncertainty analysis of 87 pesticides in mango by liquid chromatography-tandem mass spectrometry. Journal of Agricultural and Food Chemistry 57 (10), 4068–4078.

Bao, H., Chen, L., 2011. Icariin reduces mitochondrial oxidative stress injury in diabetic rat hearts. China Journal of Chinese Materia Medica 36 (11), 1503–1507.

Bays, H., 2009. From victim to ally: the kidney as an emerging target for the treatment of diabetes mellitus. Current Medical Research and Opinion 25 (3), 671–681.

Beelders, T., Brand, D.J., de Beer, D., Malherbe, C.J., Mazibuko, S.E., Muller, C.J., Joubert, E., 2014. Benzophenone C- and *O*-glucosides from *Cyclopia genistoides* (Honeybush) inhibit mammalian α-glucosidase. Journal of Natural Products 77 (12), 2694–2699.

Bharti, S.K., Krishnan, S., Kumar, A., Gupta, A.K., Ghosh, A.K., Kumar, A., 2015. Mechanism-based antidiabetic activity of fructo- and isomalto-oligosaccharides: validation by in vivo, in silico and in vitro interaction potential. Process Biochemistry 50 (2), 317–327.

Cai, M., Liu, Q., Xu, G., Lv, B., Seed, B., Roberge, J.Y., 2011. Crystalline Form of Benzylbenzene sglt2 Inhibitor. PCT Int. Appl. WO2011153953.

Cao, H., Hwang, J., Chen, X., 2011. Carbohydrate-containing natural products in medicinal chemistry. In: Tiwari, V.K. (Ed.), Opportunity, Challenge and Scope of Natural Products in Medicinal Chemistry, pp. 411–431.

Cazarolli, L.H., Pereira, D.F., Kappel, V.D., Folador, P., Figueiredo, M.D., Pizzolatti, M.G., Silva, F.R., 2013. Insulin signaling: a potential signaling pathway for the stimulatory effect of kaempferitrin on glucose uptake in skeletal muscle. European Journal of Pharmacology 712, 1–7.

Cipolla, L., Peri, F., 2011. Carbohydrate-based bioactive compounds for medicinal chemistry applications. Mini Reviews in Medicinal Chemistry 11 (1), 39–54.

Cipolla, L., Gregori, M., So, P.W., 2011. Glycans in magnetic resonance imaging: determinants of relaxivity to smart agents, and potential applications in biomedicine. Current Medicinal Chemistry 18 (1), 1002–1018.

Cipolla, L., La Ferla, B., Airoldi, C., Zona, C., Orsato, A., Shaikh, N., Russo, L., Nicotra, F., 2010. Carbohydrate mimetics and scaffolds: sweet spots in medicinal chemistry. Future Medicinal Chemistry 2 (4), 587–599.

Clar, C., Gill, J.A., Court, R., Waugh, N., 2012. Systematic review of SGLT2 receptor inhibitors in dual or triple therapy in type 2 diabetes. BMJ Open 2, e001007. http://dx.doi.org/10.1136/bmjopen-2012-001007.

Coimbra, T.C., Danni, F.F., Blotta, R.M., da Periara, C.A., Guedes, M.D., Graf, R.G., 1992. Plants employed in treatment of diabetes mellitus; results of an ethnopharmacological survey in Pordo Alegre, Brazil. Fitoterapia 63 (4), 320–322.

Complications Trial Research Group, 1993. The effect of intensive treatment of diabetes on the development and progression of long-term complications in insulin-dependent diabetes mellitus. The New England Journal of Medicine 329, 977–986.

American Diabetes Association, 2008. Diagnosis and classification of diabetes mellitus. Diabetes Care 31, S55–S60.

Espada, A., Rodriguez, J., Villaverde, M., Carmen, R., Ricardo, F., 1990. Hypoglucaemic triterpenoid saponins from *Boussingaultia baselloides*. Canadian Journal of Chemistry 68 (11), 2039–2044.

Etayo, P., Badorrey, R., Villegas, M.D., Galvez, G.A., Víu, P.L., 2010. An expeditious method for the first asymmetric synthesis of Dexoxadrol from the chiral pool. Synlett 12, 1775–1778.

Ferreira, S.B., Sodero, A.C.R., Cardoso, M.F.C., Lima, E.S., Kaiser, C.R., Silva Jr., F.P., Ferreira, V.F., 2010. Synthesis, biological activity, and molecular modeling studies of 1*H*-1,2,3-triazole derivatives of carbohydrates as α-glucosidases inhibitors. Journal of Medicinal Chemistry 53 (6), 2364–2375.

Goodwin, N.C., Mabon, R., Harrison, B.A., Shadoan, M.K., Almstead, Z.Y., Xie, Y., Healy, J., Buhring, L.M., DaCosta, C.M., Bardenhagen, J., Mseeh, F., Liu, Q., Nouraldeen, A., Wilson, A.G.E., Kimball, S.D., Powell, D.R., Rawlins, D.B., 2009. Novel l-xylose derivatives as selective sodium-dependent glucose cotransporter 2 (sglt2) inhibitors for the treatment of type 2 diabetes. Journal of Medicinal Chemistry 52 (20), 6201–6204.

Grempler, R., Thomas, L., Eckhardt, M., Himmelsbach, F., Sauer, A., Sharp, D.E., Bakker, R.A., Mark, M., Klein, T., Eickelmann, P., 2012. Empagliflozin, a novel selective sodium glucose cotransporter-2 (SGLT-2) inhibitor: characterisation and comparison with other SGLT-2 inhibitors. Diabetes, Obesity and Metabolism 14 (1), 83–90.

Hanefeld, M., Forst, T., 2010. Dapagliflozin, an SGLT2 inhibitor, for diabetes. Lancet 375, 2196–2198.

Hardman, T.C., Dubrey, S.W., 2011. Development and potential role of type-2 sodium-glucose transporter inhibitors for management of type 2 diabetes. Diabetes Therapy 2 (3), 133–145.

Hewitt, M.C., Snyder, D.A., Seeberger, P.H., 2002. Rapid synthesis of a glycosylphosphatidylinositol-based malaria vaccine using automated solid-phase oligosaccharide synthesis. Journal of American Chemical Society 124 (45), 13434–13436.

Huizinga, M.M., Rothman, R.L., 2006. Addressing the diabetes pandemic: a comprehensive approach. Indian Journal of Medical Research 124 (5), 481–484.

Ivorra, M.D., Paya, M., Villar, A., 1988. Hypoglycemic and insulin release effects of tormentic acid: a new hypoglycemic natural product. Planta Medica 54 (4), 282–286.

Jafri, M.A., Aslam, M., Javed, K., Singh, S., 2000. Effect of *Punica granatum* Linn. (flowers) on blood glucose level in normal and alloxan-induced diabetic rats. Journal of Ethnopharmacology 70 (3), 309–314.

Jenkinson, S.F., Best, D., Saville, A.W., Mui, J., Martínez, R.F., Nakagawa, S., Kunimatsu, T., Alonzi, D.S., Butters, T.D., Norez, C., Becq, F., Blériot, Y., Wilson, F.X., Weymouth-Wilson, A.C., Kato, A., Fleet, G.W.J., 2013. C-Branched iminosugars: α-glucosidase inhibition by enantiomers of isoDMDP, isoDGDP, and isoDAB–L-isoDMDP compared to miglitol and miglustat. Journal of Organic Chemistry 78 (15), 7380–7397.

Kakinuma, H., Oi, T., Hashimoto-Tsuchiya, Y., Arai, M., Kawakita, Y., Fukasawa, Y., Iida, I., Hagima, N., Takeuchi, H., Chino, Y., Asami, J., 2010. (1s)-1,5-anhydro-1-[5-(4-ethoxybenzyl)-2-methoxy-4-methylphenyl]-1-thio-d-glucitol (ts-071) is a potent, selective sodium-dependent glucose cotransporter 2 (sglt2) inhibitor for type 2 diabetes treatment. Journal of Medicinal Chemistry 53 (8), 3247–3261.

Kang, S.Y., Kim, M.J., Lee, J.S., Lee, J., 2011. Glucosides with cyclic diarylpolynoid as novel C-aryl glucoside SGLT2 inhibitors. Bioorganic and Medicinal Chemistry Letters 21 (12), 3759–3763.

Kar, A., Choudhary, B.K., Bandhopadhyay, N.G.J., 1999. Preliminary studies on the inorganic constituents of some indigenous hypoglycaemic herbs on oral glucose tolerance test. Journal of Ethnopharmacology 64, 179–184.

Katsuno, K., Fujimori, Y., Takemura, Y., Hiratochi, M., Itoh, F., Komatsu, Y., Fujikura, H., Isaji, M., 2007. Sergliflozin, a novel selective inhibitor of low-affinity sodium glucose cotransporter (SGLT2), validates the critical role of SGLT2 in renal glucose reabsorption and modulates plasma glucose level. Journal of Pharmacology and Experimental Therapeutics 320 (1), 323–330.

Kees, K.L., Caggiano, T.J., Steiner, K.E., Fitzgerald, J.J., Kates, M.J., Christos, T.E., Kulishoff, J.M., Moore, R.D., McCaleb, M.L., 1995. Studies on new acidic azoles as glucose-lowering agents in obese, diabetic db/db mice. Journal of Medicinal Chemistry 38 (4), 617–628.

Khan, A.R., Tiwari, V.K., Srivastava, A.K., Ahmad, R., Tripathi, R.P., 2004. An efficient syntheses of α-mannosylated phenolics as α-glycosidase inhibitors. Journal of Enzyme Inhibition and Medicinal Chemistry 19, 107–112.

Kim, J.H., Ryu, Y.B., Kang, N.S., Lee, B.W., Heo, J.S., Jeong, I.Y., Park, K.H., 2006. Glycosidase inhibitory flavonoids from *Sophora flavescens*. Biological and Pharmaceutical Bulletin 29 (2), 302–305.

Kim, M.J., Lee, S.H., Park, S.O., Kang, H., Lee, J.S., Lee, K.N., Jung, M.E., Kim, J., Lee, J., 2011. Novel macrocyclic *C*-aryl glucoside SGLT2 inhibitors as potential antidiabetic agents. Bioorganic and Medicinal Chemistry 19 (18), 5468–5479.

Kokil, G.R., Veedu, R.N., Ramm, G.A., Prins, J.B., Parekh, H.S., 2015. Type 2 diabetes mellitus: limitations of conventional therapies and intervention with nucleic acid-based therapeutics. Chemical Reviews 115 (11), 4719–4743.

Lakshmi, V., Kumar, R., Pandey, K., Joshi, B.S., Roy, R., Madhusudanan, K.P., Tiwari, P., Srivastava, A.K., 2009. Structure and activities of a steroidal saponin from *Chlorophytum nimonii* (Grah) Dalz. Natural Product Research 23 (10), 963–972.

Lee, J., Lee, S.H., Seo, H.J., Son, E.J., Lee, S.H., Jung, M.E., Lee, M., Han, H.K., Kim, J., Kang, J., Lee, J., 2010. Novel C-aryl glucoside SGLT2 inhibitors as potential antidiabetic agents: 1,3,4-Thiadiazolylmethylphenyl glucoside. Bioorganic and Medicinal Chemistry 18 (6), 2178–2194.

Lee, S.H., Song, K.S., Kim, J.Y., Kang, M., Lee, J.S., Cho, S.H., Park, H.J., Kim, J., Lee, J., 2011. Novel thiophenyl C-aryl glucoside SGLT2 inhibitors as potential antidiabetic agents. Bioorganic and Medicinal Chemistry 19 (19), 5813–5832.

Lee, S., Shim, S.H., Kim, J.S., Shin, K.H., Kang, S.S., 2005. Aldose reductase inhibitors from the fruiting bodies of *Ganoderma applanatum*. Biological and Pharmaceutical Bulletin 28 (6), 1103–1105.

Lemaire, S., Houpis, I.N., Xiao, T., Li, J., Digard, E., Gozlan, C., Liu, R., Gavryushin, A., Diene, C., Wang, Y., Farina, V., Knochel, P., 2012. Stereoselective C-glycosylation reactions with arylzinc reagents. Organic Letters 14, 1480–1483.

Li, Y.Q., Zhou, F.C., Gao, F., Bian, J.S., Shan, F., 2009. Comparative evaluation of quercetin, isoquercetin and rutin as inhibitors of r-glucosidase. Journal of Agricultural and Food Chemistry 57 (24), 11463–11468.

Lonngren, J., 1989. Carbohydrates and the pharmaceutical industry. Pure and Applied Chemistry 61 (7), 1313–1314.

Lv, B., Xu, B., Feng, Y., Peng, K., Xu, G., Du, J., Zhang, L., Zhang, W., Zhang, T., Zhu, L., Ding, H., Sheng, Z., Welihinda, A., Seed, B., Chen, Y., 2009. Exploration of O-spiroketal C-arylglucosides as novel and selective renal sodium-dependent glucose co-transporter 2 (SGLT2) inhibitors. Bioorganic and Medicinal Chemistry Letters 19 (24), 6877–6881.

Ma, Y.Y., Zhao, D.-G., Zhou, A.-Y., Zhang, Y., Du, Z., Zhang, K., 2015. α-Glucosidase inhibition and antihyperglycemic activity of phenolics from the flowers of *Edgeworthia gardneri*. Journal of Agricultural and Food Chemistry 63 (37), 8162–8169.

Mahmoud, A.M., Ashour, M.B., Abdel-Moneim, A., Ahmed, O.M., 2012. Hesperidin and naringin attenuate hyperglycemia-mediated oxidative stress and proinflammatory cytokine production in high fat fed/streptozotocin-induced type 2 diabetic rats. Journal of Diabetes and Its Complications 26 (6), 483–490.

Marles, R.J., Farnsworth, N.R., 1995. Antidiabetic plants and their active constituents. Phytomedicine 2, 133–189.

Mascitti, V., Maurer, T.S., Robinson, R.P., Bian, J., Boustany- Kari, C.M., Brandt, T., Collman, B.M., Kalgutkar, A.S., Klenotic, M.K., Leininger, M.T., Lowe, A., Maguire, R.J., Masterson, V.M., Miao, Z., Mukaiyama, E., Patel, J.D., Pettersen, J.C., Préville, C., Samas, B., She, L., Sobol, Z., Steppan, C.M., Stevens, B.D., Thuma, B.A., Tugnait, M., Zeng, D., Zhu, T., 2011. Discovery of a clinical candidate from the structurally unique dioxa-bicyclo[3.2.1]octane class of sodium-dependent glucose cotransporter 2 inhibitors. Journal of Medicinal Chemistry 54 (8), 2952–2960.

Maurya, R., Akanksha, Jayendra, Singh, A.B., Srivastava, A.K., 2008. Coagulanolide, a withanolide from Withania coagulans fruits and antihyperglycemic activity. Bioorganic and Medicinal Chemistry Letters 18 (24), 6534–6537.

Mishra, S., Upadhay, K., Mishra, K.B., Tripathi, R.P., Tiwari, V.K., 2016 (in press). Studies in Natural Product Chemistry. Elsevier Publication.

Mokdad, A.H., Ford, E.S., Bowman, B.A., Dietz, W.H., Vinicor, F., Bales, V.S., Marks, J.S., 2003. Prevalence of obesity, diabetes, and obesity-related health risk factors, 2001. Journal of the American Medical Association 289 (1), 76–79.

Motomura, K., Fujiwara, Y., Kiyota, N., Tsurushima, K., Takeya, M., Nohara, T., Nagai, R., Ikeda, T., 2009. Astragalosides isolated from the root of *Astragalus radix* inhibit the formation of advanced glycation end products. Journal of Agricultural and Food Chemistry 57 (17), 7666–7672.

Mukherjee, K., Ghosh, N.C., Datta, T., 1972. *Coccinia indica* Linn. as potential hypoglycaemic agent. Indian Journal of Experimental Biology 10 (5), 347–349.

Muruganandan, S., Srinivasan, K., Gupta, S., Gupta, P.K., Lal, J., 2005. Effect of mangiferin on hyperglycemia and atherogenicity in streptozotocin diabetic rats. Journal of Ethnopharmacology 97 (3), 492–501.

Nakashima, N., Kimura, I., Kimura, M., 1993. Isolation of pseudoprototimosaponin AIII from rhizomes of *Anemarrhena asphodeloides* and its hypoglycemic activity in streptozotocin-induced diabetic mice. Journal of Natural Products 56 (3), 345–350.

Narender, T., Khaliq, T., Madhur, G., 2011. Naturally occurring antihyperglycemic and antidyslipidemic agents. In: Tiwari, V.K. (Ed.), Opportunity, Challenge and Scope of Natural Products in Medicinal Chemistry, pp. 155–185.

Nauck, M.A., Del Prato, S., Meier, J.J., Durán-García, S., Rohwedder, K., Elze, M., Parikh, S.J., 2011. Dapagliflozin versus glipizide as add-on therapy in patients with type 2 diabetes who have inadequate glycemic control with metformin a randomized, 52-week, double-blind, active-controlled noninferiority trial. Diabetes Care 34, 2015–2022.

Ng, T.B., Wong, C.M., Li, W.W., Yeung, H.W., 1986. Insulin-like molecules in *Momordica charantia* seeds. Journal of Ethnopharmacology 15, 107–117.

Nogueira, C.M., Parmanhan, B.R., Farias, P.P., Corrêa, A.G., 2009. A importância crescente dos carboidratos em química medicinal. Revista Virtual de Quimica 1, 149–159.

Nomura, S., Sakamaki, S., Hongu, M., Kawanshi, E., Koga, Y., Sakamoto, T., Yamamoto, Y., Ueta, K., Kimata, H., Nakayama, K., Tsuda-Tsukimoto, M., 2010. Discovery of Canagliflozin, a novel c-glucoside with thiophene ring, as sodium-dependent glucose cotransporter 2 inhibitor for the treatment of type 2 diabetes mellitus. Journal of Medicinal Chemistry 53 (17), 6355–6360.

Nomura, S., Yamamoto, Y., Matsumura, Y., Ohba, K., Sakamaki, S., Kimata, H., Nakayama, K., Kuriyama, C., Matsushita, Y., Ueta, K., Tsuda-Tsukimoto, M., 2014. Novel indole-*N*-glucoside, TA-1887 as a sodium glucose cotransporter 2 inhibitor for treatment of type 2 diabetes. ACS Medicinal Chemistry Letters 5 (1), 51–55.

Norberg, A., Hoa, N.K., Liepinsh, E., Phan, D.V., Thuan, N.D., Joernvall, H., Sillard, R., 2004. A novel insulin-releasing substance, phanoside, from the plant *Gynostemma pentaphyllum*. Journal of Biological Chemistry 279 (40), 41361–41370.

Odetti, P.R., Borgoglio, A., De Pascale, A., Rolandi, R., Adezati, L., 1990. Prevention of diabetes-increased aging effect on rat collagen-linked fluorescence by aminoguanidine and rutin. Diabetes 39, 796–801.

Ohsumi, K., Matsueda, H., Hatanaka, T., Hirama, R., Umemura, T., Onuki, A., Ishida, N., Kageyama, Y., Maezono, K., Kondo, N., 2003. Pyrazole-*O*-glucosides as novel Na^+-glucose cotransporter (SGLT) inhibitors. Bioorganic and Medicinal Chemistry Letters 13 (14), 2269–2272.

Ohtake, Y., Sato, T., Kobayashi, T., Nishimoto, M., Taka, N., Takano, K., Yamamoto, K., Ohmori, M., Yamaguchi, M., Takami, K., Yeu, S.Y., et al., 2012. Discovery of tofogliflozin, a novel *c*-arylglucoside with an *o*-spiroketal ring system, as a highly selective sodium glucose cotransporter 2 (SGLT2) inhibitor for the treatment of type 2 diabetes. Journal of Medicinal Chemistry 55 (17), 7828–7840.

Ohtake, Y., Sato, T., Matsuoka, H., Nishimoto, M., Taka, N., Takano, K., Yamamoto, K., Ohmori, M., Higuchi, T., Murakata, M., Kobayashi, T., 2011. 5a-Carba-β-d-glucopyranose derivatives as novel sodium-dependent glucose cotransporter 2 (SGLT2) inhibitors for the treatment of type 2 diabetes. Bioorganic and Medicinal Chemistry 19 (18), 5334–5341.

Oka, M., Kato, N., 2001. Aldose reductase inhibitors. Journal of Enzyme Inhibition 16 (6), 465–473.

Ranilla, L.G., Kwon, Y.I., Genovese, M.I., Lajolo, F.M., Shetty, K., 2008. Antidiabetes and antihypertension potential of commonly consumed carbohydrate sweeteners using *in vitro* models. Journal of Medicinal Food 11 (2), 337–348.

Rao, A.V., Gurfinkel, D.M., 2000. The bioactivity of saponins: triterpenoid and steroidal glycosides. Drug Metabolism and Drug Interactions 17 (1–4), 211–236.

Saifi, A.Q., Shinde, S., Kavishwar, W.K., Gupta, S.R., 1971. Some aspects of phytochemistry and hypoglycemic actions of *Peterocarpus marsupium* (Papilionaceae). Journal of Research in Indian Medicine 6 (2), 205–207.

Santer, R., Kinner, M., Lassen, C.L., Schneppenheim, R., Eggert, P., Bald, M., Brodehl, J., Daschner, M., Ehrich, J.H., Kemper, M., Volti, S.L., Neuhaus, T., Skovby, F., Swift, P.G., Schaub, J., Klaerke, D., 2003. Molecular analysis of the SGLT2 gene in patients with renal glucosuria. Journal of the American Society of Nephrology 14 (11), 2873–2882.

Shaw, J.E., Sicree, R.A., Zimmet, P.Z., 2010. Global estimates of the prevalence of diabetes for 2010 and 2030. Diabetes Research and Clinical Practice 87 (1), 4–14.

Sheridan, C., 2012. SGLT2 inhibitors race to enter type-2 diabetes market. Nature Biotechnology 30 (10), 899–900.

Singh, N., Pandey, S.K., Tripathi, R.P., 2010. Regioselective [3+2] cycloaddition of chalcones with a sugar azide: easy access to 1-(5-deoxy-D-xylofuranos-5-yl)-4,5-disubstituted-1H-1,2,3-triazoles. Carbohydrate Research 345 (12), 1641–1648.

Song, K.S., Lee, S.H., Kim, M.J., Seo, H.J., Lee, J., Lee, S.H., Jung, M.E., Son, E.J., Lee, M., Kim, J., Lee, J., 2011. Synthesis and SAR of thiazolylmethylphenyl Glucoside as novel *C*-aryl glucoside SGLT2 inhibitors. ACS Medicinal Chemistry Letters 2 (2), 182–187.

Srinivasan, K., Kaul, C.L., Ramarao, P., 2005. Partial protective effect of rutin on multiple low dose streptozotocin-induced diabetes in mice. Indian Journal of Pharmacology 37 (5), 327–328.

Stone, P.H., Muller, J.E., Hartwell, T., York, B.J., Rutherford, J.D., Parker, C.B., Turi, Z.G., Strauss, H.W., Willerson, J.T., Robertson, T., Braunwald, E., 1989. The effect of diabetes mellitus on prognosis and serial left ventricular function after acute myocardial infarction: contribution of both coronary disease and diastolic left ventricular dysfunction to the adverse prognosis. Journal of the American College of Cardiology 14 (1), 49–57.

Strojek, K., Yoon, K.H., Hruba, V., Elze, M., Langkilde, A.M., Parikh, S., 2011. Effect of dapagliflozin in patients with type 2 diabetes who have inadequate glycaemic control with glimepiride: a randomized, 24-week, double-blind, placebo-controlled trial. Diabetes, Obesity and Metabolism 13 (10), 928–938.

Sundaram, A., Moses, C.R.A., Ilango, S., Seshiah, V., 1998. Newer antidiabetic drugs. International Journal of Diabetes in Developing Countries 18, 24–30.

Tadera, K., Minami, Y., Takamatsu, K., Matsuoka, T., 2006. Inhibition of alpha-glucosidase and alpha-amylase by flavonoids. Journal of Nutritional Science and Vitaminology 52 (2), 149–153.

Tarleton, R.L., 2005. New approaches in vaccine development for parasitic infections. Cellular Microbiology 7 (10), 1379–1386.

Taskinen, M.R., Rosenstock, J., Tamminen, I., Kubiak, R., Patel, S., Dugi, K.A., Woerle, H.J., 2011. Safety and efficacy of linagliptin as add-on therapy to metformin in patients with type 2 diabetes: a randomized, double-blind, placebo-controlled study. Diabetes, Obesity and Metabolism 13 (1), 65–74.

Tewari, N., Tiwari, V.K., Mishra, R.C., Tripathi, R.P., Srivastava, A.K., Ahmad, R., Srivastava, R., Srivastava, B.S., 2003. Synthesis and bioevaluation of glycosyl ureas as -glucosidase inhibitors and their effect on mycobacterium. Bioorganic and Medicinal Chemistry 11 (13), 2911–2922.

Tiwari, V.K., Mishra, B.B., Mishra, K.B., Mishra, N., Singh, A.S., Chen, X., 2016. Cu-catalyzed click reaction in carbohydrate chemistry. Chemical Reviews 116 (5), 3086–3240.

Tiwari, V.K., Mishra, R.C., Sharma, A., Tripathi, R.P., 2012. Carbohydrate based potential chemotherapeutic agents: recent developments and their scope in future drug discovery. Mini Reviews in Medicinal Chemistry 12, 1497–1519.

Trapero, A., Llebaria, A., 2012. A prospect for pyrrolidine iminosugars as antidiabetic α-glucosidase inhibitors. Journal of Medicinal Chemistry 55 (23), 10345–10346.

Waisundara, V.Y., Hsu, A., Huang, D.J., Tan, B.K.H., 2008. *Scutellaria baicalensis* enhances the anti-diabetic activity of metformin in streptozotocin-induced diabetic Wistar rats. The American Journal of Chinese Medicine 36 (3), 517–540.

Wan, L.S., Min, Q.X., Wang, Y.L., Yue, Y.D., Chen, J.C., 2013. Xanthone glycoside constituents of *Swertia kouitchensis* with α-glucosidase inhibitory activity. Journal of Natural Products 76 (7), 1248–1253.

Wang, H., Shi, S., Bao, B., Li, X., Wang, S., 2015. Structure characterization of an arabinogalactan from green tea and its anti-diabetic effect. Carbohydrate Polymers 124, 98–108.

Washburn, W.N., 2009a. Evolution of sodium glucose co-transporter 2 inhibitors as anti-diabetic agents. Expert Opinion on Therapeutic Patents 19, 1485–1499.

Washburn, W.N., 2009b. Development of the renal glucose reabsorption inhibitors: a new mechanism for the pharmacotherapy of diabetes mellitus type 2. Journal of Medicinal Chemistry 52 (7), 1785–1794.

Xu, G., Lv, B., Roberge, J.Y., Xu, B., Du, J., Dong, J., Chen, Y., Peng, K., Zhang, L., Tang, X., Feng, Y., Xu, M., Fu, W., Zhang, W., Zhu, L., Deng, Z., Sheng, Z., Welihinda, A., Sun, X., 2014. Design, synthesis, and biological evaluation of deuterated c-aryl glycoside as a potent and long-acting renal sodium-dependent glucose cotransporter 2 inhibitor for the treatment of type 2 diabetes. Journal of Medicinal Chemistry 57 (4), 1236–1251.

Yoshida, K., Hishida, A., Iida, O., Hosokawa, K., Kawabata, J., 2008. Flavonol caffeoylglycosides as α-glucosidase inhibitors from *Spiraea cantoniensis* flower. Journal of Agricultural and Food Chemistry 56 (12), 4367–4371.

Yoshikawa, M., Matsuda, H., Harada, E., Murakami, T., Wariishi, N., Yamahara, J., Murakami, N., 1994. Elatoside E, a new hypoglycemic principle from the root cortex of *Aralia elata* Seem.: structure-related hypoglycemic activity of oleanolic acid glycosides. Chemical and Pharmaceutical Bulletin 42, 1354–1356.

Yoshikawa, M., Murakami, T., Kadoya, M., Matsuda, H., Yamahara, J., Muraoka, O., Murakami, N., 1995a. Betavulgarosides I, II, III, IV, and V, hypoglycemic glucuronide saponins from the roots and leaves of *Beta vulgaris* L. (Sugar Beet). Heterocycles 41 (8), 1621–1626.

Yoshikawa, M., Yoshizumi, S., Ueno, T., Matsuda, H., Murakami, T., Yamahara, J., Murakami, N., 1995b. Medicinal foodstuffs. I. Hypoglycaemic constituents from a garnish foodstuff "taranome," the young shoot of Aralia elata Seem.: elatosides G, H, I, J, and K. Chemical and Pharmaceutical Bulletin 43, 1878–1882.

Zhang, W., Welihinda, A., Mechanic, J., Ding, H., Zhu, L., Lu, Y., Deng, Z., Sheng, Z., Lv, B., Chen, Y., Roberge, J.Y., Seed, B., Wang, Y.X., 2011. EGT1442, a potent and selective SGLT2 inhibitor, attenuates blood glucose and HbA(1c) levels in db/db mice and prolongs the survival of stroke-prone rats. Pharmacological Research 63, 284–293.

Zhang, X., Urbanski, M., Patel, M., Zeck, R.E., Cox, G.G., Bian, H., Conway, B.R., Beavers, M.P., Rybczynski, P.J., Demarest, K.T., 2005. Heteroaryl-O-glucosides as novel sodium glucose co-transporter 2 inhibitors. Part 1. Bioorganic and Medicinal Chemistry Letters 15 (23), 5202–5206.

Zhang, X., Urbanski, M., Patel, M., Cox, G.G., Zeck, R.E., Bian, H., Conway, B.R., Beavers, M.P., Rybczynski, P.J., Demarest, K.T., 2006. Indole-glucosides as novel sodium glucose co-transporter 2 (SGLT2) inhibitors. Part 2. Bioorganic and Medicinal Chemistry Letters 16 (6), 1696–1701.

Zhongtao, W., Guo, W., Gaoyan, L., Biao, Y., 2010. Synthesis of mangiferin, isomangiferin, and homomangiferin. Journal of Organic Chemistry 75 (16), 5725–5728.

Zhou, H., Danger, D.P., Dock, S.T., Hawley, L., Roller, S.G., Smith, C.D., Handlon, A.L., 2010. Synthesis and SAR of benzisothiazole- and indolizine-β-d-glucopyranoside inhibitors of SGLT 2. ACS Medicinal Chemistry Letters 1 (1), 19–23.

Chapter 6

Recent Developments on the Antidiabetic Sesquiterpene Lactones and Their Semisynthetic Analogues

D. Chaturvedi, P.K. Dwivedi
Amity University Uttar Pradesh, Lucknow, Uttar Pradesh, India

1. INTRODUCTION

Diabetes mellitus (DM) is a group of etiologically and clinically heterogeneous disorders resulting in metabolic defects and serious pathological consequences that eventually leads to blindness, kidney failure, gangrene, neuropathy, heart attacks, strokes, and death (Rios et al., 2015, p. 975). As per WHO report, diabetes is a multifarious group of disorders that disturbs the metabolism of carbohydrates, fats, and proteins and results in a shortage or lack of insulin secretion (type I diabetes, i.e., insulin-dependent DM) and/or reduced sensitivity of the tissue to insulin (type II diabetes, i.e., non-insulin-dependent DM). It is estimated that DM affects more than 422 million people worldwide (~8.5% of the world's population); this number would be expected to reach up to 600 million by 2030. The global prevalence of diabetes among adults over 18 years of age has risen from 4.7% in 1980 to 8.5% in 2014. Because of great expense in modern therapy, there is a clear need for scientific investigation of traditional plant-derived products for treating this disease, which may provide valuable natural lead molecules for the development of new drugs along with their therapeutic strategies.

According to Zhang et al. (2005, p. 239), Robles et al. (1995, p. 199), and their coworkers, sesquiterpene lactones (SLs) constitute a large and diverse group of biologically active plant chemicals that have been identified in several plant families such as Acanthaceae, Anacardiaceae, Apiaceae, Euphorbiaceae, Lauraceae, Magnoliaceae, Menispermaceae, Rutaceae, Winteraceae, Hepatideae, etc. Ferreira et al. (2005, p. 633) argue that these compounds are used as chemical markers in different chemosystematic studies within Asteraceae and are also characteristic of this family. According to Ohhini et al. (1997, p. 111),

Discovery and Development of Antidiabetic Agents From Natural Products.
http://dx.doi.org/10.1016/B978-0-12-809450-1.00006-5

the greatest numbers have been reported in the Compositae (Asteraceae) family with more than 3000 reported different structures. SLs are a group of secondary metabolites that represent a diverse and unique class of natural products and are important constituent of essential oils, which are formed from head-to-tail condensation of three isoprene units and subsequent cyclization and oxidative transformation to produce a *cis*- or *trans*-fused lactone (Fig. 6.1). These secondary compounds are primarily classified on the basis of their carbocyclic skeletons into pseudoguainolides, guaianolides, germacranolides, eudesmanolides, heliangolides, hyptocretenolides, etc. An important usual feature of the SLs is the presence of a γ-lactone ring (closed toward either C-6 or C-8) containing in many cases, an α-methylene group, which is also most responsible feature for biological effects and sometimes essential for the cytotoxic activity. Among other modifications, the incorporation of hydroxyls or esterified hydroxyls and epoxide ring is common.

An individual plant species generally produces one skeletal type of SLs concentrated primarily in the leaves and flower heads. The percentage of SLs per dry weight may vary from 0.01% to 8%. Losses of livestock intoxicated by

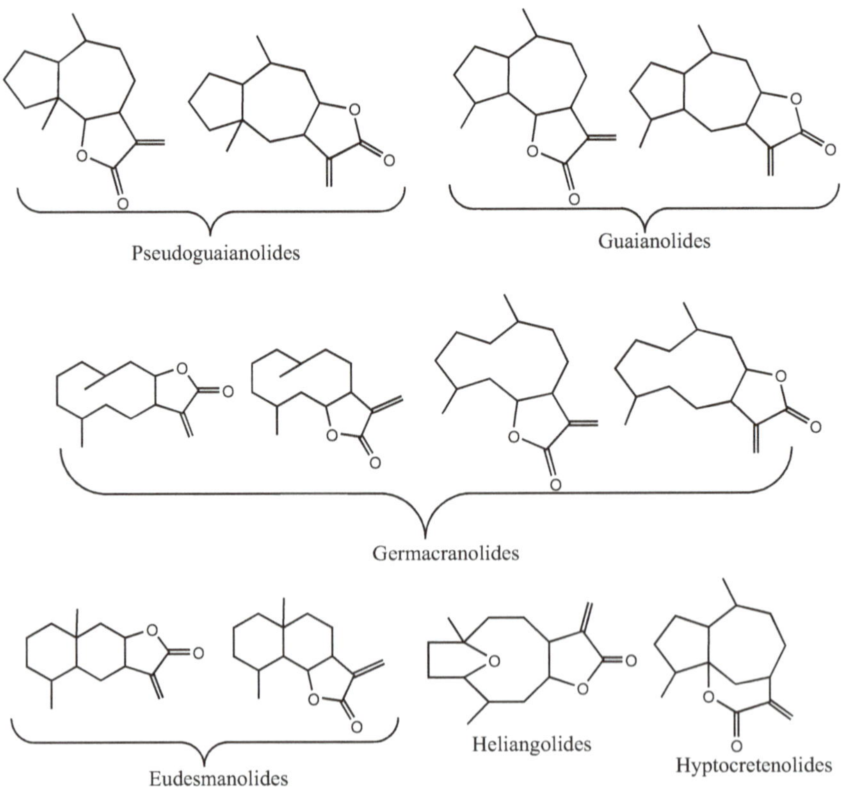

FIGURE 6.1 Basic skeletons of sesquiterpene lactones.

plants containing SLs are well known. In fact, they have been shown to exhibit a wide range of biological activities. Various biological activities exhibited by SLs include anticancer, antiinflammatory, antimalarial, antibacterial, antiviral, antidiabetic, antifungal, antidepressant, antileishmanial, antiulcer, etc. Some SLs exhibiting these activities are given in Fig. 6.2.

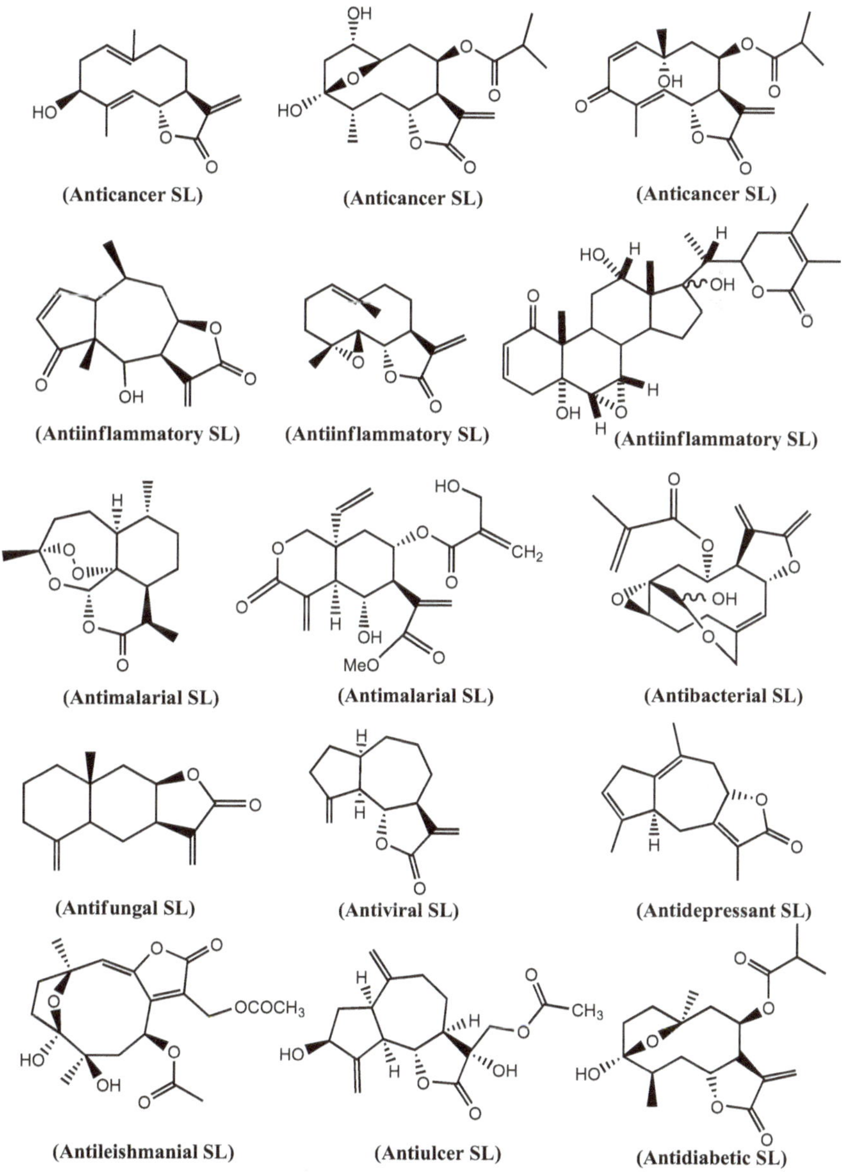

FIGURE 6.2 Some biologically potent sesquiterpene lactones (SLs).

The present chapter deals with some of the potent SL molecules from various plants having antidiabetic potential.

2. PLANT-DERIVED SESQUITERPENE LACTONES WITH ANTIDIABETIC POTENTIAL

2.1 Tirotundin and Analogues

Peroxisome proliferator-activated receptors (PPARs) behave as ligand-activated transcription factors and are members of nuclear receptor super family. In human genome, there are three types of PPAR isoforms designated as α, β/δ, and γ and encoded by different genes. Experiments on PPARs have suggested their important roles in the regulation of lipid metabolism and glucose homeostasis. In addition to this, they have shown their effects in anticancer and neuroprotection action and thus their agonist are said to play multipharmacologic functions. Because of their antidiabetic potential, PPARγ agonists such as thiazolidinediones (for example, pioglitazone, rosiglitazone, etc.) are used as the antidiabetics, thus, improving insulin resistance (IR) in hyperglycemic patients. In addition to their therapeutic effects, drug like rosiglitazone, a PPARγ synthetic agonist, also has some side effects if used for long period. It is interesting to note that pioglitazone which also belongs to same isoform, i.e., PPARγ agonist is safer than rosiglitazone but with less PPARγ agonistic property. The pharmaceutical industries started to develop PPARα/γ dual agonists to cure the diabetes because of the unwanted cardiovascular side effect exhibited by rosiglitazone. In this regard, a novel dual PPAR agonist, named as aleglitazar, has shown to lower the hyperglycemic conditions. In this way, it is thought that the natural products having PPARα/γ agonist property might be the important entity for the treatment of diabetes.

While working on *Tithonia diversifolia*, Lin (2012) isolated the tirotundin (**1**, Fig. 6.3), a SL, from the ethyl acetate (EtOAc) soluble fraction of this plant. The transient transfection reporter assay in $HepG_2$ cells was used to assess the agonistic activity of this SL against PPARs with a motto that herbal PPARα/γ agonist might play an important role in the treatment of diabetes. In this study,

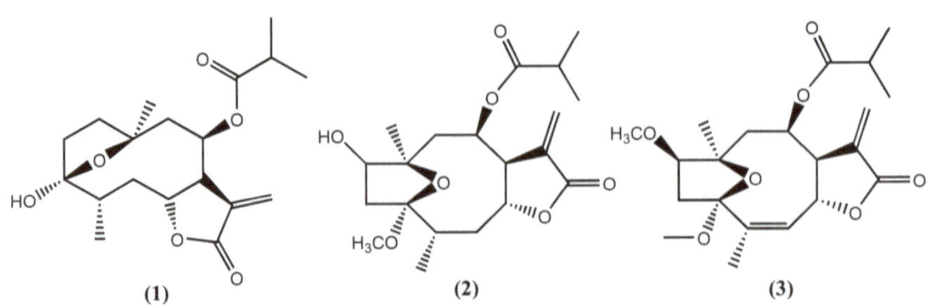

FIGURE 6.3 Structures of tirotundin and its analogues.

antidiabetic effect through PPARγ pathway by this SL was determined (Lin, 2012, p. 2954). The interaction of tirotundin (**1**) with PPARγ ligand-binding domain (LBD) was studied by computer-based modeling program to generate docking models. In this regard, Nolte et al. (1998) found that 13α-helices and four small-stranded β-strands of PPARγ LBD form a ligand-binding pocket with volume about 1300 Å. The ligandfit docking program and the crystal structure of PPARγ LBD complexed with aleglitazar were utilized to determine the binding modes of tirotundin (Nolte et al., 1998, p. 137).

From the percolated dried aerial parts of *T. diversifolia*, 1β-hydroxytirotundin-3-*O*-methyl ether (**2**, Fig. 6.3) and 1β-hydroxytirotundin-1,3-*O*-dimethyl ether (**3**, Fig. 6.3) were also isolated by Zhao et al. (2012). The antihyperglycemic activities of these compounds were tested by glucose uptake in 3T3-L1 adipocytes. For this, 96-well-plated differentiated 3T3-L1 adipocytes were preincubated with Dulbecco's Modified Eagle Medium (DMEM) containing 0.2% bovine serum albumin (BSA). The incubation of these compounds was done with various concentrations for 12 h and the amounts of glucose uptake were calculated. To monitor the cell proliferation and to adjust the glucose uptake values, MTT assay was also performed. Pioglitazone, a member of the thiazolidinedione class was used as a control compound in this study, which exerts a mechanism of action that involves binding to nuclear PPAR-γ. It was concluded that the effect of compound **2** had similar PPARγ agonist activity on glucose uptake, as pioglitazone (Zhao et al., 2012, p. 1590).

2.2 Tagitinins

Tithonia diversifolia has been used traditionally since long time in the treatment of malaria and other forms of fever or wounds in Mexico. Because of its potential biological effects, it has attracted the researchers having interest in health care phytomedical research. The plant has been shown to exhibit various biological activities including antimalarial, antidiabetic, antiinflammatory, and anticancer activity. Phytochemical investigations of *T. diversifolia* revealed the presence of SLs, chromene, and flavone as chemical constituents. Among the sesquiterpene constituents, some notable isolates are tagitinin A, C, F, G, H, and I; tirotundin 3-*O*-methyl ether; deacetylvguiestin; 1β-hydroxydiversifolin-3-*O*-methyl ether; 1β-hydroxytirotundin-3-*O*-methyl ether; 1β-hydroxytirotundin-1,3-*O*-dimethyl ether; tagitinin F-3-*O*-methyl ether; 3β-acetoxy-4α-hydroxyeduesm-11(13)-en-12-oic acid, and ilicic acid. The antihyperglycemic potential of these compounds were evaluated by glucose uptake in $3T_3$-L_1 adipocytes. Lin (2012) isolated tagitinin A (**4**, Fig. 6.4) and tagitinin C (**5**, Fig. 6.4) from aforementioned plant, which is traditionally used for the treatment of diabetes and hepatitis in Taiwan. Their structures were also identified by comparing NMR spectral data. This Mexican plant mainly includes germacrane and eudesmane types of structurally diverse chemical derivatives. By small heterodimer partner promoter-driven luciferase activity and PPARγ response

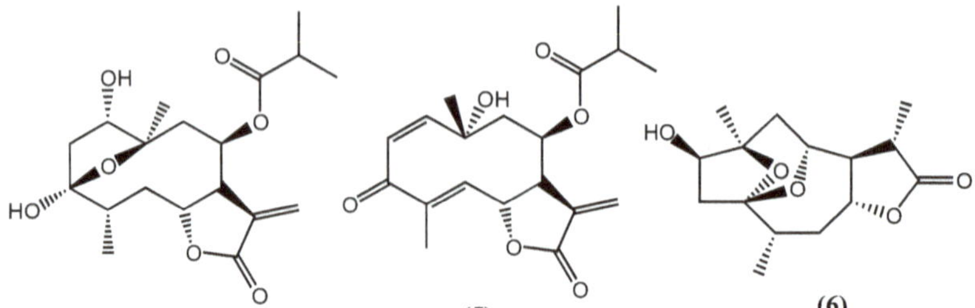

(4) (5) (6)

FIGURE 6.4 Structures of tagitinin A, C, and G.

element promoter-driven luciferase activity, the PPARγ agonistic effect was determined. To evaluate the PPARα agonistic activity of tagitinin A, the reporter plasmid-containing human SULT2A1 gene promoter was used in the transient transfection reporter assay. It was found that tagitinin A has transactivation activity as clofibrate by approximately 2.3-fold of vehicle activity at 10 μM and still exhibited 1.4-fold of relative luciferase activity at 100 nM. While working on docking studies, Nolte et al. (1998) argue that tagitinin A resided in PPARγ ligand binding pocket by leaning their lactone moiety toward to helice 3 and locating their 3-hydroxyl group nearby helice 11. Hydrophobic contact and significant hydrogen bond interaction with PPARγ ligand binding pocket have also been shown by tagitinin A. Thus, it was also confirmed that these compounds not only had similar docking score but also occupied in PPARγ ligand binding pocket in the same way as in the case of tirotundin (Nolte et al., 1998, p. 137; Lin, 2012, p. 2954).

Zhao et al. (2012) obtained the crystals of tagitinin G (**6**, Fig. 6.4) by slow evaporation from EtOAc extract of aforementined plant. The compound was evaluated for its antihyperglycemic activity. The 10 μg/mL of this compound was found to increase the glucose uptake of 3T3-L1 adipocytes by 1.2-($p < .05$) times as compared with the basal level. It was also concluded that the effect on glucose uptake was not due to a nonspecific cytotoxicity, although there may be some cellular toxicity that did not result in cell death during these experiments (Zhao et al., 2012, p. 1590).

2.3 Costunolide

Costunolide (**7**, Fig. 6.5), a plant-derived active principle, is known to possess a wide range of biological activities, which attracted its possible use in the treatment of type 2 diabetes. According to Chen et al. (1995), costunolide (**7**) is an active component from the crude extracts of a traditional Chinese medicinal herb, *Saussurea lappa* Clarks (Chen et al., 1995, p. 99). In addition to various biological actions, it also possesses important immunological actions. For example, costunolide also exhibited preventive effects on carcinogenesis like some other sesquiterpene compounds, which induced apoptosis in

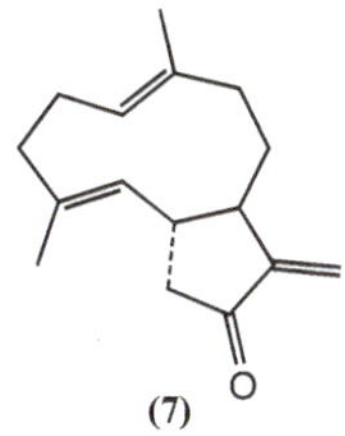

(7)

FIGURE 6.5 Structure of costunolide.

cancer cells. Besides these effects, costunolide isolated from the leaves of *Laurus nobilis* has shown the potential inhibition of blood ethanol increase in ethanol-loaded rats. The hypoglycemic effect of *Costus speciosus* root crude extracts has already been reported. In addition to anticancer potential as reported by Park et al. (2001, p. 342) and Choi et al. (2002, p. 480), Kassuyaa et al. (2009, p. 369) have also claimed its antiinflammatory property. Recently, considerable efforts have been devoted to understand its potential against the diabetes. *Costus speciosus* is widely known for its therapeutic potential including treatment against diabetes in Indian medicine. In this sequence, Eliza et al. (2009a,b) isolated costunolide from the hexane extract of roots of this plant collected from southern region of India. To demonstrate the antidiabetic potential, they used the male Wistar strain rats weighing about 190–200 gm. All the animals were maintained under laboratory conditions and DM was induced. On 0, 7, 15, and 30 days of treatment with costunolide, plasma glucose level was recorded in normal and experimental rats. The workers claimed that there occurred a significant reduction in the plasma glucose level at the dose of 20 mg/kg b.w. (body weight) on oral administration of costunolide. In addition to this, significant increased body weight (22.5%), plasma insulin (41.6%) and decreased HbA1c (56.8%) was observed on oral administration of costunolide as compared to untreated diabetic rats. It was realized that the normoglycemic action of costunolide was due to potentiation of insulin release from the existing beta cells of islets of Langerhans (Eliza et al., 2009a,b, p. 329). Li et al. (2004, p. 1) argue that the increased sensitivity of insulin to uptake glucose by costunolide might be due to stimulation of the beta islets to secrete insulin.

2.4 Eremanthin

Costus speciosus, also known as spiral or wild ginger, belongs to the family Costaceae. The use of *C. speciosus* rhizomes in many human and veterinary medicines has already been documented. The anticholinesterase activity of this plant alkaloids has been shown by both in vitro and in vivo experiments. The hypoglycemic effects of an aqueous extract of this plant showed its important role lowering the glucose level when administered orally. The crude extract of the roots of the aforementioned plant is said to exert hypoglycemic effect. In search of plant-derived active natural products for the treatment of diabetes,

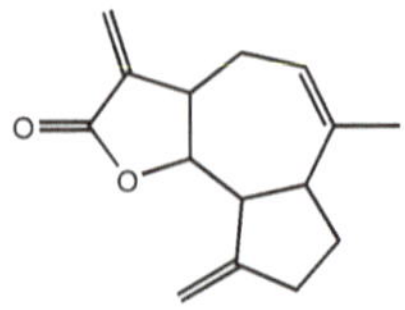

(8)

FIGURE 6.6 Structure of eremanthin.

eremanthin (**8**), a SL has been extracted from *Pterodon pubescens* and *Eremanthus eleaegnus*. It is also found to be present in *n*-hexane leaves extract of *L. nobilis*. Eremanthin (**8**) possess immunomodulating property by inhibiting the killing function of cytotoxic T lymphocytes (CTL) and the induction of intercellular adhesion molecule-1 (ICAM-1). To investigate its hypoglycemic potential, the compound (**8**, Fig. 6.6) was isolated from Indian medicinal plant, *C. speciosus*. The presence of eremanthin has also been reported from the extract of *L. nobilis, P. pubescens* and *Eremanthus eleaegnus*. Eliza et al. (2009a,b) evaluated the hypoglycemic activity on plasma glucose levels in normal and diabetic rats, by administering it to streptozotocin (STZ)-induced diabetic male Wistar rats at different doses for a period of 60 days. They found the significant ($p < .05$) reduction in plasma glucose level in a dose-dependent manner as compared to the control. It was also reported that the oral administration of eremanthin (at 20 mg/kg b.w.) markedly increased the plasma insulin. It was realized that the primary action of eremanthin is via stimulation of insulin release (Eliza et al., 2009a,b, p. 67). Tian et al. (1998, p. 1722) thought that the action of eremanthin seems to be similar to that of glibenclamide. This study indicated that eremanthin possessed hypoglycemic activities and thus could be used as a drug for treatment of diabetes.

2.5 Enhydrin

Aybar et al. (2001) found that there was a significant reduction in plasma glucose levels in normal rats by a 10% decoction of yacon (*Smallanthus sonchifolius*), an herbaceous South American plant belonging to the family Asteraceae. It was also documented that a significant hypoglycemic effect in STZ-induced diabetic rats can be produced by its daily administration of 2% infusion for 30 days (Aybar et al., 2001, p. 125).

Inoue et al. (1995, p. 845) and Lin et al. (2003, p. 2154) worked on the phytochemical investigations of yacon leaves, where they confirmed the presence of a biologically potent SL, enhydrin (**9**, Fig. 6.7). During the last few years, the antidiabetic potential of enhydrin, the major lactone component isolated from yacon leaves has attracted the researchers around the globe. Although the inclusion of enhydrin by Kawashima et al. (2001) in a patented antidiabetic pharmaceutical formulation has been done, but there was no experimental data to prove its antidiabetic activity (Schorr and Da Costa, 2005, p. 161; Kawashima

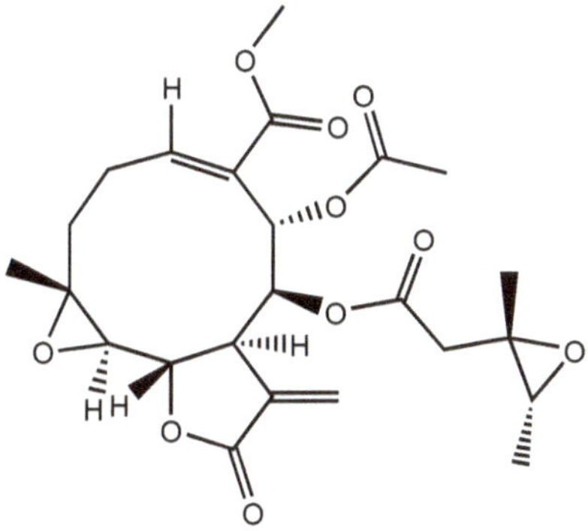

(9)

FIGURE 6.7 Structure of enhydrin.

et al., 2001; Patent). Later on, glucose lowering effect of enhydrin obtained from *S. sonchifolius* leaves, on STZ-induced diabetic rats was investigated. The experiments were carried out to examine the long-term antidiabetic potential of methanol extract, chloroform extract, *n*-butanol extract, and ethyl acetate extract of yacon leaves and pure enhydrin in STZ-induced diabetic rats. It was found that methanol extract at a dose of 50 mg/kg b.w./day was able to decrease blood glucose levels significantly, and this effect was observed from the starting of the treatment, the highest activity being detected at day 10. With chloroform extract or *n*-butanol extract at 20 and 10 mg/kg b.w./day, respectively, the same results were obtained. It was also found at daily administration of pure enhydrin (0.8 mg/kg b.w.) or ethanol extract (20 mg/kg b.w.) for 8 weeks resulted in a reduction in fasting blood glucose in diabetic animals. However, the reduction in blood glucose level was slow and significantly lower than that of methanol, chloroform, and *n*-butanol extract. The data obtained in this study reveal that this SL decreases blood glucose levels in both normal and diabetic rats. Interestingly, it was found to be effective in the prevention of the hyperglycemic peak after food ingestion. During a period of 60 days study, this compound was noticed to control the blood glucose levels in diabetic animals associated with a late increase in plasma insulin level. It was suggested that by a decrease in the intestinal index of glucose absorption and/or by stimulation of insulin release only at high blood glucose concentrations, enhydrin would exert its biological action. Thus the in vivo activity of enhydrin as demonstrated by Genta et al. (2010, p. 143) proved its effectiveness in reduction of postprandial glucose levels and usefulness in the treatment of diabetes without risk of hypoglycemia.

2.6 Polymatin A

Biologically active natural products produced by plants are stored in various organs including leaves, stems, and roots. It is found that the glandular trichomes on the surface of leaf are specialized secretory structures that might be the sites of production or accumulation of various bioactive secondary metabolites. These natural compounds may have various functions, such as plant defense, and could be of interest as pharmaceutical products for the treatment of many pathological conditions. From the genus *Smallanthus,* various secondary metabolites belonging to category of melampolide-type SLs are produced in glandular trichomes of the foliar surface. Gum from *Smallanthus siegesbeckia* and *Smallanthus maculatus* var. *maculatus* is an important source of polymatin A, a type of SL. Although, enhydrin is more abundant SL but there are 19 melampolide-type SLs, which have been isolated and identified in *S. sonchifolius* leaves. With regard to the chemical constituents of *Smallanthus macroscyphus* leaves, within the mixture of SLs, polymatin A was the major one, which has been isolated and identified using 1H NMR spectrometry and MS analysis in this species.

Serra-Barcellona et al. (2014) isolated polymatin A (**10**, Fig. 6.8) from *S. macroscyphus* (wild yacon) leaves to analyze its hypoglycemic effects by performing the oral glucose tolerance test in vivo. The minimum effective dose of polymatin A was 14mg crystalline powder/kg b.w. in this experiment (Serra-Barcellona et al., 2014, p. 35). Earlier, Cabrera et al. (2007, p. 311) have also demonstrated the hypoglycemic effect by the aqueous extracts of dried wild yacon leaves on rats.

Later on, to study the hypoglycemic effects of polymatin A (**10**; Fig. 6.8), a single intraperitoneal injection of freshly prepared STZ solution in sodium citrate buffer was given to induce the diabetes while only citrate buffer was given to control rats. Induction of diabetes was determined by measuring the fasting blood glucose levels and glucosuria. In this experiment, an effective glycemic control in diabetic animals was achieved after daily administration of polymatin A for 4weeks. The reduced urinary glucose excretion and improved fasting blood glucose/insulin ratio were also claimed. In addition to this, no deaths or signs of toxicity was observed after oral administration of polymatin A within the tested doses of this study. The measurement of blood glucose was based on glucose dye oxidoreductase mediator reaction and the determination of urine glucose was done by a reagent-based glucose oxidase–peroxidase enzymatic method. Enzyme-linked immunosorbent

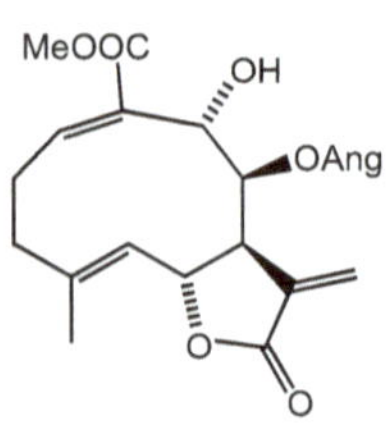

(10)

FIGURE 6.8 Structure of polymatin A.

assay was applied for the determination of plasma insulin. In this study, it was concluded that polymatin A has significant hypoglycemic and antidiabetic potential.

2.7 β-Caryophyllene

Various Indian and Chinese spices possess essential oils, which are traditionally used against various diseases. A SL present in these spices is β-caryophyllene (**11**, Fig. 6.9). Some important sources of biologically potent β-caryophyllene discussed by Ghalib et al. (2012, p. 601) and Bento et al. (2011, p. 1153) include cloves (*Syzygium aromaticum*), oregano (*Origanum vulgare*), thyme (*Thymus vulgaris*), black pepper (*Piper nigrum*), cinnamon (*Cinnamomum cassia*) and *Bupleurum bicaule*. Hidhayath Basha and Sankaranarayanan (2014, p. 9) argue that the antidiabetic potential of β-caryophyllene studied in STZ-induced diabetic rats exhibits a dose-dependent response in glycemic control. The experiment to assess the effect of β-caryophyllene on blood glucose and insulin was also carried out by Basha and Sankaranarayanan (2015), which concludes that the oral administration of β-caryophyllene reduced the blood glucose and increased plasma insulin to near normal at dose of 200 mg/kg b.w. in diabetic rats for 45 days. It was also found that there was no alteration in plasma glucose and insulin levels of normal rats when β-caryophyllene was administered (Basha and Sankaranarayanan, 2015, p. 9).

Prajna et al. (2013, p. 2435) hypothesized that since the metabolism of sialic acid is drastically changed in diabetic condition hence may lead to various complications such as retinopathy, nephropathy, and neuropathy, Therefore, Pari and Rajarajeswari (2010) became interested to carry out the study regarding this and it was observed that oral administration of β-caryophyllene to diabetic treated rats significantly restored the sialic acid levels in plasma, hepatic, and renal tissues to near normal. It was also observed that β-caryophyllene significantly reduces the elevated level of fucose, which is increased in case of diabetes (Pari and Rajarajeswari, 2010, p. 61).

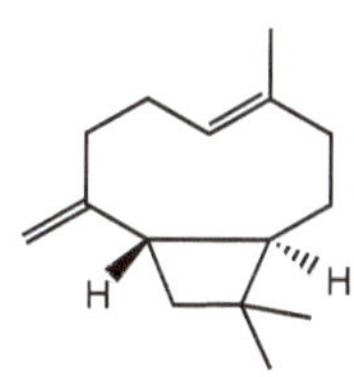

(11)

FIGURE 6.9 Structure of β-caryophyllene.

2.8 Lactucains

Lactuca indica Linn. is widely distributed in Asia and belongs to the family Compositae; the plant species is reported to contain terpenoids as one of the main chemical constituents. The isolation of α-amyrin, β-amyrin, lupeol, pseudotaraxasterol, taraxasterol, and germanicol has been reported from this herb. Recently, experiments have been carried out to assess biological activity of the methanol extract of *L. indica* and observed that the plant extract significantly lowers the serum levels of total cholesterol and low-density lipoprotein (LDL) cholesterol. From the active extract of this plant, three novel SLs and one new lignan glycoside have been isolated and characterized. After structural elucidation of these SL compounds using spectroscopic and chemical methods, they were identified as lactucain A, B, and C (Hou et al., 2003, p. 625).

Lactucain A, B, and C (**12**, **13**, and **14**, Fig. 6.10) belong to the guianolide category of SLs. These compounds were isolated from fresh herb of *L. indica* Linn. In preliminary investigations, it was found that the aqueous acetone extract of this fresh herb has significant antidiabetic potential. Hou et al. (2003) isolated the

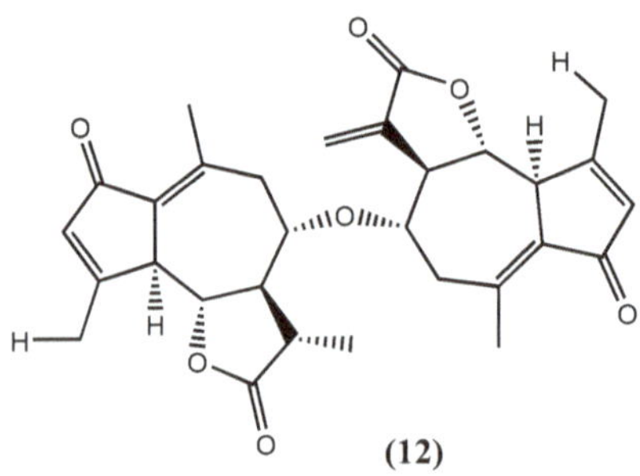
(12)

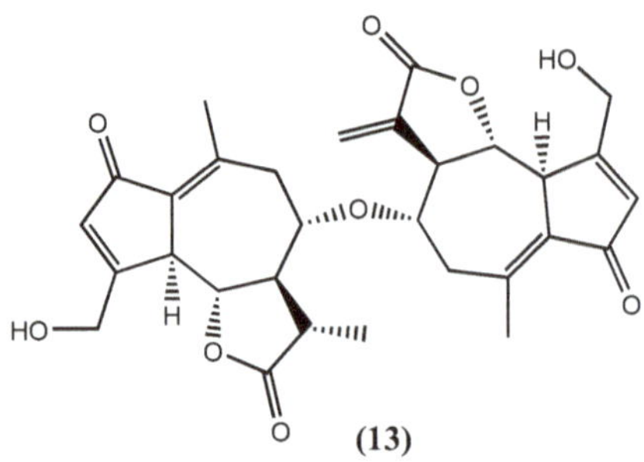
(13)

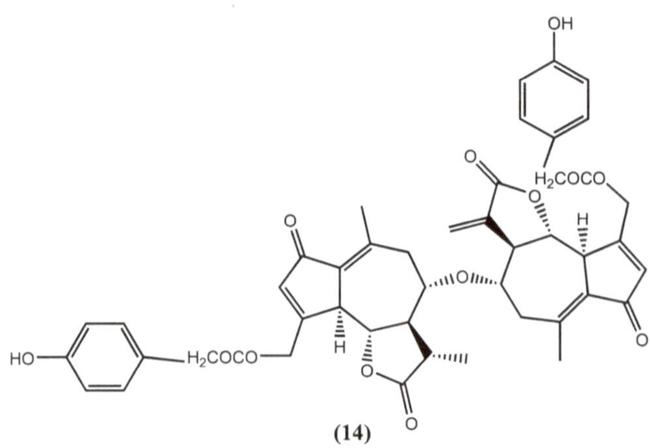
(14)

FIGURE 6.10 Structures of lactucain A, B, and C.

aforementioned three novel SLs by the chromatographic fractionation of the active extract. Lactucain C, which was obtained as an amorphous powder, had shown the promising antidiabetic activity among these compounds (Hou et al., 2003, p. 625).

2.9 Caleins

Brickellia cavanillesii (Asteraceae) has been used in Mexican folk medicine since long time. In the sequence for the search of antidiabetic agent, first, an aqueous extract of the aerial parts of this plant was tested using acute hypoglycemic as well as oral glucose and sucrose tolerance tests in two sets of animals namely normal and diabetic. In the experiment, type 2 DM was achieved by treatment of mice with STZ, 100 mg/kg 15 min after an injection of β-nicotinamide adenine dinucleotide (NAD, 40 mg/kg). This preliminary treatment with NAD creates a diabetic syndrome similar to type 2 DM by stimulating partial protection against the cytotoxic action of STZ by scavenging free radicals and only small damage to pancreatic β-cell mass. It was observed that in the acute preliminary experiments, the aqueous extract lowered ($p < .05$) the blood glucose levels in both normal and NAD–STZ diabetic mice at doses of 100 and 316 mg/kg, the effect similar to that have been produced by glibenclamide at the dose of 10 mg/kg.

Bioassay-guided fractionation of the active extract of *B. cavanillesii* led to the isolation of two SLs, namely calein B (**15**, Fig. 6.11) and calein C (**16**, Fig. 6.11). It was demonstrated that calein C inhibited the activity of yeast α-glucosidase with IC_{50} value 0.28 mM versus 1.7 mM for acarbose. Escandon-Rivera et al. (2012) postulated that the nature of the acyl residue at C-8 has an impact on the enzymatic activity that is why the best activity was found for calein C in comparison with calein B. In the analysis of enzyme inhibition properties of calein C using Lineweaver–Burk and Dixon plots, it was found that this SL shows a mixed type of inhibition against *Saccharomyces cerevisiae* α-glucosidase (Escandon-Rivera et al., 2012, p. 968). In addition to this, a molecular docking study using the program AutoDock4 as mentioned by Morris et al. (1998, p. 1639) was performed to investigate the binding model of calein C with α-glycosidase. It was

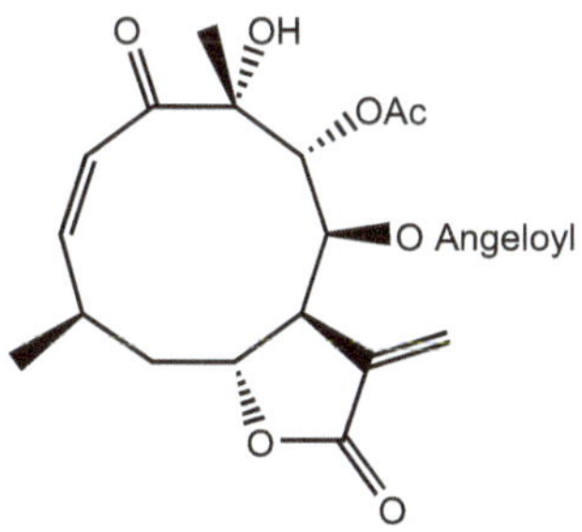

(15)

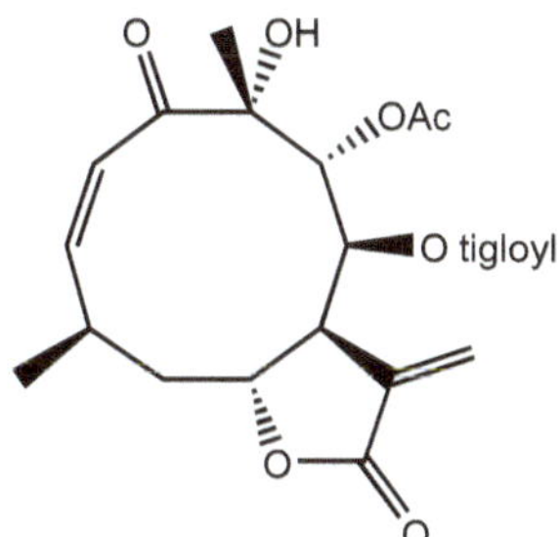

(16)

FIGURE 6.11 Structures of caleins B and C.

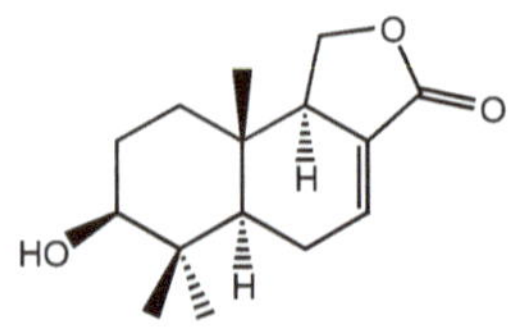

(17)

FIGURE 6.12 Structure of hydroxycinnamolide.

found that this compound bound close to the catalytic site of the enzyme with a Ki value (the inhibition constant for enzyme) of 0.30 μM.

2.10 Drimane Sesquiterpenoids

To manage type 2 diabetes and its associated problems, α-glucosidase inhibitors (AGIs) are extensively used. This class of oral hypoglycemic agents includes the drugs such as acarbose, miglitol, and voglibose. The drawbacks of these AGIs are the strict and repetitive dosing schedule and a high incidence of gastrointestinal disturbances, which include abdominal distension, flatulence, and diarrhea. Hence, AGIs with fewer side effects and better patient compliance are major points, which must be considered during the development of antidiabetic drugs.

In this context, since last one decade *Inonotus obliquus*, a medicinal mushroom (Hymenochaetaceae), has drawn attention because of its antihyperglycemic potential. In this regard, Cha et al. (2006, p. 122), Noh et al. (2004, Patent 1020040079662), Sun et al. (2008, p. 7), Xu et al. (2010, p. 235), Lu et al. (2009, p. 276) and their coworkers carried out the antihyperglycemic activity of the mycelia powder or the crude (mostly water-soluble) extracts of *I. obliquus*. Further research in this context, Ying et al. (2014) isolated three compounds bearing drimane sesquiterpenoid skeleton from the chloroform extract of the submerged culture of this mushroom. The α-glucosidase inhibitory assay was performed on these compounds. The α-glucosidase (from *S. cerevisiae*) inhibitory activity was determined spectrophotometrically in a 96-well microtiter plate with *p*-nitrophenyl-α-D-glucopyranoside (PNPG) as a substrate. After incubation, it was found that 3β-hydroxycinnamolide (**17**, Fig. 6.12) exhibited moderate activity ($IC_{50,}$ 3.39 ± 0.07 mM) (Ying et al., 2014, p. 171).

3. PLANT EXTRACTS CONTAINING SESQUITERPENE LACTONE MOLECULES

From the ethanolic extract of *Vernonia cinerea* L. (Asteraceae), an annual herbaceous Indian plant, a brownish semisolid compound having SL structure was obtained. After structural determination of this compound, acute toxicity study was carried out on this plant extract and the isolated compound using female and male Swiss albino mice. Choudhary et al. (2013) observed that oral administration of the isolated compound and crude ethanol extract (300 and 500 mg/kg) significantly decreased ($p < .05$) the blood glucose level after measurement of the blood sugar levels in alloxan-induced diabetic rates. It was hypothesized that the potentiation of the insulin effect may be increasing either due to the pancreatic secretion of insulin from β-cells of the islets of Langerhans or its release from the bound form, similar to the mechanism by which many plants have been found to exert their hypoglycemic effects (Choudhary et al., 2013, p. 35).

The antidiabetic property of the roots of *Sphaeranthus indicus* Linn. (Asteraceae), a multibranched aromatic Indian medicinal herb, was studied in STZ-induced type 1 diabetic rats. Ramachandran et al. (2011) found that the ethanolic extract 100 and 200 mg/kg had significant effect on reduction in blood glucose when compared with diabetic control rats. The plant is said to possess sesquiterpenoids, SLs, sesquiterpene acids along with other secondary metabolites (Ramachandran et al., 2011, p. 1). Significant lowering in blood glucose levels and increases in hepatic glycogen and plasma insulin levels was recorded by Prabhu et al. (2008) after oral administration of alcoholic extract of this aromatic herb for 15 days. In addition to this, there was a significant improvement in oral glucose tolerance test. The experiment was carried out with glibenclamide as a reference standard (Prabhu et al., 2008, p. 909).

Pareek et al. (2010) used the isolated rat hemidiaphragm in vitro model to investigate the antidiabetic potential of ethanol extract of aerial part of the same Indian herb using the glucose uptake. It was concluded that there is a significant increase ($p < .01$) in the uptake of glucose by medicinal herb and more effective than insulin (Pareek et al., 2010, p. 2284). To study the effect of the methanol extract of the same herb in dexamethasone-induced IR, the mice were treated with dexamethasone for 22 days. It was found that there was significant lowering in plasma glucose at doses, of 400 mg/kg p.o. by its extract. It was realized that its effect in the treatment of type 2 DM may be useful as the extract significantly restored dexamethasone-induced body weight loss (Ghaisas et al., 2010, p. 1371). The screening for biological activity against alloxan-induced hypoglycemia was performed for the dried petroleum ether (60–80°C) extract of flower head of *S. indicus* in Wistar rats by Jha et al. (2010). The results showed that there was a significant reduction in blood glucose after oral administration of flower head extract at doses of 200 mg/kg (Jha et al., 2010, p. 16). The antidiabetic effect of MES in alloxan-induced diabetic rabbits was studied in comparison with 80 mg/kg of diamicron standard by Muhammad et al. (2011). A significant reduction in blood glucose level was reported by them at the dose of 300 mg/kg b.w. for the extract (Muhammad et al., 2011, p. 378).

Achillea santolina (Asteraceae), mainly distributed in the Northern Hemisphere had been used traditionally against diabetes and other pathological conditions. The chemical constituents of *Achillea* plants have many terpenoids, especially monoterpenes, sesquiterpenes, diterpenes, and triterpenes along with other compounds. As discussed in the conference of SRC Baghdad, Hatam argues that SLs are one of the main chemical constituents found in *A. santolina*(Hatam et al., 1988; SRC Baghdad). Various pharmacological activities and therapeutic effects including antidiabetic potential have been highlighted by researchers for this plant. Al-Awwadi (2013) argues that the aqueous extract of the aforementioned plant possess hypoglycemic effects at acute administration of oral dose of 150 and 250 mg/kg b.w. It was observed that the serum glucose level become reduced in STZ-induced diabetic rats after its administration. In addition to this, chronic administration of the same extract with dose of 250 mg/kg orally for 28 days also resulted in hypoglycemic effects in STZ-induced diabetic rats when compared with diabetic control group (Al-Awwadi, 2013, p.129). The experiment regarding elevated levels of liver malondialdehyde and protein carbonyls had also been performed by Ardestani and Yazdanparast (2006) on STZ-induced diabetic rats. The data showed that the extract of *A. santolina* (ethanol–water, 7:3 v/v) at a dose of 100 mg/kg of b.w./day for 30 consecutive days resulted in lowering of the elevated levels of liver malondialdehyde and protein carbonyls in STZ-induced diabetic rats (Ardestani and Yazdanparast, 2006, p. 298).

The antidiabetic effects of *T. diversifolia* (Asteraceae) extract (water and 80% ethanol) have been established by Miura et al. (2002, 2005) in an animal model of type 2 diabetes (Miura et al., 2005, p. 2152; Miura et al., 2002, p. 81). The results showed that the extracts have hypoglycemic effect along with improved IR in peripheral tissues in type 2 diabetes. It was also found that the extracts have their prominent effect on lowering blood glucose after 7 h when administered with a single dose and has no such effect on blood glucose in normal mice. It was realized after collection of experimental data that these extracts show their effect only after metabolic processes. It was concluded that *T. diversifolia* appears to be useful for the treatment of type 2 diabetes, which also explains its use as a medicinal plant for treating diabetes. Takahashi (1998, p. 814) has filed a patent about preparations processes and usage of same of extracts for diabetic treatment.

Centratherum anthelminticum (Asteraceae), an ethnomedicinal plant of India and Southeast Asia, had shown a range of biological activities including antidiabetic potential. Liu et al. (2010, p. 6584) reported the bioactive SLs in the extract of this plant, which includes vernodalol, vernodalidimers A, vernodalidimers B, and vernodalin. Ani and Naidu (2008, p. 897) documented the antihyperglycemic potential of aqueous methanol–acetone extract of this Indian plant. Bhatia et al. (2008, p. 326) found that water extract of seeds of the same plant at the dose of 200 and 500 mg/kg significantly lowered the blood sugar level after 1 week (Bhatia et al., 2008, p. 326).

With more than 100 species, the genus *Inula* (Asteraceae) is commonly found in Europe and Asia. The Greek and Roman ancient physicians have first documented the uses of this genus as herbal medicines. Various types of SLs have been investigated from different species of *Inula*. Mainly eudesmanolides, guaianolides, pseudoguaianolides, germacranolides, xanthanolides, and dimeric sesquiterpenes have been reported from *Inula* species (Zeggwagh et al., 2006). The aqueous extract of aerial parts of this plant showed the hypoglycemic effect at a dose of 20 mg/kg by reducing blood glucose levels (Zeggwagh et al., 2006, p. 223). Shan et al. (2006) have shown that the antihyperglycemic effect of aqueous extract of flowers of *Inula japonica* has the better activity than metformin and gliclazide, which were used as positive controls in the experiment (Shan et al., 2006, p. 455). The methanolic extract of roots of *Inula racemosa*, was tested on alloxan-treated animals to see the effect. Ajani et al. (2009, p. 118) tract has significant effect on lowering of blood sugar level in alloxan-induced hyperglycemia status as compared to alone alloxan-treated animals.

Various parts of *Elephantopus scaber* Linn. (Asteraceae), a small herb found in the Neotropics, Europe, Asia, Africa, and Australia have been used traditionally due to its therapeutic effects against many diseases. Among the species of *Elephantopus* plants, *E. scaber* has been widely grown in India. SLs are one of the main chemical constituents of this plant. From whole plant and roots, sesquiterpenoid lactones such as scabertopin, isoscabertopin, deoxyelephantopin, isodeoxyelephantopin, elescaberin, 17,19-dihydrodeoxyelephantopin, and iso-17,19-dihydrodeoxyelephantopin have been isolated so far. Methanol, hexane, and ethyl acetate extracts of *E. scaber* leaves were tested in STZ-induced diabetic rats to assess their respective hypoglycemic activity with effective dosage at 250 mg/kg b.w./day. Daisy et al. (2009, p. 304) have shown that these extracts not only improved the biochemical parameter of the study but also brought about regeneration of β-cells of pancreas with improved insulin levels. Lowering of blood glucose levels in STZ-induced diabetic rats by acetone extract of *E. scaber* plant has been reported by Daisy et al. (2007). It was hypothesized that by increasing either the pancreatic secretion of insulin from the β-cells of islets of Langerhans or its release from bound insulin, the improvement in the insulin effect of plasma results (Daisy et al., 2007, p. 433). Daisy and Jasmine (2008) proved that the methanol extract of the root of this herb was more effective than its leaf in the treatment of various metabolic disorders. The adverse effect on glucose oxidation in liver and skeleton muscles of diabetic rats due to the deficiency of insulin was also corrected by this herb (Daisy and Jasmine, 2008, p. 22). The comparative study between the aqueous extract of *E. scaber* root and leaf against alloxan-induced hyperglycemic rats has also been done. It was observed that the aqueous root extract of the same herb possesses better hypoglycemic activity than leaf extract of same plant by reducing the blood glucose level. Reduced serum glucose and the activity of gluconeogenic enzyme glucose-6-phosphatase along with increased serum insulin and the activity of glycolytic enzyme glucokinase has been reported by Modilal and Daisy (2011, p. 98) upon oral administration of aqueous extract of leaves and roots into alloxan-induced hyperglycemic rats.

Oran and Al-eisawi (1998, p.84) argued that in Jordan, various indigenous plants have been used since long time for the treatment of diabetes in folk medicine. Abdelgaleil et al. (2011, p. 3385), Mahmoud et al. (1999, p. 575), El-Ela et al. (1990, p. 2704), Vukovic et al. (2006, p. 303) and their coworkers supported that infusion of *Ambrosia maritima* L. and *Anthemis pseudocotula* Boiss (flowering head and leaves) (Asteraceae) have been used for this purpose. The phytochemical investigation of these plants reveals the presence of SL molecules.

4. STRUCTURE–ACTIVITY RELATIONSHIP OF SESQUITERPENE LACTONES

It is generally believed that the bioactivity of SLs is mediated by alkylation of nucleophiles through their α-, β-, or α,β,γ-unsaturated carbonyl structures, such as α-methylene-γ-lactones or α,β-unsaturated cyclopentenones. These structure elements react with nucleophiles, especially the cysteine sulfhydryl groups by Michael-type addition. Therefore, it is widely accepted that thiol groups such as cysteine residues in proteins, as well as the free intracellular GSH, serve as the major targets of SLs. In essence, the interaction between SLs and protein thiol groups or GSH leads to reduction of enzyme activity or causes the disruption of GSH metabolism and vitally important intracellular cell redox balance.

The relationship between chemical structure and bioactivity of SLs has been studied in several systems, especially with regard to cytotoxicity, antiinflammatory, and antitumor activity (Fig. 6.13). It is believed that the exomethylene group on the lactone is essential for cytotoxicity because structural modifications such as saturation or addition to the methylene group resulted in the loss of cytotoxicity and tumor inhibition. However, it has also been shown that the factor responsible for the cytotoxicity of SLs might be the presence of the O=C—C=CH_2 system, regardless of lactone or cyclopentenone. It was later demonstrated that the presence of additional alkylating groups greatly enhanced the cytotoxicity of SLs. Furthermore, it was established that the α-methylene-γ-lactones and α,β-unsaturated cyclopentenone ring (or α-epoxycyclopentenone) present in SLs essential for their in vivo antitumor activity. It has been confirmed through various published reports that the various kinds of biological activities displayed by SLs is due to the presence of either α-methylene-γ-lactones and α,β-unsaturated cyclopentenone ring.

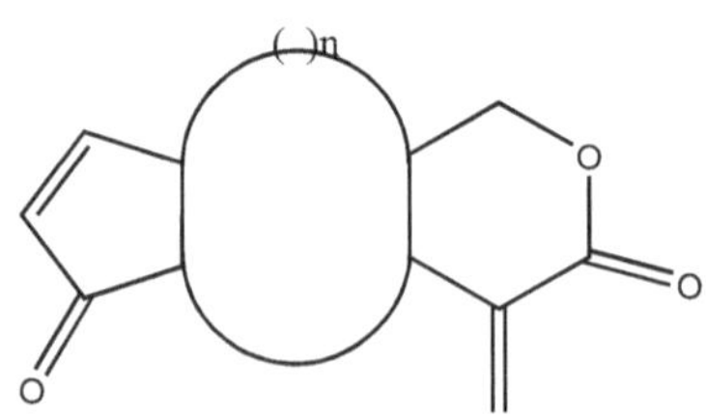

FIGURE 6.13 General structure of sesquiterpene lactones.

In summary, the differences in activity among individual SLs may be explained by differences in the number of alkylating elements, lipophilicity, molecular geometry, and the chemical environment of the target sulfhydryl group.

5. CONCLUDING REMARKS

SLs are an important group of natural products obtained from many species of medicinal plants. Their structural diversity and diverse potential biological activities such as anticancer, antiinflammatory, antitumor, antimalarial, antiviral, antibacterial, antifungal, etc., have made further interest among the chemists to the drug discovery research. Although the exact mechanism of action of SLs is not well known but it has been documented through various published reports that the biological activity displayed by majority of SLs is due to the presence of α-methylene-γ-lactones and α,β-unsaturated cyclopentenone ring. The present chapter deals with an overview on the antidiabetic potential of structurally diverse SLs, which may be useful for the chemists/pharmacologists working in the area of drug discovery of the relevant subject.

LIST OF ABBREVIATIONS

AGIs α-Glucosidase inhibitors
BSA Bovine serum albumin
CTL Cytotoxic T lymphocytes
DM Diabetes mellitus
DMEM Dulbecco's Modified Eagle Medium
EtOAc Ethyl acetate
GC–MS Gas chromatography–mass spectrometry
HbA1c Hemoglobin A1c
ICAM-1 Intercellular adhesion molecule-1
IR Insulin resistance
LBD Ligand-binding domain
LDL Low-density lipoprotein
NAD β-nicotinamide adenine dinucleotide
NMR Nuclear magnetic resonance
PNPG *p*-Nitrophenyl-α-D-glucopyranoside
PPARs Peroxisome proliferator-activated receptors
SAR Structure–activity relationship
SLs Sesquiterpene lactones
STZ Streptozotocin

ACKNOWLEDGMENTS

We wish to thank Pro-Vice Chancellor and Dean, Research (Science and Technology), Amity University Uttar Pradesh, Lucknow Campus, Lucknow, for his constant encouragement and support for research. Financial support from the Department of Science and Technology (DST), Government of India (Grant No.: SR/FT/CS-147/2010) is duly acknowledged. We are also grateful to Professor (Dr.) Goutam Brahmachari, Editor of the book for his kind invitation and fruitful suggestions during the preparation of this chapter.

REFERENCES

Abdelgaleil, S.A.M., Badawy, M.E.I., Suganuma, T., Kitahara, K., 2011. Antifungal and biochemical effects of pseudoguaianolide sesquiterpenes isolated from *Ambrosia maritima* L. African Journal of Microbial Research 5, 3385–3393.

Ajani, H.B., Patel, H.P., Shah, G.B., Acharya, S.R., Shah, S.K., 2009. Evaluation of antidiabetic effect of methanolic extract of *Inula racemosa* root in rats. Pharmacologyonline 3, 118–129.

Al-Awwadi, N.A.J., 2013. Acute and chronic hypoglycaemic effect of *Achillea santolina* aqueous leaves extract. International Journal of Medicinal Plant Research 2, 129–134.

Ani, V., Naidu, K.A., 2008. Antihyperglycemic activity of polyphenolic components of black/bitter cumin, *Centratherum anthelminticum* Kuntze seeds. European Food Research and Technology 226, 897–903.

Ardestani, A., Yazdanparast, R., 2006. *Achillea santolina* reduced oxidative stress in the liver of streptozotocin-induced diabetic rats. Pharmacologyonline 3, 298–308.

Aybar, M.J., Sánchez Riera, A.N., Grau, A., Sánchez, S.S., 2001. Hypoglycemic effect of the water extract of *Smallanthus sonchifolius* (yacon) leaves in normal and diabetic rats. Journal of Ethnopharmacology 74, 125–132.

Basha, R.H., Sankaranarayanan, C., 2015. Protective role of β-caryophyllene, a sesquiterpene lactone on plasma and tissue glycoprotein components in streptozotocin-induced hyperglycemic rats. Journal of Acute Medicine 5, 9–14.

Bento, A.F., Marcon, R., Dutra, R.C., Claudino, R.F., Cola, M., Leite, D.F.P., Calixto, J.B., 2011. β-Caryophyllene inhibits dextran sulfate sodium-induced colitis in mice through CB2 receptor activation and PPARγ pathway. American Journal of Pathology 178, 1153–1166.

Bhatia, D., Gupta, M.A., Gupta, A., Singh, M., Kaithwas, G., 2008. Pharmacognostical studies on seeds of *Centratherum anthelminticum* Kuntze. Natural Product Radiance 7, 326–329.

Cabrera, W., De Pedro, A., Perotti, M., Grau, A., Catalán, C., Genta, S., Sánchez, S.S., 2007. Hypoglycemic effect of *Smallanthus macroscyphus* leaves. Identification of the active principle. Biocell 31, 311.

Cha, J.Y., Jun, B.S., Yoo, K.S., 2006. Fermented chaga mushroom (*Inonotus obliquus*) effects on hypolipidemia and hepatoprotection in Otsuka Long-Evans Tokushima fatty (OLETF) rats. Food Science and Biotechnology 15, 122–127.

Chen, H.C., Chou, C.K., Lee, S.D., Wang, J.C., Yeh, S.F., 1995. Active compounds from *Saussurea lappa* Clarks that suppress hepatitis B virus surface antigen gene expression in human hepatoma cells. Antiviral Research 27, 99–109.

Choi, J.H., Seo, B.R., Seo, S.H., Lee, K.T., Park, J.H., Park, H.J., Choi, J.W., Itoh, Y., Miyamoto, K., 2002. Costunolide induces differentiation of human leukemia HL-60 cells. Archives of Pharmacal Research 25, 480–484.

Choudhary, S., Sharma, M., Tripathi, J., Mishra, P., 2013. Antihyperglycemic activity of *Vernonia cinerea* L. on alloxan-induced diabetic mice. International Journal of Advanced Research 1, 35–42.

Daisy, P., Jasmine, R., 2008. Role of *Elephantopus scaber* on the glucose oxidation in liver and skeletal muscles of streptozotocin induced diabetic adult male rats. Research Journal of Medicinal Plants 2, 22–27.

Daisy, P., Rayan, N.A., Rajathi, D., 2007. Hypoglycemic and other related effects of *Elephantopus scaber* extracts on alloxan induced diabetic rats. Journal of Biological Science 7, 433–437.

Daisy, P., Vargese, L., Priya, C.E., 2009. Comparative studies on the different leaf extracts of *Elephantopus scaber* on stz-induced diabetic rats. European Journal of Scientific Research 32, 304–313.

El-Ela, M.A., Jakupovic, J., Bohlmann, F., Ahmed, A.A., Seif El-Din, A., Khafagi, S., Sabri, N., EL-Ghazouly, M., 1990. Seco-germacranolides from *Anthemis pseudocotula*. Phytochemistry 29, 2704–2706.

Eliza, J., Daisy, P., Ignacimuthu, S., Duraipandiyan, V., 2009a. Antidiabetic and antilipidemic effect of eremanthin from *Costus speciosus* (Koen.) Sm., in STZ-induced diabetic rats. Chemico-Biological Interactions 182, 67–72.

Eliza, J., Daisya, P., Ignacimuthub, S., Duraipandiyan, V., 2009b. Normo-glycemic and hypolipidemic effect of costunolide isolated from *Costus speciosus* (Koen ex. Retz.) Sm. in streptozotocin-induced diabetic rats. Chemico-Biological Interaction 179, 329–334.

Escandon-Rivera, S., Gonzalez-Andrade, M., Bye, R., Linares, E., Navarrete, A., Mata, R., 2012. α-Glucosidase inhibitors from *Brickellia cavanillesii*. Journal of Natural Products 75, 968–974.

Ferreira, M.J.P., Brant, A.J.C., Alvarenga, S.A.V., Emerenciano, V.P., 2005. Neural networks in chemosystematic studies of Asteraceae: a classification based on a dichotomic approach. Chemistry & Biodiversity 2, 633–644.

Genta, S.B., Cabrera, W.M., Mercado, M.I., Grau, A., Catalán, C.A., Sánchez, S.S., 2010. Hypoglycemic activity of leaf organic extracts from *Smallanthus sonchifolius*: constituents of the most active fractions. Chemico-Biological Interaction 185, 143–152.

Ghaisas, M., Zope, V., Takawale, A., Navghare, V., Tanwar, M., Deshpande, A., 2010. Preventive effect of *Sphaeranthus indicus* during progression of glucocorticoid-induced insulin resistance in mice. Pharmaceutical Biology 48, 1371–1375.

Ghalib, R.M., Hashim, R., Sulaiman, O., Mehdi, S.H., Valkonen, A., Rissanen, K., Trifunovic, R., Ahamed, M.B.K., Majid, A.M.S.A., Kawamura, F.A., 2012. Novel caryophyllene type sesquiterpene lactone from *Asparagus falcatus* (Linn.); Structure elucidation and anti-angiogenic activity on HUVECs. European Journal of Medicinal Chemistry 47, 601–607.

Hatam, N.A.R., Hamad, M.N., Nadir, M.T., 1988. The Constituents of *Achillea Santolina*: Phytochemical and Antimicrobial Studies. Abstracts of the Fifth Scientific Conference. Scientific Research Council in Baghdad.

Hidhayath Basha, R., Sankaranarayanan, C., 2014. β-Caryophyllene, a natural sesquiterpene, modulates carbohydrate metabolism in streptozotocin induced diabetic rats. Acta Histochemica 5, 9–14.

Hou, C., Lin, S., Cheng, J., Hsu, F., 2003. Antidiabetic dimeric guianolides and a lignan glycoside from *Lactuca indica*. Journal of Natural Products 66, 625–629.

Inoue, A., Tamogami, S., Kato, H., Nakazato, Y., Akiyama, M., Kodama, O., Akatsuka, T., Hashidoko, Y., 1995. Antifungal melampolides from leaf extracts of *Smallanthus sonchifolius*. Phytochemistry 39, 845–848.

Jha, R.K., Mangilal Bhandari, A., Nema, R.K., 2010. Antidiabetic activity of flower head petroleum ether extracts of *Sphaeranthus indicus* Linn. Asian Journal of Pharmaceutical and Clinical Research 3, 16–19.

Kassuyaa, C.A.L., Cremonezea, A., Barrosb, L.F.L., Simasa, A.S., Lapaa, F.R., Mello-Silvac, R., Stefanellob, M.E.A., Zampronio, A.R., 2009. Antipyretic and anti-inflammatory properties of the ethanolic extract, dichloromethane fraction and costunolide from *Magnolia ovata* (Magnoliaceae). Journal of Ethnopharmacology 124, 369–376.

Kawashima, S., Sakato, M., Terada, S., Koide, Y., Tanaka, M., Ko, M., 2001. Anti-diabetic agents containing lactones. Japanese Kokai Tokkyo Koho Patent no. JP 2001247461.

Li, W.L., Zheng, H.C., Bukuru, J., De Kimpe, N., 2004. Natural medicines used in the traditional Chinese medical system for therapy of diabetes mellitus. Journal of Ethnopharmacology 92, 1–21.

Lin, F., Hasegawa, M., Kodama, O., 2003. Purification and identification of antimicrobial sesquiterpene lactones fromyacon (*Smallanthus sonchifolius*) leaves. Bioscience, Biotechnology & Biochemistry 67, 2154–2159.

Lin, H., 2012. Sesquiterpene lactones from *Tithonia diversifolia* act as peroxisome proliferator-activated receptor agonists. Bioorganic and Medicinal Chemistry Letters 22, 2954–2958.

Liu, Y., Nugroho, A.E., Hirasawa, Y., Nakata, A., Kaneda, T., 2010. Vernodalidimers A and B novel orthoester elmanololide dimers from the seed of *Vernonia anthelmintica*. Tetrahedron Letters 51, 6584–6587.

Lu, X.M., Chen, H.X., Dong, P., Fu, L.L., Zhang, X., 2009. Phytochemical characteristics and hypoglycaemic activity of fraction from mushroom *Inonotus obliquus*. Journal of the Science of Food and Agriculture 90, 276–280.

Mahmoud, A.A., Ahmed, A.A., El Bassuony, A.A., 1999. A new chlorosesquiterpene lactone from *Ambrosia maritima*. Fitoterapia 70, 575–578.

Miura, T., Furuta, K., Yasuda, A., Iwamoto, N., Kato, M., Ishihara, E., Ishida, T., Tanigawa, K., 2002. Antidiabetic effect of nitobegiku in KK-Ay diabetic mice. American Journal of Chinese Medicine 30, 81–86.

Miura, T., Nosaka, K., Ishii, H., Ishida, T., 2005. Antidiabetic effect of Nitobegiku, the herb *Tithonia diversifolia*, in KK-Ay diabetic mice. Biological and Pharmaceutical Bulletin 28, 2152–2154.

Modilal, M.R.D., Daisy, P., 2011. Hypoglyceamic effects of *Elephatopus scaber* in alloxan-induced diabetic rats. Indian Journal of Novel Drug Delivery 3, 98–103.

Morris, G.M., Goodsell, D.S., Halliday, R.S., Huey, R., Hart, W.E., Belew, R.K.J., 1998. Automated docking using a Lamarckian genetic algorithm and an empirical binding free energy function. Journal of Computational Chemistry 19, 1639–1662.

Muhammad, R.T., Malik, S., Ghulam, M., Waseem, H., 2011. Determination of the anti-diabetic effect of methanolic extract of *Sphaeranthus indicus* L. on alloxan induced diabetic rabbits. Latin American Journal of Pharmacy 30, 378–382.

Noh, J.Y., Yoon, B.G., Yoon, J.J., 2004. Composition for Prevention and Treatment of Type 1 and Type 2 Diabetes, Containing Water Extract of *Cordyceps Militaris* and *Inonotus Obliquus* as Effective Ingredient. Korean Patent, Publication No. 1020040079662.

Nolte, R.T., Wisely, G.B., Westin, S., Cobb, J.E., Lambert, M.H., Kurokawa, R., Rosenfeld, M.G., Willson, T.M., Glass, C.K., Milburn, M.V., 1998. Ligand binding and co-activator assembly of the peroxisome proliferator-activated receptor-gamma. Nature 395, 137–143.

Ohhini, M., Yoshimi, N., Kawamori, T., Ino, N., Hirose, Y., Tanaka, T., Yamahara, J., Miyata, H., Mori, H., 1997. Inhibitory effects of dietary protocatechuic acid and costunolide on 7,12 dimethylbenz[a]anthracene induced hamster cheek pouch carcinogenesis. Japanese Journal of Cancer Researh 88, 111.

Oran, S.A., Al-Eisawi, D.M., 1998. Check list of medicinal plants in Jordan. Dirasat. Medical and Biological Sciences 25, 84–112.

Pareek, A., Suthar, M., Bansal, V.K., Chauhan, R.K., Batra, N., 2010. In-vitro study of *Sphaeranthus indicus* on glucose uptake by isolated rat hemi-diaphragm. Journal of Pharmacy Research 3, 2284–2285.

Pari, L., Rajarajeswari, N., 2010. Protective role of coumarin on plasma and tissue glycoprotein components in streptozotocin- nicotinamide induced hyperglycemic rats. International Journal of Biological and Medical Research 1, 61–65.

Park, H.J., Kwon, S.H., Han, Y.N., Choi, J.W., Miyamoto, K., Lee, S.H., Lee, K.T., 2001. Apoptosis-Inducing costunolide and a novel acyclic monoterpene from the stem bark of *Magnolia sieboldii*. Archives of Pharmacal Research 24, 342–348.

Prabhu, K.S., Lobo, R., Shirwaikar, A., 2008. Antidiabetic properties of the alcoholic extract of *Sphaeranthus indicus* in streptozotocin-nicotinamide diabetic rats. Journal of Pharmacy and Pharmacology 60, 909–916.

Prajna, K., Kumar, A.J., Rai, S., Shetty, S.K., Rai, T., Shrinidhi, Begum, M., Shashikala, 2013. Predictive value of serum sialic acid in type-2 diabetes mellitus and its complication (Nephropathy). Journal of Clinical and Diagnostic Research 7, 2435–2437.

Ramachandran, S., Asokkumar, K., Maheswari, M.U., Ravi, T.K., Sivashanmugam, A.T., Saravanan, S., 2011. Investigation of antidiabetic, antihyperlipidemic and in-vivo antioxidant properties of *Sphaeranthus indicus* Linn. in type 1 diabetic rats: an identification of possible biomarkers. Evidence-Based Complementary and Alternative Medicine 2011, 1–8.

Rios, J.L., Franchini, F., Schinella, G.R., 2015. Natural products for the treatment of type 2 diabetes mellitus. Planta Medica 82, 975–994.

Robles, M., Aregullin, M., West, J., Rodriguez, E., 1995. Recent studies on the zoopharmacognosy, pharmacology and neurotoxicology of sesquiterpene lactones. Planta Medica 61, 199–203.

Schorr, K., Da Costa, F.B., 2005. Quantitative determination of enhydrin in leaf rinse extracts and in glandular trichomes of *Smallanthus sonchifolius* (Asteraceae) by reversed-phase high-performance liquid chromatography. Phytochemical Analysis 16, 161–165.

Serra-Barcellona, C., Aráoz, M.V.C., Cabrera, W.M., Habib, N.C., Honoré, S.M., Catalán, C.A.N., Grau, A., Genta, S.B., Sánchez, S.S., 2014. *Smallanthus macroscyphus*: a new source of antidiabetic compounds. Chemico-Biological Interactions 209, 35–47.

Shan, J.J., Yang, M., Ren, J.W., 2006. Anti-diabetic and hypolipidemic effects of aqueous-extract from the flower of *Inula japonica* in alloxan-induced diabetic mice. Biological and Pharmaceutical Bulletin 29, 455–459.

Sun, J.E., Ao, Z.H., Lu, Z.M., 2008. Antihyperglycemic and antilipidperoxidative effects of dry matter of culture broth of *Inonotus obliquus* in submerged culture on normal and alloxan-diabetes mice. Journal of Ethnopharmacology 118, 7–13.

Takahashi, M., 1998. Compositions for Curing Diabetes Mellitus, Processes for the Preparation of Same, and Usage of Same. US Pat. 5,773,004 (Fukuoka 1998, 814–01).

Tian, Y.M., Johnson, G., Ashcroft, J.H., 1998. Sulfonylureas enhance exocytosis from pancreatic β-cells by a mechanism that does not involve direct activation of protein kinase C. Diabetes 47, 1722–1726.

Vukovic, I., Vujisic, L., Vajs, V., Tesevic, V., Macura, S., Janackovic, P., Milosavljevic, S., 2006. Sesquiterpene lactones from the aerial parts of *Anthemis arvensis* L. Biochemical Systematics and Ecology 34, 303–309.

Xu, X., Pang, C., Yang, C.J., Zheng, Y.T., Xu, H.Y., Lu, Z.M., Xu, Z.H., 2010. Antihyperglycemic and antilipidperoxidative effects of polysaccharides extracted from medicinal mushroom Chaga, *Inonotus obliquus* (Pers.: Fr.) Pilát (Aphyllophoromycetideae) on alloxan-diabetes mice. International Journal of Medicinal Mushrooms 12, 235–244.

Ying, Y., Zhang, L., Zhang, X., Bai, H., Liang, D., Ma, L., Shan, W., Zhan, Z., 2014. Terpenoids with alpha-glucosidase inhibitory activity from the submerged culture of *Inonotus obliquus*. Phytochemistry 108, 171–176.

Zeggwagh, N.A., Ouahidi, M.L., Lemhadri, A., Eddouks, M., 2006. Study of hypogly-caemic and hypolidemic effects of *Inula viscosa* L. aqueous extract in normal and diabetic rats. Journal of Ethnopharmacology 108, 223–227.

Zhang, Y., Won, Y.K., Ong, C.N., Shen, H.M., 2005. Anti-cancer potential of sesquiterpene lactones: bioactivity and molecular mechanisms. Current Medicinal Chemistry – Anti-Cancer Agents 5, 239–249.

Zhao, G., Li, X., Chen, W., Xi, Z., Sun, L., 2012. Three new sesquiterpenes from *Tithonia diversifolia* and their anti-hyperglycemic activity. Fitoterapia 83, 1590–1597.

Chapter 7

Prediction of Prediabetes and Its Prevention by Functional Food Compounds

S.J. Chen, T. Matsui
Graduate School of Kyushu University, Fukuoka, Japan

1. INTRODUCTION

Metabolic diseases usually occur for years before becoming clinically apparent. Diabetes, a complex group of metabolic disorders which are improperly regulated by insulin, is projected as one of the leading causes of death worldwide (Wild et al., 2004, p. 1047). The long-term manifestation of these metabolic disorders results primarily in elevated fasting and postprandial blood glucose levels (BGL). Moreover, severe diabetic complications, like diabetic vascular complication, diabetic nephropathy, and diabetic retinopathy, continue to puzzle the patients and physicians (Forbes and Cooper, 2013, p. 137). Prediabetes, typically defined as impaired glucose tolerance and/or impaired fasting blood glucose, is a high risk state of developing type 2 diabetes. It is predicted that more than 470 million people will have prediabetes by 2030 (Tabák et al., 2012, p. 2279). People with prediabetes will progress diabetes or convert back to normal stage with the same proportion per year. For prediabetic individuals, the onset of type 2 diabetes can be prevented or delayed by early interventions such as life style change and therapeutic intervention with evidence of a 40–70% relative risk reduction (Nathan et al., 2007, p. 753, 2009, p. 17). Given the availability of effective intervention for preventing increased burden of diabetes worldwide, earlier prediction of individuals at risk is particularly crucial (Knowler et al., 2002, p. 393; Kitabchi et al., 2005, p. 2404). Current clinical indicators, such as fasting, postprandial blood glucose, and hemoglobin A1c (HbA1c), can be helpful to reduce the risks of diabetes, however, more sensitive and specific predictors are needed for early prediction of prediabetes (American Diabetes Association, 2014, p. S8).

Previous studies developed several approaches for the management of diabetes by reducing fasting and postprandial BGL rise (Tahrani et al., 2011, p. 182). Since multiple and variable pathophysiological changes were associated

Discovery and Development of Antidiabetic Agents From Natural Products.
http://dx.doi.org/10.1016/B978-0-12-809450-1.00007-7

with type 2 diabetes, pharmacological compounds targeting different aspects of the disease are needed for each stages of diabetes. Natural products are regularly used as alternative medicinal approach, owing to their multiple pharmacological potentials and fewer side effects, such as hypoglycemia, weight gain, gastrointestinal disturbances, etc. (Tiwari and Madhusudana Rao, 2002, p. 30; Leiherer et al., 2013, p. 3). Despite the great efforts that have been made in understanding pathophysiology and management of diabetes, preventive strategy of this disease and its complications are still insufficient. Thereafter, early prediction of prediabetes and therapeutic intervention at prediabetic stage by natural products should be paid more attention. This chapter illustrates analytical prediction of prediabetes by monitoring levels of advanced glycation end products (AGEs), typical diabetes-related metabolites, and presents an overview of preventive strategy of prediabetes by natural products through the inhibition of AGEs accumulation.

2. ADVANCED GLYCATION END PRODUCTS

Altered metabolite levels may reflect the pathogenic progress of diabetes and serve as diagnostic predictors. Recent studies allow the investigation of metabolites involved in diseases by monitoring metabolite levels. AGEs, mainly formed at the advanced stage of protein glycation, has been revealed to exert considerable effects in diabetes and its complications (Goldin, 2006, p. 597). Moreover, increased levels of AGEs have been found in diabetic patients, compared with health subjects (Ahmed et al., 2005b, p. 2465). Therefore, AGEs are considered as potential predictor of prediabetes.

2.1 Protein Glycation

Protein glycation, a major cause of spontaneous damage to proteins in physiological system, is a series of nonenzymatic reactions of protein or lipid molecules with reducing saccharides and saccharide derivatives (Wolff et al., 1991, p. 339; Lis and Sharoon, 1993, p. 1). With time, it became clear that glycation involves a series of sequential and parallel reactions, often termed as the Maillard reaction, named after the leading pioneer of glycation research, Louis Camille Maillard. Protein glycation is accelerated in diabetes as a consequence of chronic hyperglycemia and increased oxidative stress in blood (Goh and Cooper, 2008, p. 1143). Glycation adducts in organs and fluids in vivo are usually classified into two kinds: (1) early glycation adducts including Schiff's base and Amadori products formed at early stages of glycation processes, (2) AGEs, which initially considered to be formed in later or advanced stages of the Maillard reaction but are rather believed to be formed in both early and advanced stages (Rabbani and Thornalley, 2012, p. 1087).

Both intracellular and extracellular proteins may be glycated and oxidized through the Maillard reaction. AGEs appear to play a vital role in the process of protein damage in vivo. As reviewed by Brownlee (1995, p. 223), AGEs may produce reactive oxygen species (ROS), bind to specific cell surface receptors, form cross links, and consequently alter the functional properties of extracellular matrix or modify the action of hormones and intracellular proteins. Moreover, increasing evidence has shown that AGEs play a pivotal role in diabetes, microvascular diabetic complications, and vascular cell dysfunction (Goh and Cooper, 2008, p. 1143; Ahmed and Thornalley, 2007, p. 233; Brownlee, 2001, p. 813).

2.2 Formation and Metabolism of AGEs

There are several processes involved in the formation of AGEs in physiological system. The major AGE formation process begins from the Schiff's base produced by the reaction of carbonyl group of glucose with free amino groups of proteins. According to the Amadori rearrangement, highly reactive intermediates, known as dicarbonyls or α-oxoaldehydes, including 3-deoxyglucosone (3-DG), glyoxal (GO), and methylglyoxal (MG), are formed and accumulated (Thornalley et al., 1999, p. 109). The dicarbonyls are powerful glycating agents, which react with lysine or arginine residues in proteins, leading to the formation of irreversible AGE residues, then release AGE-free adducts by proteolysis. The Schiff's base adduct may also degrade to dicarbonyls via the non-Amadori rearrangement reaction. In addition, dicarbonyls can also be formed by the degradation of glycolytic intermediates and lipid peroxidation (Rabbani and Thornalley, 2012, p. 1087; Thornalley, 2005, p. 111). Limiting factors for the formation of AGEs include the rate of turnover of proteins (Verzijl et al., 2000, p. 39027), the degree of hyperglycemia, and the extent of oxidant stress (Yan et al., 1994, p. 9889) (Fig. 7.1). Except for endogenously originated products, AGEs can

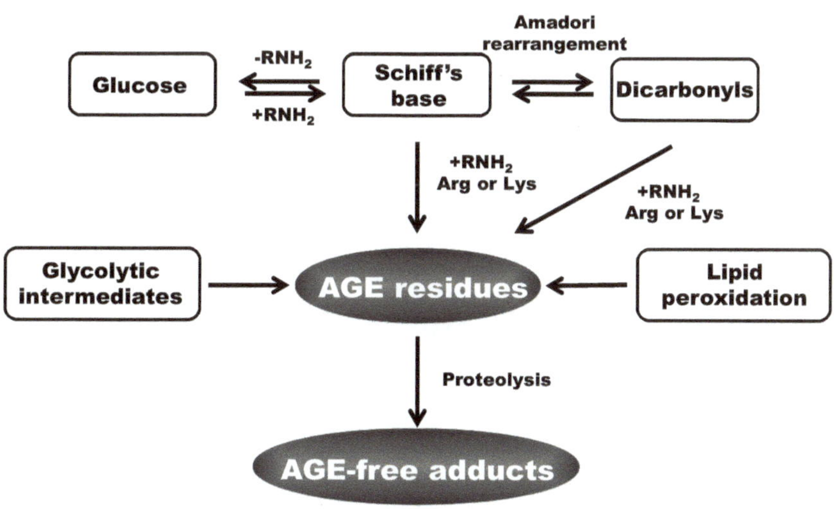

FIGURE 7.1 Formation of AGEs in physiological system.

also be formed from exogenous sources. Foods high in protein and fat, such as meat, egg yolk, and dairy products, are rich in AGEs (Goldberg et al., 2004, p. 1287). Food processing, particularly prolonged heating or frying, promotes the generation of AGEs. A large portion of ingested AGEs are absorbed with food. Diet-derived AGEs are considered as another risk factor for several chronic disorders (Uribarri et al., 2010, p. 911; Poulsen et al., 2013, p. 10).

AGE adducts of proteins are termed "AGE residues," in keeping with the nomenclature used for amino acid residues of proteins. "AGE-free adducts" are glycated amino acids formed by the degradation of glycated proteins by cellular proteolysis (Ahmed and Thornalley, 2007, p. 233). A scheme of formation, gastrointestinal tract absorption of AGE residues, proteolytic processing in body, and renal excretion in AGE-free adduct forms is given in Fig. 7.2.

Glycation of proteins was initially considered as a posttranslational modification and gave rise to AGEs that remain in tissues and body fluids throughout life (Lis and Sharon, 1993, p. 1; Brownlee et al., 1988, p. 1315). It is now acceptable that most AGEs are chemically stable and formed on long-lived proteins, the others formed on short-lived proteins (2–6 weeks) in physiological system. Their levels depend on the rate of formation and decomposition (Goldberg, 2003, p. 895). Moreover, the turnover of glycated proteins by proteolysis releases AGE-free adducts (Thornalley et al., 2003, p. 581). AGE-free adducts, from digested proteins in ingested foods, are also absorbed through the gastrointestinal tract. AGE-free adducts in blood are filtered through renal glomeruli and can be reabsorbed by organic anion and cation transporter in the renal tubular. Active secretion of AGE-free adducts may also occur from the renal circulation (Bohlender et al., 2005, p. F645).

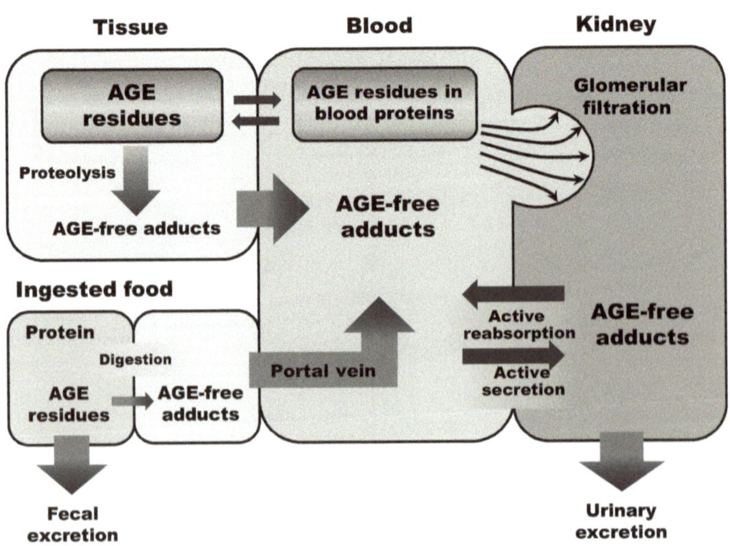

FIGURE 7.2 A scheme of formation, proteolytic processing, and excretion of AGEs.

2.3 Association of AGEs With Diabetes and Its Complications

AGEs accumulate in various tissues that are damaged in diabetes, while the rate of accumulation varies with hyperglycemic conditions. It has been revealed that AGEs are associated with intensive glycemic control in experimental models of diabetic nephropathy (For the Diabetes, T.W.T., 2003, p. 2159). The association of AGEs with indicators of glycemic control has been wildly investigated according to the credible AGEs quantification. N^{ε}-carboxymethyl lysine (CML) and pentosidine residues in plasma protein or hemoglobin are associated with fasting plasma glucose. A similar correlation with HbA1c was found for CML residues of plasma protein (Ahmed et al., 2005b, p. 2465). CML and pentosidine residues in skin collagen may also serve as indicators of glycemic control (Verzijl et al., 2000, p. 39027). Conversely, some AGE residues, e.g., N^{ε}-(1-carboxyethyl)-lysine (CEL), were not linked with the markers of glycemic control (Ahmed et al., 2005b, p. 2465). Plasma AGEs were even found to be negatively related to 2-h BGL after oral glucose load in health subjects, the association likely to be involved in the modulation of the receptor of AGEs (RAGE) (Forbes et al., 2014, p. 321). Taken together, although some AGE residues have the potential to serve as surrogate indicators of glycemic control, it seems to be hard to come to a conclusion that AGE levels can reflect the glycemic control independently.

General mechanisms through which AGEs contribute to diabetic complications include the following: (1) the formation of cross-links between molecules in the basement membrane of extracellular matrix, altering cellular structure (Brownlee, 1995, p. 223; Sell and Monnier, 1989, p. 21597), and (2) interaction of AGEs with RAGE on cell membranes, altering cellular function (Tanji et al., 2000, p. 1656). Evidence for this emerges not only from an increased number of in vitro and in vivo studies exploring the role of AGEs in different pathologies (Thornalley et al., 2003, p. 31; Ahmed et al., 2005a, p. 1590; Vlassara et al., 1992, p. 12043; Cai et al., 2012, p. 15888) but also from studies demonstrating significant improvement of features of diabetic complications by inhibiting AGE accumulation. Various inhibitors of dicarbonyls, major precursors of AGEs, have been investigated for reducing AGEs. Aminoguanidine, a well-known inhibitor of AGE formation, is an efficient scavenger of MG (Brownlee et al., 1986, p. 1629), being effective to prevent a range of diabetic vascular complications in animal studies (Hammes et al., 1994, p. 32), while it is unlikely to serve as clinical agent for its toxicity (Thornalley, 2003, p. 31). Pyridoxamine, a derivative of vitamin B_6, prevents the degradation of the Amadori intermediates to AGE residues (Voziyan and Hudson, 2005, p. 807), consequently ameliorates early renal disease and dyslipidemia in diabetic rat (Degenhardt et al., 2002, p. 939). Benfotiamine, a lipid-soluble thiamine derivative, inhibits the formation pathway of AGE (Hammes et al., 2003, p. 294). OPB-9195, a thiazolidine derivative, was shown to prevent the progression of glomerular sclerosis(Nakamura et al., 1997, p. 895) and diabetic neuropathy(Wada et al., 2001, p. 513) in animal models. These studies are not only helpful to understand the pathogenesis of diabetic complications but also are encouraging, since beneficial effects may be achieved by therapeutic intervention to reduce the accumulation of AGEs.

2.4 Methylglyoxal-Derived Hydroimidazolone Residue—A Novel Predictor of Prediabetes

Clinical studies revealed that an increase of glycation metabolites was related to the progression of diabetes and its complications (Goh and Cooper, 2008, p. 1143; Ahmed and Thornalley, 2007, p. 233). Our previous research hypothesized that the variation of certain AGEs may reflect the progression of prediabetes, consequently can be served as a diagnostic predictor of prediabetes, and enable preventive interventions before the onset of diabetes (Chen et al., 2015b, p. 12,477). In order to investigate AGE production behavior with high sensitivity and specificity, we proposed an LC-MS/MS method in combination with 2,4,6-trinitrobenzene sulfonate (TNBS) derivatization technique for reliable quantitation of AGEs (Hashimoto et al., 2013; Chen et al., 2015a). The TNBS derivatization with amines or AGEs allows the enhanced MS detection of AGEs due to hydrophobic trinitrobenzene moiety in derivatives. With the aid of proposed TNBS-MS method, we got the first insight of metabolic behavior of some typical diabetes-related glycation metabolites during the pathogenic progress of diabetes in spontaneously diabetic Torii (SDT) rats. The study also proposed MG-derived hydroimidazolone (MG-H1) residue as a potent predictor of prediabetes, according to the monitored AGE levels (Chen et al., 2015b, p. 12477). The results (Fig. 7.3) showed that MG-H1 residue levels in plasma protein were increased significantly since prediabetic stage (16-week-old) of SDT

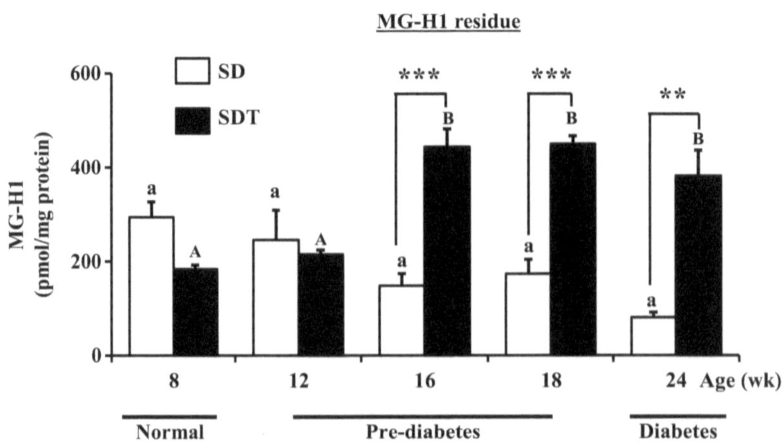

FIGURE 7.3 Prediction of prediabetes by monitoring MG-H1 residue levels in SDT rats.

rats, compared with age-matched Sprague–Dawley (SD) rats (SD: 148.2 ± 25.9 versus SDT: 444.2 ± 37.5 pmol/mg protein, $p < .001$). Besides, it is also given an over twofold increase, with respect to the baseline of SDT rats at the age of 8 weeks (186.1 ± 8.6 pmol/mg protein, $p < .001$).

Data are mean ± SEM ($n = 4$). Key: hollow bars, control SD rats; solid bars, SDT rats. Significance: statistical differences within group were evaluated using one-way ANOVA, different letters represent the statistical differences at $p < .05$ among the means in individual group by Dunnett's test for multiple comparisons. * indicates significance of difference with respect to age-matched SD control rats by Student's t-test, where two and three symbols reflect $p < .01$, $p < .001$, respectively.

MG-H1 is formed by the reaction of MG with the side chain of arginine residues. It has been revealed that arginine residues have a high frequency of occurrence in the active sites of enzymes and other proteins (Gallet et al., 2000, p. 917). Moreover, MG-H1 was identified as the most abundant AGEs in vivo in body fluids and cellular proteins (Thornalley, 2005, p. 111). Therefore, it is reasonable to speculate that the high reactivity of MG and the wide distribution of arginine may contribute to excessive MG-H1 response to the early progression of diabetes. The increase of MG-H1 residue might also arise in other species, especially in prediabetic human. Previous reports revealed that some AGEs (e.g., MG-H1, CEL residue of plasma protein) were increased both in diabetic patients (Ahmed et al., 2005b, p. 2465) and in streptozotocin (STZ)-induced diabetic rats (Karachalias et al., 2010, p. 1506). In this sense, although no study has been reported regarding the increase of MG-H1 at prediabetic stages, it should be reasonable for us to consider that the increase of MG-H1 residue of plasma protein at the prediabetic stages, followed by the apparent increase of plasma AGE-free adducts at diabetic stages, possibly can be a common phenomenon for any species.

3. INHIBITION OF AGES ACCUMULATION BY NATURAL PRODUCTS

Prevention of protein damages caused by AGE accumulation is expected to contribute to the preventive therapy of life-style related diseases, especially diabetic complications (Peyroux and Sternberg, 2006, p. 405). Recent studies have highlighted the beneficial role of daily intake of natural products with anti-glycation properties in the management of life-style related diseases (Kim and Kim, 2003, p. 1586; Jariyapamornkoon et al., 2013, p. 171). Therefore, studies on screening natural products as potential inhibitors of AGE formation attract lots of concern (Wu et al., 2011, p. 224).

3.1 In vitro Studies

The inhibition of AGE formation by natural products has been widely investigated in the in vitro assay. In an in vitro assay, glucose/fructose-mediated bovine serum albumin (BSA) protein glycation model is commonly used to evaluate antiglycation properties (Yamaguchi et al., 2000, p. 180). The formation of fluorescent AGEs is the major indicator of the protein damage. Antiglycation properties of some natural products are of a comparable order with that of medical inhibitor (Jang et al., 2010, p. 329). The antiglycation properties mainly involves in the antioxidant activity (Deetae et al., 2012, p. 953), the interaction with reactive dicarbonyl compounds (Wang et al., 2011, p. 2666), and the radical scavenging activity of natural products (Muthenna et al., 2012, p. 221).

Polyphenols, most abundant antioxidants existed in leaf tissue, the epidermis, bark layers, flowers, and fruits, are mainly classified into flavonoids (e.g., quercetin) and phenolic acids (e.g., caffeic acid) (Manach et al., 2004, p. 727). Evidence for their role in the prevention of life style-related diseases is emerging. Flavonoids have been demonstrated to inhibit glycoxidation effectively. Some typical flavonoids, including luteolin (**5**) (Schröder et al., 2004, p. 213; Kontogianni et al., 2013, p. 817), catechin (**15**) (Wang et al., 2011, p. 2666; Peng et al., 2008, p. 475), and quercetin (**20**) (Ferchichi et al., 2012, p. 98; Li et al., 2014, p. 12,152) were found to inhibit glycoxidation induced by sugars or dicarbonyls. Phenolic acids are a kind of polyphenol derived from plants. Some studies revealed that phenolic acids were a candidate that responsible for antiglycation effects of natural products in vitro (Silván et al., 2011, p. 208).

In vitro studies indicate that active compounds derived from natural products have potential anti-glycation properties. However, some foundational flaws in the experimental design make the results of little use from an in vivo point of view. In most assays, BSA glucose model cannot represent the real condition of protein glycation in vivo. In addition, AGEs with fluorescence only represent limited kinds of AGEs, therefore the inhibitory properties of natural products evaluated in vitro are inadequate. In any case, the earlier results indicate the potential of dietary consumption of polyphenols for the inhibition of AGE accumulation in vivo.

3.2 Studies on Animal Models

Some data regarding to the inhibition of AGE accumulation by natural products can be found in the literature. Most studies are carried out in diabetic rat models. The studies available suggested the beneficial potential of dietary supplementation of natural products, such as turmeric (Sajithlal et al., 1998, p. 1607) and garlic (Ola-Mudathir et al., 2008, p. 3604). The main outcomes include not only hypoglycemic activity but also decreased AGE accumulation in body fluids or tissues (Kim et al., 2007, p. 8491). Vitamin E (**22**), a typical antioxidant, has been proven to suppress AGE formation in rats (Aoki et al., 1992, p. 913). Supplement of the diet containing 5% caffeic acid for 12 weeks showed the protective effect of kidney in diabetic mice, including the reduction of renal CML and urinary glycated albumin (Chao et al., 2010, p. 388). Although some animal studies seem to indicate the antiglycation properties, however, the dosages are far higher than those for nutritional approach.

In addition, beneficial effects of natural products were mainly evaluated under the diabetic status due to the lack of appropriate prediabetic animal models, which made the results of little value concerning the prevention of diabetes. The aforementioned SDT rats, which possess a polygenic basis and develop severe diabetic complications resemble human type 2 diabetes, have been proven to be a suitable model for the prediabetic studies. More efforts should be made to develop alternative strategy to prevent the burden of diabetes in prediabetic subjects.

Examples of more detailed results in vitro and animal studies are shown in Tables 7.1 and 7.2 (structures are given in Fig. 7.4), respectively.

4. THE PREVENTION OF PREDIABETES BY NATURAL PRODUCTS

Although the extended investigation of prediabetes and its prevention is limited due to the lack of appropriate animal models, scientists have made great efforts to provide valuable evidences concerning this issue. Some clinical drug treatment strategies were proved to offer beneficial effects on prediabetic subjects, including troglitazone (Higa et al., 1999, p. 11,513), sitagliptin (Chen et al., 2011, p. 309), as well as the combination treatment of alogliptin and voglibose (Moritoh et al., 2010, p. 224).

The use of natural products to ameliorate prediabetic syndrome is also promising. The dietary supplement of black chokeberry extract improved the antioxidant status and leads to intestinal changes in a rat model (Jurgoński et al., 2008, p. 176). Zhang et al. (2014, p. 221) discovered that *Lactobacillus casei* ameliorated impaired glucose tolerance in hyperinsulinemia rats. Alves et al. (2015, p. 102) demonstrated that white tea consumption improves cardiac glycolytic and oxidative profile of prediabetic rats.

5. CONCLUDING REMARKS

Diabetes with multifactorial pathophysiology demands multimodal therapeutic approach. Natural products have been proven to have different pharmacological benefits to treat life style-related diseases. However, well-organized evidence and the mechanisms of action are not adequately available. Future studies need to be carefully designed so that the results can provide reliable evidence to understanding the mechanisms underlying the beneficial potentials. In addition, novel strategies, particularly targeted the prediabetic patients are needed to complement the current strategy to management diabetes.

TABLE 7.1 In vitro Inhibition Studies on AGE Accumulation by Natural Products

Source	Assay	Active Ingredient	Concentration/ Duration	AGE Measurement	Main Outcome	Ref(s).
Garcinia indica	BSA-fructose	Garcinol **(1)**	0.01, 0.1, 1 mM/5 days	Fluorescence, SDS-PAGE	↓ AGE formation↓cross-link formation	Yamaguchi et al. (2000)
Green tea (*Camellia sinensis*)	AAPH-BSA-glucose	EGCg **(2)**	0.25 μM/2 weeks	Fluorescence	↓ AGE formation	Nakagawa et al. (2002)
Gramineae (*Zea mays*)	Lysozyme-fructose	*C*-glycoside **(3)**	1 mM/1 week	Fluorescence	↓ AGE formation	Suzuki et al. (2003)
Garlic (*Allium sativum*)	Low-density lipoprotein-glucose	Organosulfur compounds	5, 10, 15 μM/6 days	Fluorescence	↑ Oxidative-delaying effects↓LDL glycation	Huang et al. (2004)
Tomato paste	BSA-glucose/BSA-fructose	Rutin **(4)**	7–49 days	Fluorescence, ELISA, Western blotting	↓ Fructosamines formation ↓ AGE formation	Kiho et al. (2004)
–	BSA-glucose/BSA-MG/ G.K. peptide-ribose	Flavonoids [e.g., luteolin **(5)**]	0–300 μM/7 days	Fluorescence, SDS-PAGE	↓ Protein glycation at each stage (early, middle, later)	Schröder et al. (2004)
Aged garlic extract (*A. sativum*)	Lysozyme-fructose	*S*-allylcysteine **(6)**	0–80 mM/35 days	PAGE-image analysis	↓ AGE formation	Ahmad and Ahmed (2006)
Rosemary (Rosmarinus officinalis)	Low-density lipoprotein-glucose	Carnosic acid **(7)**, rosmarinic acid **(8)**	9 days	Fluorescence	↓ AGE formation in LDL and collagen	Hsieh et al. (2007)
Vanilla	Glycation in apoptotic Neuro-2A cells	Vanillic acid **(9)**	–	Western blotting	↓ CML formation↓ROS production	Huang et al. (2005)
Mung bean (*Vigna radiata*)	BSA-glucose	Vitexin **(10)**, isovitexin **(11)**	100 μM/7 days	Fluorescence	↓ AGE formation	Peng et al. (2008)
Astragali radix (*Astragalus membranaceus*)	BSA-ribose	Astragaloside V **(12)**	0.03–1.0 mM/7 days	HPLC, ELISA	↓ CML formation↓pentosidine formation	Motomura et al. (2009)

–	Histone H1-MG/ADP-ribose	Rutin metabolites (e.g., 3,4-dihydroxy phenylacetic acid **(13)**)	25–1600 μM/12 h	Fluorescence, Western blotting	↓ AGE formation↓CML and argpyrimidine formation	Pashikanti et al. (2010)
Cinnamon bark (*Cinnamomum zeylanicum*)	BSA-glucose/ribose	Proanthocyanidins (e.g., procyanidin B2 **(14)**)	100 μM/7 days	Fluorescence, ELISA, HPLC	↓ Pentosidine formation↓CML formation↓Cysteine-Derived AGEs	Peng et al. (2010)
Berries	BSA-fructose/BSA-MG/Arginine-MG	Catechin **(15)**	0.05, 0.5 mg/mL; 6 days	Fluorescence	↑ MG scavenging↓AGE formation	Wang et al. (2011)
Cranberry (*Vaccinium macrocarpon*)	Human hemoglobin-glucose/Human serum albumin (HSA)-MG	Procyanidins (e.g., Epicatechin **(16)**)	4 days	Fluorescence	↑ MG scavenging↓AGE formation	Liu et al. (2011)
Calophyllum flavoramulum	BSA-ribose	Quercitrin **(17)**3,4-dihydroxybenzoic acid**(18)**	24 h	Fluorescence	↓ AGE formation	Ferchichi et al. (2012)
Olive leaves (*Olea europaea*)	BSA-ribose	Luteolin **(5)**, luteolin-4′-*O*-*β*-D-glucopyranoside **(19)**	10, 100 μM/21 days	Fluorescence	↓ Pentosidine formation	Kontogianni et al. (2013)
Lotus seedpod	BSA-glucose	Oligomeric procyanidins	0.01, 1.0 mg/mL; 35 days	Fluorescence	↑ MG scavenging↓AGE formation	Wu et al. (2013)
Red grape (*Vitis vinifera*)	BSA-fructose	–	0.03–0.5 mg/mL; 1–4 weeks	Fluorescence, ELISA	↑ Radical scavenging↓fructosamines formation↓fluorescent AGEs and CML formation	Jariyapamornkoon et al. (2013)
–	BSA-MG	Quercetin **(20)**	1.5 mM/0–720 h	Fluorescence	↑ MG and GO trapping↓AGE formation	Li et al. (2014)
Garlic (*A. sativum*)	Lysine-glucose	C-glutamylcysteine derivatives (e.g., γ-glutamylmethylcysteine **(21)**)	0.02 g/mL; 70°C, 60 min	HPLC, ELISA	↓ Lysine radicals↓AGE formation	Shi et al. (2016)

AGE, advanced glycation end product; *BSA*, bovine serum albumin; *CML*, Nε-carboxymethyl-lysine; *GO*, glyoxal; *LDL*, low density lipoprotein; *MG*, methylglyoxal; *ROS*, reactive oxygen species.

TABLE 7.2 In vivo Inhibition Studies on AGE Accumulation by Natural Products

Source	Animal	Active Ingredient	Intervention	AGE Measurement	Main Outcome	Ref(s).
–	Streptozotocin (STZ)-induced diabetic rats	Vitamin E **(22)**	Fed a vitamin E supplemented diet for 4 weeks	Collagen-linked fluorescence	↓ Serum lipid peroxide levels ↓ Fluorescence and thermal rupture time of collagen	Aoki et al. (1992)
Turmeric (*Curcuma longa*)	STZ-induced diabetic rats	Curcumin **(23)**	Supply curcumin (200 mg/kg/day) for 8 weeks	Competitive ELISA	↓ Serum lipid peroxidation products↓AGE-collagen	Sajithlal et al.(1998)
American ginseng (*Panax quinquefolium*)	STZ-induced diabetic rats	–	Treat with extract (100 mg/kg/day) for 20 days	Western blotting	↓ Serum glucose↓CML protein in renal↓ RAGE protein in renal	Kim et al. (2007)
Turmeric (*C. longa*)	STZ-induced diabetic rats	Curcumin, tetrahydrocurcumin (THC) **(24)**	Supply curcumin or THC (80 mg/kg/day) for 45 days	Collagen-linked fluorescence	↓ AGE formation↓cross-linking of collagen	Pari and Murugan (2007)
Green tea (*Camellia sinensis*)	STZ-induced diabetic rats	EGC **(25)**, EGCg **(2)**	Treat with green tea extract (300 mg/kg/day) for 6 weeks	Collagen-linked fluorescence	↓ Blood glucose↓AGE formation↓Cross-linking of collagen	Babu et al. (2008)
Garlic and onion (*Allium*)	STZ-induced diabetic mice	*S*-ethyl cysteine **(26)**, *S*-propyl cysteine **(27)**	Supply in drinking water (0, 1, 2 g/L) for 12 weeks	HPLC	↓ Plasma glucose↓Renal CML and pentosidine	Lin and Yin (2008)
Jeera (*Cuminum cyminum*)	STZ-induced diabetic rats	–	Treat with extract (200–600 mg/kg/day) for 28 days	Fluorescence	↓ Serum glucose↓Renal AGEs↑Radical scavenging	Jagtap and Patil (2010)
Japanese chestnut (*Castanea crenata*)	Otsuka Long-Evans Tokushima fatty (OLETF) rats	–	Supply 1% extract with powder diet for 23 weeks	Competitive ELISA	↓ Plasma and renal cortex CML	Mizutani et al. (2014)
Helianthus annuus	Methylglyoxal-induced hyperglycemic rats	Scopoletin **(28)**	Supply scopoletin (10 mg/kg/day) for 12 weeks by gavage	Immunohistochemistry (IHC) Stain	↓ Hepatic AGEs level↓Protein tyrosinephosphatase 1B	Chang et al. (2015)

AGE, advanced glycation end product; *CML*, *Nε*-carboxymethyl-lysine; *RAGE*, receptor of AGEs.

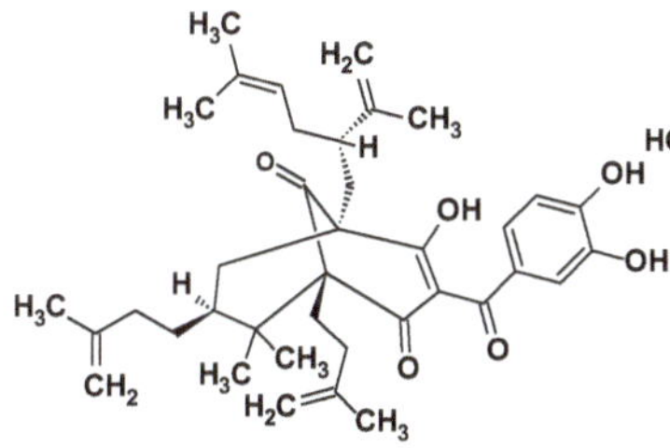

Garcinol (1)

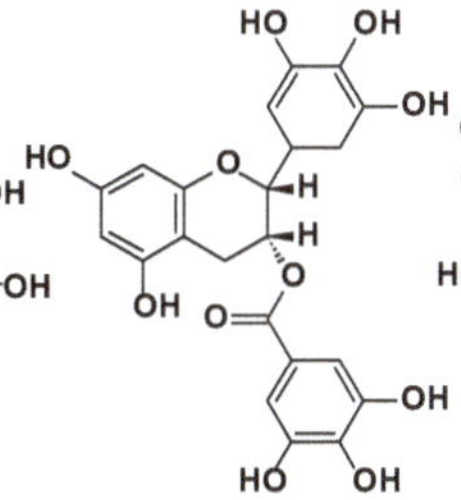

EGCg (2)

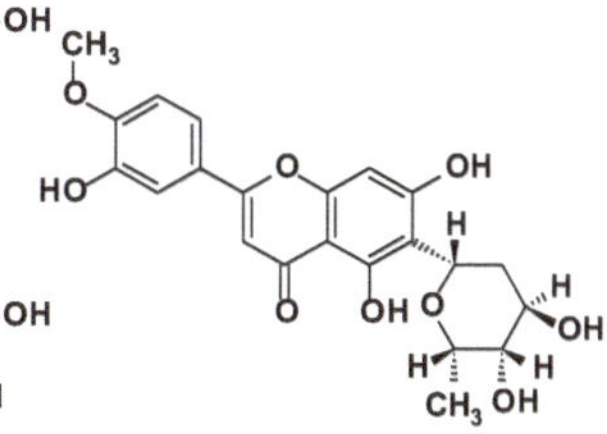

C-Glycoside (3)

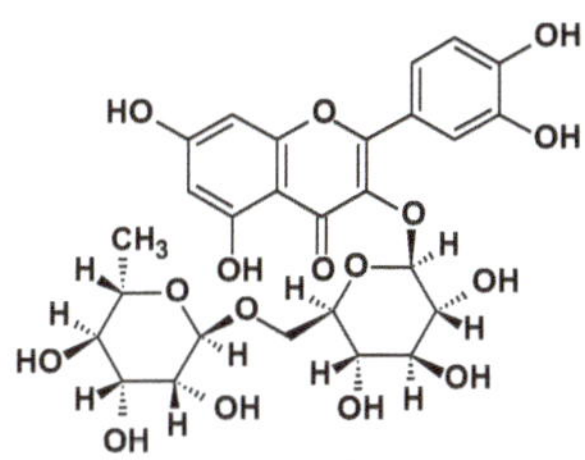

Rutin (4)

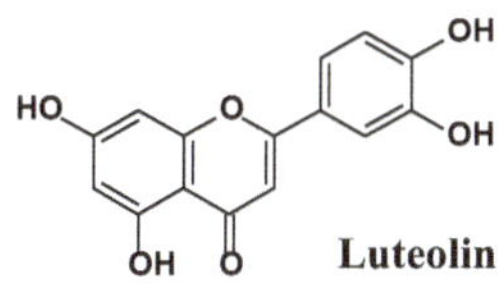

Luteolin (5)

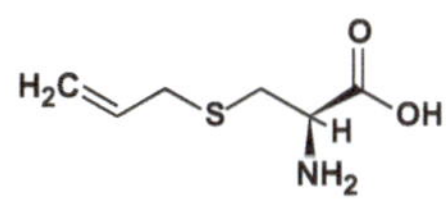

S-Allylcysteine (6)

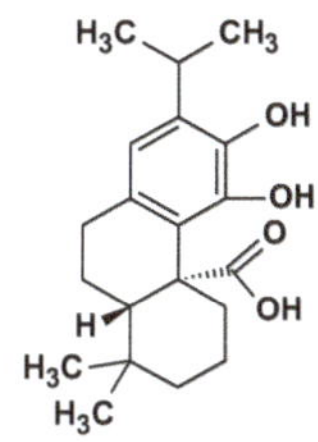

Carnosic acid (7)

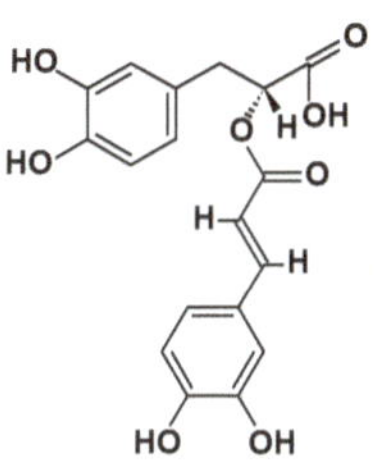

Rosmarinic acid (8)

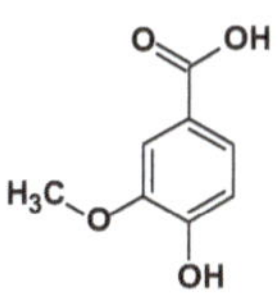

Vanillic acid (9)

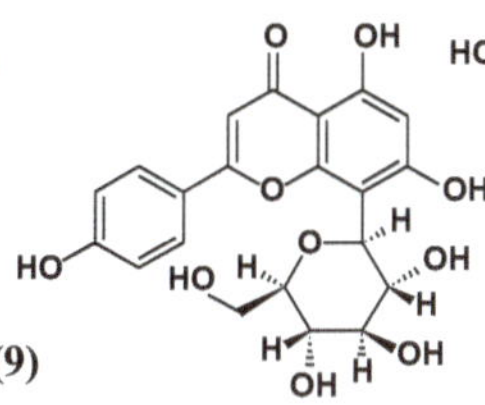

Vitexin (10)

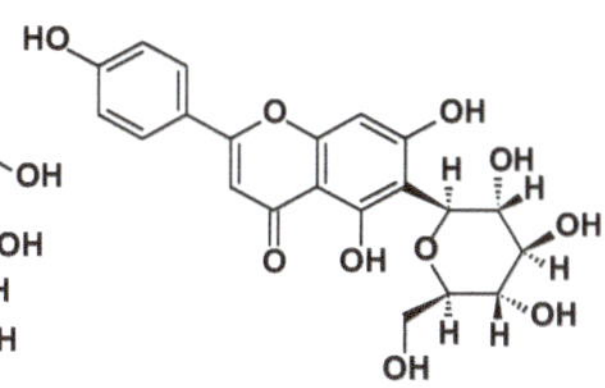

Isovitexin (11)

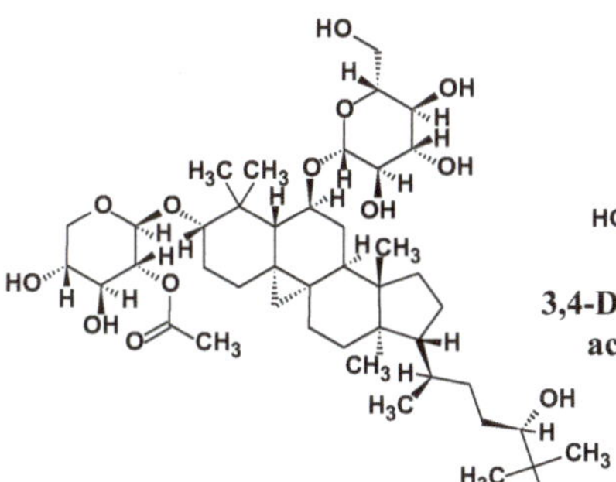

Astragaloside V (12)

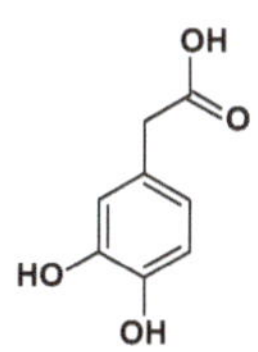

3,4-Dihydroxyphenyl-acetic acid (13)

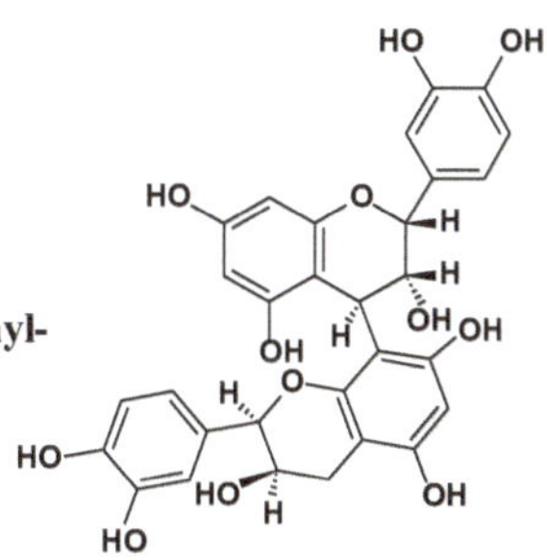

Proanthocyanidin B2 (14)

FIGURE 7.4 Structures of candidates for antidiabetic effect.

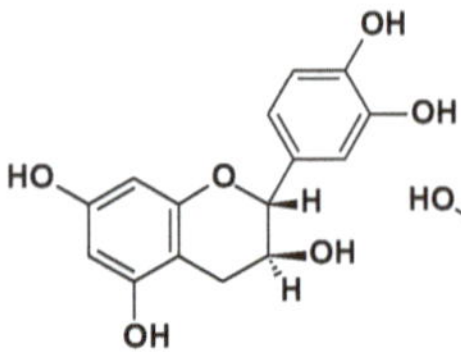

Catechin (15)

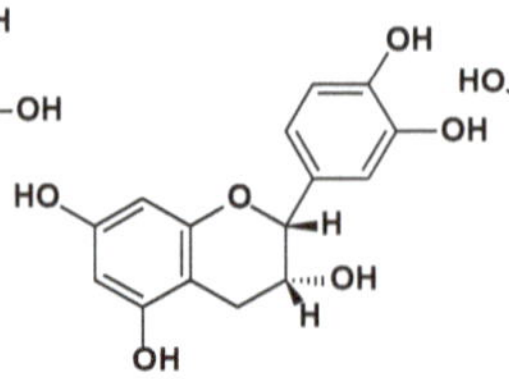

Epicatechin (16)

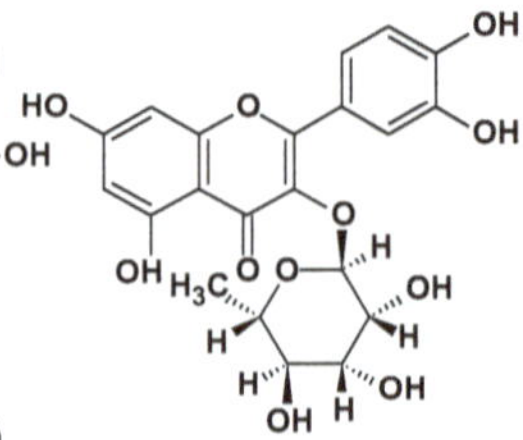

Quercitrin (17)

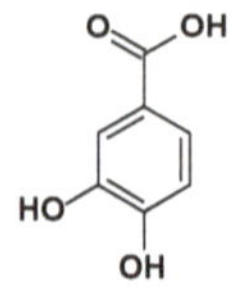

3,4-Dihydroxy-benzoic acid (18)

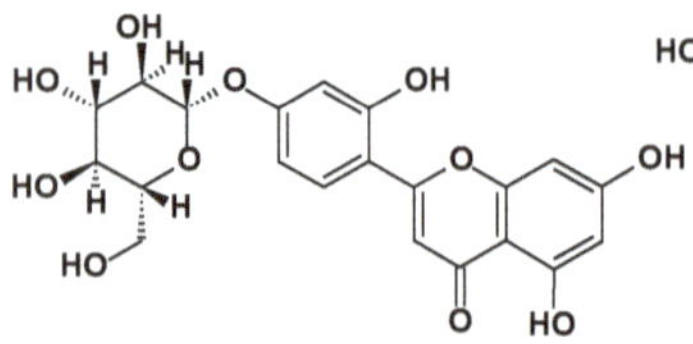

Luteolin-4'-*O*-*β*-D-glucopyranoside (19)

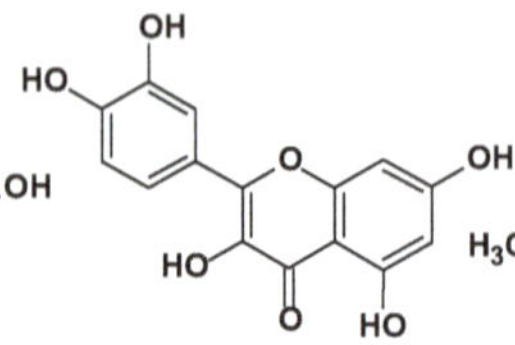

Quercetin (20)

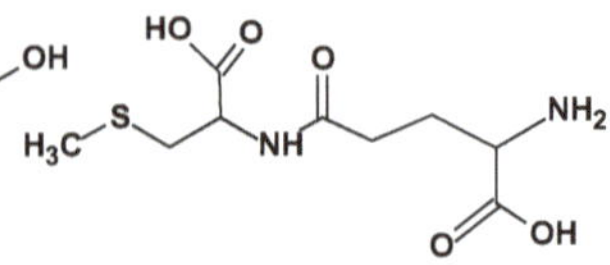

γ-glutamylmethylcysteine (21)

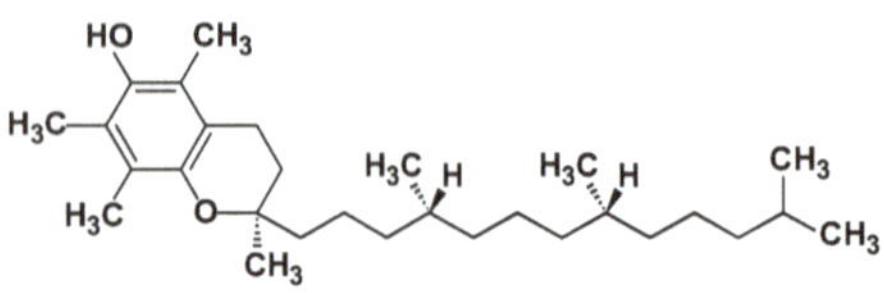

Vitamin E (22)

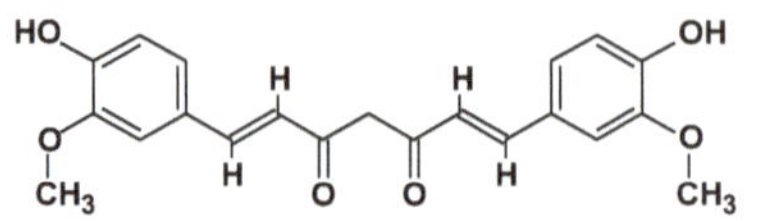

Curcumin (23)

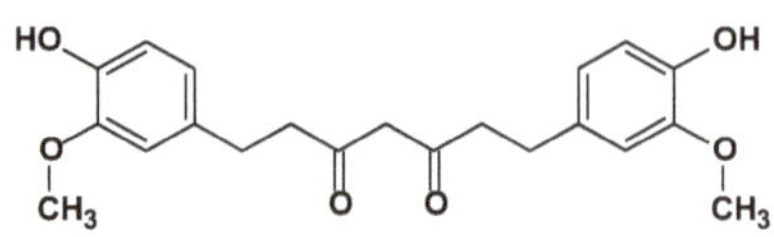

Tetrahydrocurcumin (24)

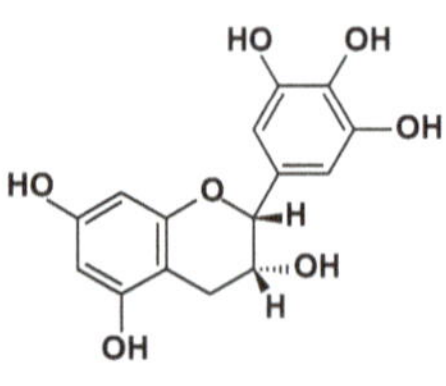

EGC (25)

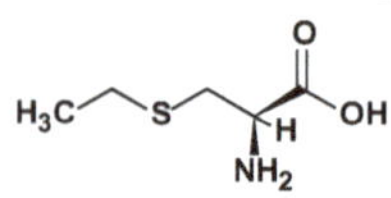

***S*-Ethyl cysteine (26)**

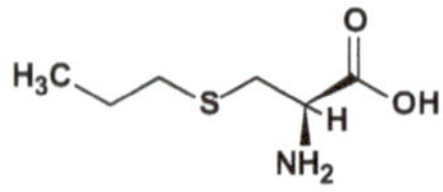

***S*-Propyl cysteine (27)**

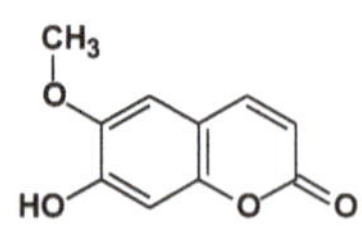

Scopoletin (28)

FIGURE 7.4 Cont'd.

LIST OF ABBREVIATIONS

3-DG 3-Deoxyglucosone
AGE Advanced glycation end product
BGL Blood glucose level
BSA Bovine serum albumin
CEL N^{ε}-(1-carboxyethyl)-lysine
CML N^{ε}-carboxymethyl-lysine

GO Glyoxal
HbA1c Hemoglobin A1c
MG Methylglyoxal
MG-H1 Methylglyoxal-derived hydroimidazolone
RAGE Receptor of AGEs
ROS Reactive oxygen species
SD Sprague–Dawley rat
SDT Spontaneously diabetic Torii rat
STZ Streptozotocin

REFERENCES

Ahmad, M.S., Ahmed, N., 2006. Antiglycation properties of aged garlic extract: possible role in prevention of diabetic complications. Journal of Nutrition 136 (3), 796S–799S.

Ahmed, N., Babaei-Jadidi, R., Howell, S.K., Beisswenger, P.J., Thornalley, P.J., 2005a. Degradation products of proteins damaged by glycation, oxidation and nitration in clinical type 1 diabetes. Diabetologia 48 (8), 1590–1603.

Ahmed, N., Babaei-Jadidi, R., Howell, S.K., Thornalley, P.J., Beisswenger, P.J., 2005b. Glycated and oxidized protein degradation products are indicators of fasting and postprandial hyperglycemia in diabetes. Diabetes Care 28 (10), 2465–2471.

Ahmed, N., Thornalley, P.J., 2007. Advanced glycation endproducts: what is their relevance to diabetic complications? Diabetes, Obesity and Metabolism 9 (3), 233–245.

Alves, M.G., Martins, A.D., Teixeira, N.F., Rato, L., Oliveira, P.F., Silva, B.M., 2015. White tea consumption improves cardiac glycolytic and oxidative profile of prediabetic rats. Journal of Functional Foods 14, 102–110.

American Diabetes Association, 2014. (2) Classification and diagnosis of diabetes. Diabetes Care 38 (Suppl. 1), S8–S16.

Aoki, Y., Yanagisawa, Y., Yazaki, K., Oguchi, H., Kiyosawa, K., Furuta, S., 1992. Protective effect of vitamin E supplementation on increased thermal stability of collagen in diabetic rats. Diabetologia 35 (10), 913–916.

Babu, P.V.A., Sabitha, K.E., Shyamaladevi, C.S., 2008. Effect of green tea extract on advanced glycation and cross-linking of tail tendon collagen in streptozotocin induced diabetic rats. Food and Chemical Toxicology 46 (1), 280–285.

Bohlender, J.M., Franke, S., Stein, G., Wolf, G., 2005. Advanced glycation end products and the kidney. American Journal of Physiology-Renal Physiology 289 (4), F645–F659.

Brownlee, M., 1995. Advanced protein glycosylation in diabetes and aging. Annual Review of Medicine 46, 223–234.

Brownlee, M., 2001. Biochemicstry and molecular cell biology of diabetic complications. Nature 414 (6865), 813–820.

Brownlee, M., Cerami, A., Vlassara, H., 1988. Advanced glycosylation end products in tissue and the biochemical basis of diabetic complications. The New England Journal of Medicine 318 (20), 1315–1321.

Brownlee, M., Vlassara, H., Kooney, A., Ulrich, P., Cerami, A., 1986. Aminoguanidine prevents diabetes-induced arterial wall protein cross-linking. Science 232 (4758), 1629–1632.

Cai, W., Ramdas, M., Zhu, L., Chen, X., Striker, G.E., Vlassara, H., 2012. Oral advanced glycation endproducts (AGEs) promote insulin resistance and diabetes by depleting the antioxidant defenses AGE receptor-1 and sirtuin 1. Proceedings of the National Academy of Sciences of the United States of America 109 (39), 15888–15893.

Chang, W.C., Wu, S.-C., Xu, K.-D., Liao, B.-C., Wu, J.-F., Cheng, A.-S., 2015. Scopoletin protects against methylglyoxal-induced hyperglycemia and insulin resistance mediated by suppression of advanced glycation endproducts (AGEs) generation and antiglycation. Molecules 20 (2), 2786–2801.

Chao, C., Mong, M., Chan, K., Yin, M., 2010. Anti-glycative and anti-inflammatory effects of caffeic acid and ellagic acid in kidney of diabetic mice. Molecular Nutrition & Food Research 54 (3), 388–395.

Chen, B., Moore, A., Escobedo, L.V.S., Koletsky, M.S., Hou, D., Koletsky, R.J., Ernsberger, P., 2011. Sitagliptin lowers glucagon and improves glucose tolerance in prediabetic obese SHROB rats. Experimental Biology and Medicine 236 (3), 309–314.

Chen, S.J., Aikawa, C., Matsui, T., 2015a. Quantitative analysis of methylglyoxal, glyoxal and free advanced glycation end-products in the plasma of Wistar rats during the oral glucose tolerance test. Biological & Pharmaceutical Bulletin 38 (2), 336–339.

Chen, S.J., Aikawa, C., Yoshida, R., Matsui, T., 2015b. Methylglyoxal- derived hydroimidazolone residue of plasma protein can behave as a predictor of prediabetes in Spontaneously Diabetic Torii rats. Physiological Reports 3 (8), e12477.

Deetae, P., Parichanon, P., Trakunleewatthana, P., Chanseetis, C., Lertsiri, S., 2012. Antioxidant and anti-glycation properties of Thai herbal teas in comparison with conventional teas. Food Chemistry 133 (3), 953–959.

Degenhardt, T.P., Alderson, N.L., Arrington, D.D., Beattie, R.J., Basgen, J.M., Steffes, M.W., Thorpe, S.R., Baynes, J.W., 2002. Pyridoxamine inhibits early renal disease and dyslipidemia in the streptozotocin-diabetic rat. Kidney International 61 (3), 939–950.

Ferchichi, L., Derbré, S., Mahmood, K., Touré, K., Guilet, D., Litaudon, M., Awang, K., Hadi, A.H., Le Ray, A.M., Richomme, P., 2012. Bioguided fractionation and isolation of natural inhibitors of advanced glycation end-products (AGEs) from *Calophyllum flavoramulum*. Phytochemistry 78, 98–106.

Forbes, J.M., Cooper, M.E., 2013. Mechanisms of diabetic complications. Physiological Reviews 93 (1), 137–188.

Forbes, J.M., Sourris, K.C., de Courten, M.P., Dougherty, S.L., Chand, V., Lyons, J.G., Bertovic, D., Coughlan, M.T., Schlaich, M.P., Soldatos, G., Cooper, M.E., Straznicky, N.E., Kingwell, B.A., de Courten, B., 2014. Advanced glycation end products (AGEs) are cross-sectionally associated with insulin secretion in healthy subjects. Amino Acids 46 (2), 321–326.

For the Diabetes, T. W. T., & Control and Complications Trial/Epidemiology of Diabetes Interventions and Complications Research Group, 2003. Sustained effect of intensive treatment of type 1 diabetes mellitus on development and progression of diabetic nephropathy: the Epidemiology of Diabetes Interventions and Complications (EDIC) study. JAMA: The Journal of the American Medical Association 290 (16), 2159–2167.

Gallet, X., Charloteaux, B., Thomas, A., Brasseur, R., 2000. A fast method to predict protein interaction sites from sequences. Journal of Molecular Biology 302 (4), 917–926.

Goh, S.-Y., Cooper, M.E., 2008. The role of advanced glycation end products in progression and complications of diabetes. The Journal of Clinical Endocrinology & Metabolism 93 (4), 1143–1152.

Goldberg, A.L., 2003. Protein degradation and protection against misfolded or damaged proteins. Nature 426 (6968), 895–899.

Goldberg, T., Cai, W., Peppa, M., Dardaine, V., Baliga, B.S., Uribrri, J., Vlassara, H., 2004. Advanced glycoxidation end products in commonly consumed foods. Journal of the American Dietetic Association 104, 1287–1291.

Goldin, A., 2006. Advanced glycation end products: sparking the development of diabetic vascular injury. Circulation 114 (6), 597–605.

Hammes, H.-P., Brownlee, M., Edelstein, D., Saleck, M., Martin, S., Federlin, K., 1994. Aminoguanidine inhibits the development of accelerated diabetic retinopathy in the spontaneous hypertensive rat. Diabetologia 37 (1), 32–35.

Hammes, H.-P., Du, X., Edelstein, D., Taguchi, T., Matsumura, T., Ju, Q., Lin, J., Bierhaus, A., Nawroth, P., Hannak, D., Neumaier, M., Bergfeld, R., Giardino, I., Brownlee, M., 2003. Benfotiamine blocks three major pathways of hyperglycemic damage and prevents experimental diabetic retinopathy. Nature Medicine 9 (3), 294–299.

Hashimoto, C., Iwaihara, Y., Chen, S.J., Tanaka, M., Watanabe, T., Matsui, T., 2013. Highly-sensitive detection of free advanced glycation end-products by liquid chromatography-electrospray ionization-tandem mass spectrometry with 2,4,6-trinitrobenzene sulfonate derivatization. Analytical Chemistry 85 (9), 4289–4295.

Higa, M., Zhou, Y.-T., Ravazzola, M., Baetens, D., Orci, L., Unger, R.H., 1999. Troglitazone prevents mitochondrial alterations, beta cell destruction, and diabetes in obese prediabetic rats. Proceedings of the National Academy of Sciences of the United States of America 96 (20), 11513–11518.

Hsieh, C.-L., Peng, C.-H., Chyau, C.-C., Lin, Y.-C., Wang, H.-E., Peng, R.Y., 2007. Low-density lipoprotein, collagen, and thrombin models reveal that Rosemarinus officinalis L. exhibits potent antiglycative effects. Journal of Agricultural and Food Chemistry 55 (8), 2884–2891.

Huang, C.-N., Horng, J.-S., Yin, M.-C., 2004. Antioxidative and antiglycative effects of six organosulfur compounds in low-density lipoprotein and plasma. Journal of Agricultural and Food Chemistry 52 (11), 3674–3678.

Huang, T.H.W., Peng, G., Kota, B.P., Li, G.Q., Yamahara, J., Roufogalis, B.D., Li, Y., 2005. Antidiabetic action of Punica granatum flower extract: activation of PPAR-gamma and identification of an active component. Toxicology and Applied Pharmacology 207 (2), 160–169.

Jagtap, A.G., Patil, P.B., 2010. Antihyperglycemic activity and inhibition of advanced glycation end product formation by Cuminum cyminum in streptozotocin induced diabetic rats. Food and Chemical Toxicology 48 (8–9), 2030–2036.

Jang, D.S., Yoo, N.H., Kim, N.H., Lee, Y.M., Kim, C.-S., Kim, J., Kim, J.-H., Kim, J.S., 2010. 3,5-Di-O-caffeoyl-epi-quinic acid from the leaves and stems of erigeron annuus inhibits protein glycation, aldose reductase, and cataractogenesis. Biological & Pharmaceutical Bulletin 33 (2), 329–333.

Jariyapamornkoon, N., Yibchok-Anun, S., Adisakwattana, S., 2013. Inhibition of advanced glycation end products by red grape skin extract and its antioxidant activity. BMC Complementary and Alternative Medicine 13 (1), 171.

Jurgoński, A., Juśkiewicz, J., Zduńczyk, Z., 2008. Ingestion of black chokeberry fruit extract leads to intestinal and systemic changes in a rat model of prediabetes and hyperlipidemia. Plant Foods for Human Nutrition 63 (4), 176–182.

Karachalias, N., Babaei-Jadidi, R., Rabbani, N., Thornalley, P.J., 2010. Increased protein damage in renal glomeruli, retina, nerve, plasma and urine and its prevention by thiamine and benfotiamine therapy in a rat model of diabetes. Diabetologia 53 (7), 1506–1516.

Kiho, T., Usui, S., Hirano, K., Aizawa, K., Inakuma, T., 2004. Tomato paste fraction inhibiting the formation of advanced glycation end-products. Bioscience, Biotechnology, and Biochemistry 68 (1), 200–205.

Kim, H.Y., Kang, K.S., Yamabe, N., Nagai, R., Yokozawa, T., 2007. Protective effect of heat-processed American ginseng against diabetic renal damage in rats. Journal of Agricultural and Food Chemistry 55 (21), 8491–8497.

Kim, H.Y., Kim, K., 2003. Protein glycation inhibitory and antioxidative activities of some plant extracts in vitro. Journal of Agricultural and Food Chemistry 51 (6), 1586–1591.

Kitabchi, A.E., Temprosa, M., Knowler, W.C., Kahn, S.E., Fowler, S.E., Haffner, S.M., Andres, R., Saudek, C., Edelstein, S.L., Arakaki, R., Murphy, M.B., Shamoon, H., 2005. Role of insulin secretion and sensitivity in the evolution of type 2 diabetes in the diabetes prevention program: effects of lifestyle intervention and metformin. Diabetes 54 (8), 2404–2414.

Knowler, W.C., Barrett-Connor, E., Fowler, S.E., Hamman, R.F., Lachin, J.M., Walker, E.A., Nathan, D.M., 2002. Reduction in the incidence of type 2 diabetes with lifestyle intervention or metformin. The New England Journal of Medicine 346 (6), 393–403.

Kontogianni, V.G., Charisiadis, P., Margianni, E., Lamari, F.N., Gerothanassis, I.P., Tzakos, A.G., 2013. Olive leaf extracts are a natural source of advanced glycation end product inhibitors. Journal of Medicinal Food 16 (9), 817–822.

Leiherer, A., Mündlein, A., Drexel, H., 2013. Phytochemicals and their impact on adipose tissue inflammation and diabetes. Vascular Pharmacology 58 (1–2), 3–20.

Li, X., Zheng, T., Sang, S., Lv, L., 2014. Quercetin inhibits advanced glycation end product formation by trapping methylglyoxal and glyoxal. Journal of Agricultural and Food Chemistry 62 (50), 12152–12158.

Lin, C.C., Yin, M.C., 2008. Effects of cysteine-containing compounds on biosynthesis of triacylglycerol and cholesterol and anti-oxidative protection in liver from mice consuming a high-fat diet. British Journal of Nutrition 99 (1), 37–43.

Lis, H., Sharon, N., 1993. Protein glycosylation. Structural and functional aspects. European Journal of Biochemistry 218 (1), 1–27.

Liu, H., Liu, Ha., Wang, W., Khoo, C., Taylor, J., Gu, L., 2011. Cranberry phytochemicals inhibit glycation of human hemoglobin and serum albumin by scavenging reactive carbonyls. Food and Function 2 (8), 475–482.

Manach, C., Scalbert, A., Morand, C., Remesy, C., Jimenez, L., 2004. Polyphenols: food sources and bioavailability. The American Journal of Clinical Nutrition 79 (5), 727–747.

Mizutani, T., Shizuka, F., Matsuzawa, T., Amano, Y., Arikawa, Y., 2014. Anti-glycation activity of Japanese chestnut (Castanea crenata) inner skin extract is beneficial for type 2 diabetes in a rat model. Anti-Aging Medicine 10, 112–119.

Motomura, K., Fujiwara, Y., Kiyota, N., Tsurushima, K., Takeya, M., Nohara, T., Nagai, R., Ikeda, T., 2009. Astragalosides isolated from the root of astragalus radix inhibit the formation of advanced glycation end products. Journal of Agricultural and Food Chemistry 57 (17), 7666–7672.

Moritoh, Y., Takeuchi, K., Hazama, M., 2010. Combination treatment with alogliptin and voglibose increases active GLP-1 circulation, prevents the development of diabetes and preserves pancreatic beta-cells in prediabetic db/db mice. Diabetes, Obesity & Metabolism 12 (3), 224–233.

Muthenna, P., Akileshwari, C., Reddy, G.B., 2012. Ellagic acid, a new antiglycating agent: its inhibition of Nε-(carboxymethyl) lysine. Biochemical Journal 442 (1), 221–230.

Nakagawa, T., Yokozawa, T., Terasawa, K., Shu, S., Juneja, L.R., 2002. Protective activity of green tea against free radical- and glucose-mediated protein damage. Journal of Agricultural and Food Chemistry 50 (8), 2418–2422.

Nakamura, S., Makita, Z., Ishikawa, S., Yasumura, K., Fujii, W., Yanagisawa, K., Kawata, T., Koike, T., 1997. Progression of nephropathy in spontaneous diabetic rats is prevented by OPB-9195, a novel inhibitor of advanced glycation. Diabetes 46 (5), 895–899.

Nathan, D.M., Buse, J.B., Davidson, M.B., Ferrannini, E., Holman, R.R., Sherwin, R., Zinman, B., 2009. Medical management of hyperglycaemia in type 2 diabetes mellitus: a consensus algorithm for the initiation and adjustment of therapy: a consensus statement from the American Diabetes Association and the European Association for the Study of Diabetes. Diabetologia 52 (1), 17–30.

Nathan, D.M., Davidson, M.B., Defronzo, R.A., Heine, R.J., Henry, R.R., Pratley, R., Zinman, B., 2007. Impaired fasting glucose and impaired glucose tolerance: implications for care. Diabetes Care 30 (3), 753–759.

Ola-Mudathir, K.F., Suru, S.M., Fafunso, M.A., Obioha, U.E., Faremi, T.Y., 2008. Protective roles of onion and garlic extracts on cadmium-induced changes in sperm characteristics and testicular oxidative damage in rats. Food and Chemical Toxicology 46 (12), 3604–3611.

Pari, L., Murugan, P., 2007. Influence of tetrahydrocurcumin on tail tendon collagen contents and its properties in rats with streptozotocin-nicotinamide-induced type 2 diabetes. Fundamental Clinical Pharmacology 21 (6), 665–671.

Pashikanti, S., de Alba, D.R., Boissonneault, G.A., Cervantes-Laurean, D., 2010. Rutin metabolites: novel inhibitors of nonoxidative advanced glycation end products. Free Radical Biology and Medicine 48 (5), 656–663.

Peng, X., Ma, J., Chao, J., Sun, Z., Chang, R.C.-C., Tse, I., Li, E.T.S., Chen, F., Wang, M., 2010. Beneficial effects of cinnamon proanthocyanidins on the formation of specific advanced glycation endproducts and methylglyoxal- induced impairment on glucose consumption. Journal of Agricultural and Food Chemistry 58 (11), 6692–6696.

Peng, X., Zheng, Z., Cheng, K.-W., Shan, F., Ren, G.-X., Chen, F., Wang, M., 2008. Inhibitory effect of mung bean extract and its constituents vitexin and isovitexin on the formation of advanced glycation end products. Food Chemistry 106 (2), 475–481.

Peyroux, J., Sternberg, M., 2006. Advanced glycation endproducts (AGEs): pharmacological inhibition in diabetes. Pathologie-Biologie 54 (7), 405–419.

Poulsen, M.W., Hedegaard, R.V., Andersen, J.M., de Courten, B., Bügel, S., Nielsen, J., Skibsted, L.H., Dragsted, L.O., 2013. Advanced glycation endproducts in food and their effects on health. Food and Chemical Toxicology 60, 10–37.

Rabbani, N., Thornalley, P.J., 2012. Glycation research in amino acids: a place to call home. Amino Acids 42 (4), 1087–1096.

Sajithlal, G., Chithra, P., Chandrakasan, G., 1998. Effect of curcumin on the advanced glycation and cross-linking of collagen in diabetic rats. Biochemical Pharmacology 56 (12), 1607–1614.

Schröder, L., Schmitz, C., Bachert, P., 2004. Molecular dynamics and information on possible sites of interaction of intramyocellular metabolites in vivo from resolved dipolar couplings in localized 1H NMR spectra. Journal of Magnetic Resonance 171 (2), 213–224.

Sell, D., Monnier, V., 1989. Structure elucidation of a senescence cross-link from human extracellular matrix. Implication of pentoses in the aging process. Journal of Biological Chemistry 264 (36), 21597–21602.

Shi, F., Bai, B., Ma, S., Ji, S., Liu, L., 2016. The inhibitory effects of γ-glutamylcysteine derivatives from fresh garlic on glycation radical formation. Food Chemistry 194, 538–544.

Silván, J.M., Assar, S.H., Srey, C., Dolores Del Castillo, M., Ames, J.M., 2011. Control of the Maillard reaction by ferulic acid. Food Chemistry 128 ((1), 208–213.

Suzuki, R., Okada, Y., Okuyama, T., 2003. Two flavone C-glycosides from the style of Zea mays with glycation inhibitory activity. Journal of Natural Products 66 (4), 564–565.

Tabák, A.G., Herder, C., Rathmann, W., Brunner, E.J., Kivimaki, M., 2012. Prediabetes: a high-risk state for diabetes development. The Lancet 379 (9833), 2279–2290.

Tahrani, A.A., Bailey, C.J., Del Prato, S., Barnett, A.H., 2011. Management of type 2 diabetes: new and future developments in treatment. The Lancet 378 (9786), 182–197.

Tanji, N., Markowitz, G.S., Fu, C., Kislinger, T., Taguchi, A., Pischetsrieder, M., Stern, D., Schmidt, A.M., D'agati, V.D., 2000. Expression of advanced glycation end products and their cellular receptor RAGE in diabetic nephropathy and nondiabetic renal disease. Journal of the American Society of Nephrology 11 (9), 1656–1666.

Thornalley, P.J., 2005. Dicarbonyl intermediates in the maillard reaction. Annals of the New York Academy of Sciences 1043 (1), 111–117.
Thornalley, P.J., 2003. Use of aminoguanidine (Pimagedine) to prevent the formation of advanced glycation endproducts. Archives of Biochemistry and Biophysics 419 (1), 31–40.
Thornalley, P.J., Battah, S., Ahmed, N., Karachalias, N., Agalou, S., Babaei-Jadidi, R., Dawnay, A., 2003. Quantitative screening of advanced glycation endproducts in cellular and extracellular proteins by tandem mass spectrometry. The Biochemical Journal 375 (2), 581–592.
Thornalley, P.J., Langborg, A., Minhas, H.S., 1999. Formation of glyoxal, methylglyoxal and 3-deoxyglucosone in the glycation of proteins by glucose. Biochemical Journal 344 (1), 109–116.
Tiwari, A.K., Madhusudana Rao, J., 2002. Diabetes mellitus and multiple therapeutic approaches of phytochemicals: present status and future prospects. Current Science 83 (1), 30–38.
Uribarri, J., Woodruff, S., Goodman, S., Cai, W., Chen, X., Pyzik, R., Yong, A., Striker, G.E., Vlassara, H., 2010. Advanced glycation end products in foods and a practical guide to their reduction in the diet. Journal of the American Dietetic Association 110 (6), 911–916.
Verzijl, N., Degroot, J., Thorpe, S.R., Bank, R.A., Shaw, J.N., Lyons, T.J., Bijlsma, J.W.J., Lafeber, F.P.J.G., Baynes, J.W., Tekoppele, J.M., 2000. Effect of collagen turnover on the accumulation of advanced glycation end products. Journal of Biological Chemistry 275 (50), 39027–39031.
Vlassara, H., Fuh, H., Makita, Z., Krungkrai, S., Cerami, A., Bucala, R., 1992. Exogenous advanced glycosylation end products induce complex vascular dysfunction in normal animals: a model for diabetic and aging complications. Proceedings of the National Academy of Sciences of the United States of America 89 (24), 12043–12047.
Voziyan, P.A., Hudson, B.G., 2005. Pyridoxamine: the many virtues of a maillard reaction inhibitor. Annals of the New York Academy of Sciences 1043 (1), 807–816.
Wada, R., Nishizawa, Y., Yagihashi, N., Takeuchi, M., Ishikawa, Y., Yasumura, K., Nakano, M., Yagihashi, S., 2001. Effects of OPB-9195, anti-glycation agent, on experimental diabetic neuropathy. European Journal of Clinical Investigation 31 (6), 513–520.
Wang, W., Yagiz, Y., Buran, T.J., Nunes, C.N., Gu, L., 2011. Phytochemicals from berries and grapes inhibited the formation of advanced glycation end-products by scavenging reactive carbonyls. Food Research International 44 (9), 2666–2673.
Wild, S., Roglic, G., Green, A., Sicree, R., King, H., 2004. Global prevalence of diabetes: estimates for the year 2000 and projections for 2030. Diabetes Care 27 (5), 1047–1053.
Wolff, S.P., Jiang, Z.Y., Hunt, J.V., 1991. Protein glycation and oxidative stress in diabetes mellitus and ageing. Free Radical Biology and Medicine 10 (5), 339–352.
Wu, Q., Chen, H., Lv, Z., Li, S., Hu, B., Guan, Y., Xie, B., Sun, Z., 2013. Oligomeric procyanidins of lotus seedpod inhibits the formation of advanced glycation end-products by scavenging reactive carbonyls. Food Chemistry 138 (2–3), 1493–1502.
Wu, C.-H., Huang, S.-M., Lin, J.-A., Yen, G.-C., 2011. Inhibition of advanced glycation endproduct formation by foodstuffs. Food & Function 2 (5), 224.
Yamaguchi, F., Ariga, T., Yoshimura, Y., Nakazawa, H., 2000. Antioxidative and anti-glycation activity of garcinol from garcinia indica fruit rind. Journal of Agricultural and Food Chemistry 48 (2), 180–185.
Yan, S.D., Schmidt, A.M., Anderson, G.M., Zhang, J.H., Brett, J., Zou, Y.S., Pinsky, D., Stern, D., 1994. Enhanced cellular oxidant stress by the interaction of advanced glycation end-products with their receptors binding-proteins. Journal of Biological Chemistry 269 (13), 9889–9897.
Zhang, Y., Wang, L., Zhang, J., Li, Y., He, Q., Li, H., Guo, X., Guo, J., Zhang, H., 2014. Probiotic *Lactobacillus casei* Zhang ameliorates high-fructose-induced impaired glucose tolerance in hyperinsulinemia rats. European Journal of Nutrition 53 (1), 221–232.

Chapter 8

Immunomodulators in Prophylaxis and Therapy of Type-1 Diabetes

V. Sravanthi, H.M. Sampath Kumar
Vaccine Immunology Lab, Natural Products Chemistry Division CSIR-Indian Institute of Chemical Technology, Hyderabad, India

1. INTRODUCTION

Autoimmune diseases are majorly caused due to the improper recognition of immune system between self- and nonself-entities. Many reasons lead triggering of autoimmune reactions in body which mainly include but are not limited to bacteria/virus, drugs, chemical irritants, and environmental factors. Type 1 diabetes (T1D), an autoimmune disease is outlined by hyperglycemia due to substantial loss of pancreatic insulin-secreting cells (β-cells). Destruction of pancreatic β-cells is due to progressive infiltration of cells into pancreatic islets, which are destroyed by T cells leading to an inflammatory condition known as *insulitis* (Bardsley and Want, 2004). Adverse immune response is caused and aggravated by many genetic and environmental factors over a period of time; progression of T1D is illustrated in Fig. 8.1. The major genetic factor involved in this autoimmune disease is human leukocyte antigen (HLA-DQ) alleles and many unknown environmental factors (Lernmark and Helena, 2013). This autoimmune disease is considered as juvenile diabetes and formerly as insulin-dependent diabetes, but later it was found that not all suffering with T1D are neither juvenile nor insulin dependent. Replacing the depleted insulin does not prevent marginal complications, which are major cause of mortality (James et al., 2002). According to a survey by International Diabetes Federation (IDF) worldwide, prevalence of diabetes mellitus was about 366 million in 2011 and is estimated to increase up to 522 million by 2030 (Diabetes federation, 2013). T1D represents about 10% of the population majorly of European heritage (EURODIAB ACE, 2000; Onkamo et al., 1999). β-Cell destruction by immune system is caused by many of its components like cluster differentiation (CD)4$^+$, CD8$^+$, B cells, natural killer (NK) cells, invariant natural killer cells (iNKT), dendritic cells (DCs), and macrophages. Studies on humans and

Discovery and Development of Antidiabetic Agents From Natural Products.
http://dx.doi.org/10.1016/B978-0-12-809450-1.00008-9

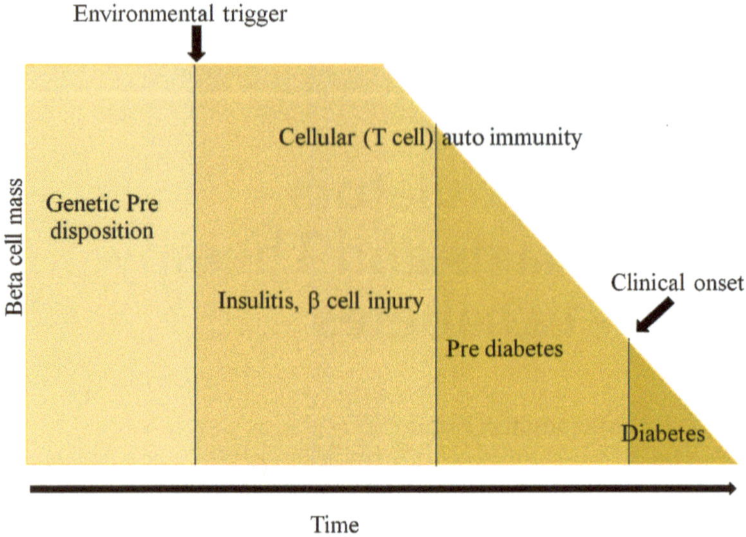

FIGURE 8.1 Progression of T1D.

animal models have revealed that disease pathogenesis involves both innate and adaptive immune responses (Bakay et al., 2013).

1.1 Epidemiology

Despite aforementioned controversies, T1D is one of the most common chronic diseases prevailing in children (Karvonen et al., 2000). T1D occurs as two peaks in childhood once at age of 5–7 years and other at puberty (Harjutsalo et al., 2008). As many other autoimmune diseases, T1D affects both males and females equally and in addition, incidence of T1D is higher in autumn and winter and lower in the summer (Moltchanova et al., 2009). According to recent surveys, T1D is mostly observed in Finland, Sardinia (Patterson et al., 2009) but contrastingly very few cases are observed in India and China (Atkinson et al., 2015).

1.2 Autoantigens of T1D

Most of the autoantigens associated with T1D are related to β-cell components like insulin, insulin and glutamic acid decarboxylase 65 (GAD65) (Baekkeskov et al., 1990), insulinoma antigen 2 (IA-2), heat shock protein (HSP) (Birk et al., 1996), islet-specific glucose 6 phosphatase catalytic protein-related subunit (IGRP) (Yang et al., 2006), imogen 38, pancreatic duodenal homeobox factor 1 (PDX1), chromogranin A (CHGA), islet amyloid polypeptide (IAPP) (Wenzlau et al., 2010; Delong et al., 2011; Wenzlau et al., 2007) and in recent times ZnT-8-β-cell-specific zinc transporter (Michael and Morran, 2008). Preclinical stage of the disease is associated with autoantibodies to these antigens and detection of these antibodies in serum is indicative of onset of T1D (Bingley et al., 1999; Achenbach et al., 2005). In addition, this stage also includes activation of self-reactive lymphocytes that enter pancreas and cause silent damage of β-cells. By onset of clinical stage, about 80% of β-cells would be destroyed.

1.3 Symptoms

Symptoms of T1D include polydipsia (increase in thirst), polyuria (frequent urination), polyphagia (extreme hunger), irritability and other mood changes, fatigue and weakness, blurred vision, and vaginal yeast infection in females.

1.4 Treatment

Due to complex intervention of immune and endocrine systems, very slow progress has been made in research and development of drugs used to treat T1D. At present, insulin is the only approved drug in market (Waldron-Lynch and Herold, 2011), despite many valuable phytoconstituents available for treatment of T1D very less effort is being made in this direction. A major reason for overlooking this area of green pharmacy is due to the requirement of tedious efforts in plant species identification, extraction of active molecule(s) from the plant, and their purifications and characterizations. Despite all the aforementioned efforts, only minimal amount of the active ingredients are usually obtained. At present, WHO has listed 21,000 plants, which are medicinally used worldwide, and among them 2500 species belong to India and hence, India is nicely called as "*botanical garden of the world*" (Seth and Sharma, 2004). Dating from ages, Indians have been using many herbs in the traditional system of medicine "Ayurveda." One good example is metformin—an oral hypoglycemic drug, which is prototype derived from *Galega officinalis* rich in guanidine is widely used in treating diabetes (Modak et al., 2007).

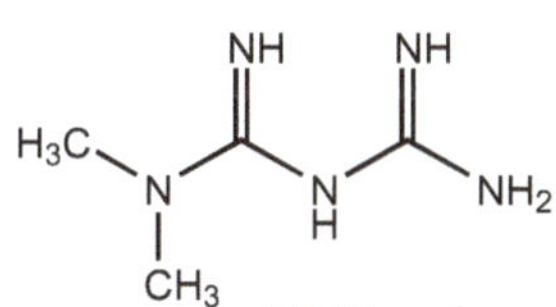

Metformin

Till date about 400 medicinal plants have been used and are reported for diabetes treatment and WHO expert committee has recommended for further investigation of these medicinal herbs.

The main obstacle in merging herbal medicine and modern day medicine is lack of scientific and clinical data for the former providing proof of their efficacy and safety. To some extent, potentiality of these phytoconstituents is being assessed using mice models like nonobese diabetic (NOD), streptozotocin (STZ)-induced diabetic but there is still requirement to develop simple bioassays for standardization and toxicological evaluation of herbal medicines (Modak et al., 2007).

Many options for treating T1D are available, which include drugs and immune modulators from plants. Immune modulators are gaining interest at present due to their capacity to alter immune response and avoid adverse interactions between immune system and pancreatic cells not only in initial stages but also after transplantation. In this chapter, we will discuss in detail about natural immunomodulators for prophylaxis and treatment of T1D.

2. IMMUNE MODULATORS

Immunomodulators are molecules that stimulate the immune system either as agonists or antagonists. Many natural phytochemicals act as immune system modulators in T1D majorly by two mechanisms, i.e., firstly, by regulating growth, differentiation of β-cells, and protecting them from invading T cells, and secondly, by regulating immune cells and modulating the Th1 and Th2 cytokine responses. Some phytoconstituents act both by protecting β-cells and also modulating the immune cells.

2.1 Prophylaxis

Prophylaxis for T1D has come into picture very recently due to the advancement of research in the areas of pathogenesis and factors responsible for its cause. A successful prophylaxis includes antigen-specific immunization or preserving the β-cells mass and its functions from invading immune cells. Apart from the previously mentioned prophylactic measures, avoidance of environmental stimuli to islet of pancreas such as cow milk, gluten, etc., are also effective ways to prevent this autoimmune disease. Metabolic modifications such as weight loss, increased physical activity are also among the suggested remedies (Rewers and Gottlieb, 2009).

2.1.1 Preserving β-Cells Mass

A booming approach for prevention of T1D is preserving β-cells mass, which holds good for not only existing cells but also to the regenerated ones. The major requirements for achieving the aforementioned goal are (1) controlling the immune system both in autoimmune reactions and also in allorejection, (2) protecting β-cells mass by immune stimulation, (3) using pharmacologically active ingredients for restimulating growth and differentiation of β-cells mass (Mineo et al., 2008). The aforesaid targets can be achieved by using antiinflammatory agents, immunomodulatory agents such as anti-CD3, anti-CD20 monoclonal antibodies.

2.1.2 Antigen-Specific Vaccination

Vaccination with autoantigens, which lead to β-cell destruction, is considered as another important parameter for prophylaxis of T1D. The first antigen considered for vaccination is pancreatic β-cell-specific antigen—insulin, whereas other antigens are expressed more widely in later time (Narendran et al., 2003). In infants who are at genetic predisposition to T1D, insulin is the first marker for preclinical disease, preceding autoantibodies to GAD65 and other antigens (Krishnamurthy et al., 2006). Second major antigen to be considered for vaccinating in T1D is GAD65. These antigen antibodies are sensitive and specific markers for T1D and are also present in NOD mice model (De Aizpurua et al., 1994; Tisch et al., 1993). Systemic administration of GAD65 or plasmid DNA-encoding gene has been reported to reduce the incidence of T1D (Petersen et al., 1994; Tisch et al., 2001).

2.2 Treatment

Many options are available for treatment of T1D, which include phytoconstituents that aid in rescuing β-cell from invading T cells and also which promote the regeneration and development of β-cell in pancreas. These constituents, depending on their mechanism of action, can be classified as follows.

2.2.1 Plant Constituents That Protect β-Cells

β-Cells in islets of pancreas are destroyed by T cells in the body due to inflammation of pancreas caused due to infiltration (insulitis). In this section, we will discuss about phytoconstituents that aid in protecting the β-cells from invading immune cells.

2.2.1.1 Alkaloids

An isoquinoline alkaloid, berberine obtained from many plants like *Hydrastis canadensis* (goldenseal), *Rhizoma coptidis* (Huanglian), *Berberis vulgaris* (barberry), has potential antiinflammatory effect by increasing glucagon like peptide-1 secretion and biosynthesis. Reports suggest that berberine protected NOD mice from T1D by significantly increasing population in the islet cells in a dose-dependent manner (Chueh and Lin, 2011). Researchers have also suggested that other alkaloid constituents of ginseng root extract show protective effect against pancreatic β-cell apoptosis in vitro (Yuan et al., 2012). Recent reports on kinsenoside, a phytochemical from *Anoectochilus roxburghii* plant, showed partial hypoglycemic activity in STZ mice, which may be attributed to β-cell repair or regeneration in addition to exert facilitation toward the glucose and lipid metabolism (Zhang et al., 2007, 2015).

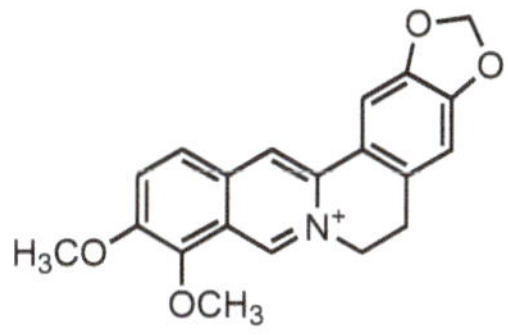

Berberine

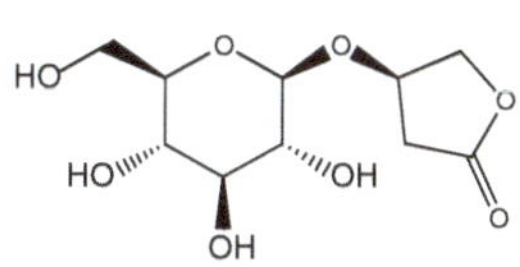

Kinsenoside

2.2.1.2 Polyphenols

Polyphenols are group of naturally occurring molecules with multiple phenol units. They are abundantly present in many plants. Most of phenolic compounds have been identified as active components for reducing blood glucose levels. (-)-Epigallocatechin-3 gallate ((-)-EGCG), one of the flavanols in tea has shown potentiality to delay T1D in NOD mice by protecting from β-cell death mediated by IAPP (Fu et al., 2011; Meng et al., 2010). Reports state that EGCG is also active in protecting β-cells from toxicity induced by proinflammatory cytokines and apoptosis as well by regulating the expression of Bcl-2 (B cell lymphoma 2) and suppressing Bax translocation to mitochondria (Babu et al., 2013). Next to EGCG, naringin isolated from many citrus fruits has a profound effect in STZ-induced animal models by increasing plasma insulin and leptin concentrations. Many researchers have also proved that antidiabetic effect of naringin is mainly due to lowering the hepatic glucose output (Babu et al., 2013).

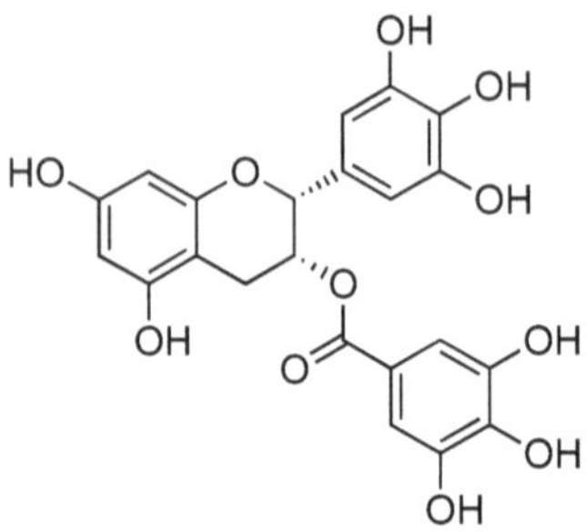

(-)-Epigallocatechin gallate (EGCG)

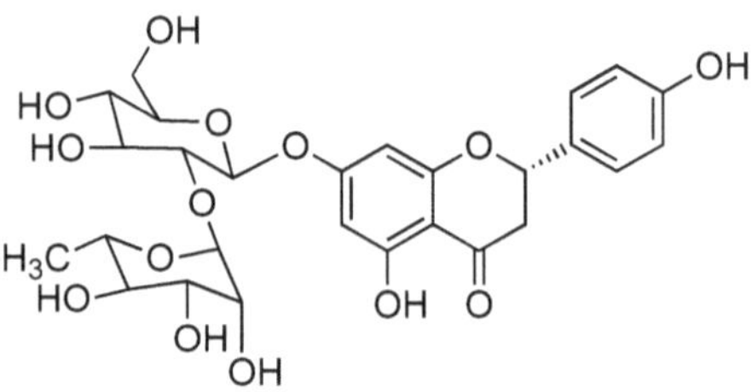

Naringin

2.2.2 *Plant Constituents That Promote Growth, Differentiation, and Maturation of β-Cells*

The second mechanism involved in treatment of T1D is using of phytoconstituents that promote the β-cells growth in pancreas to substitute the destroyed cells and also aid in their differentiation and maturation. Few promising such that phytoconstituents are mentioned herein.

2.2.2.1 Conophylline

Conophylline, an alkaloid derived from the plant *Tabernaemontana divaricata* or *Ervatamia microphylla*, was evaluated in generating and differentiating of pancreatic cells in vitro and in vivo (Kawakami et al., 2010; Ogata et al., 2004). This natural compound was found to reduce pancreatic islet cells fibrosis and inhibited cystic structure formation (Saito et al., 2012). Researchers proved that conophylline along with nicotinamide enhanced differentiation of insulin-producing cell monolayer and helped in insulin secretion (Kawakami et al., 2010). Hence, this alkaloid can generate and differentiate the pancreatic cells, acting in a dual manner.

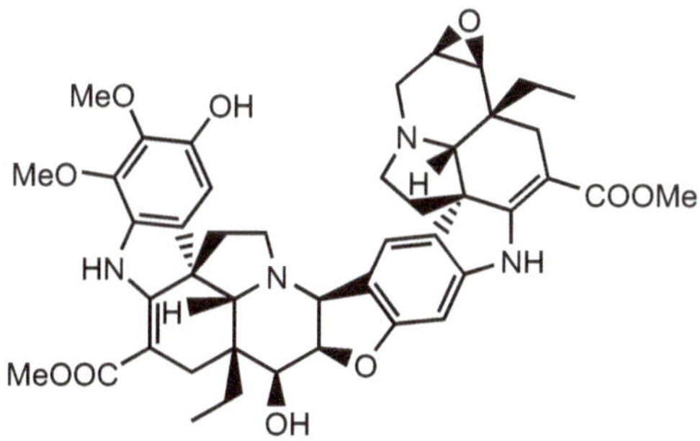

Conophylline

2.2.2.2 Silymarin

Flavonoids in general are plant metabolites that possess many health benefits. Silymarin, a combination of silybin A, silybin B, isosilybin A, isosilybin B, silychristin A, silydianin, and taxifolin, isolated from milk thistle (*Silybum marianum* L. Gaertn., Asteraceae) has shown protective action of β-cell in alloxan-treated rats. Researchers also reported that this drug enhances the serum insulin levels and reduces the raise of antioxidant enzymes and glutathione. Reports also suggest that silymarin enhanced NKx6.1 and insulin gene expression in pancreas, in addition to β-cell neogenesis in male Wistar rats. It also aided in raising the serum insulin levels and normalized the blood glucose levels in treated groups (Soto et al., 2014). In addition, this plant also has antioxidant and hepatoprotective activities (Soto et al., 2004).

2.2.2.3 Nicotinamide

Amide of nicotinic acid (niacin/vitamin B3) is abundantly found in plants, animals, and yeast cells. Nicotinamide exerts a beneficial effect on human β-cells and also aids in their growth and differentiation in mice, thus protecting from T1D (Sandler et al., 1993). Many reports also suggest that nicotinamide elicits a protective action on islets cells from inflammatory cytokines and prevents destruction of β-cells (Vague et al., 1989). Human trials in combination with insulin were performed and nicotinamide has shown similar response in C-type peptide secretion and residual β-cell function (Crino et al., 2005; Visalli et al., 1999).

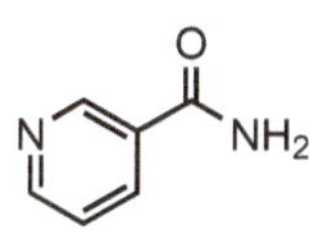

Nicotinamide

2.2.2.4 Other Plant Constituents

Carica papaya and *Pandanus amaryfollius* ethanolic extracts have shown reduction in blood glucose level through β-cell regeneration. Specific active components of these particular plants were not identified till data, but it is hypothesized that alkaloids, saponins, and tannins in both plants have bioactive phytochemicals (Sasidharan et al., 2011).

2.2.3 Phytochemical Ingredients That Regulate Immune Cells

Herein, we discuss the immunomodulatory compounds from plants that show effect on T1D condition in both animal and human models. Majority of saponins, glucosides, alkaloids, flavanols, steroids, anthranoids, polyynes, and stilbene class of compounds are used for treating T1D. Details of active components and their mechanism of action are described as follows.

2.2.3.1 Anthraquinones

Anthraquinones are naturally available functional chemicals with anthracene structure. Among all the anthracenes, *Fallopia japonica* has demonstrated preventive capacity of T1D in NOD mice by inhibiting leukocyte migration with help of CXCR4. Two of its active compounds viz., emodin and physcion suppress inflammation by inhibiting leukocyte movement and in turn preventing β-cell destruction (Shen et al., 2011). Other anthraquinones such as canterin, rhein, and cascarin also demonstrated β-cell protection by inhibiting CXCR4 and CCR5 receptors. These compounds also protect islet cells from destruction and diabetes in NOD mice by leukocyte migration inhibition (Shen et al., 2012).

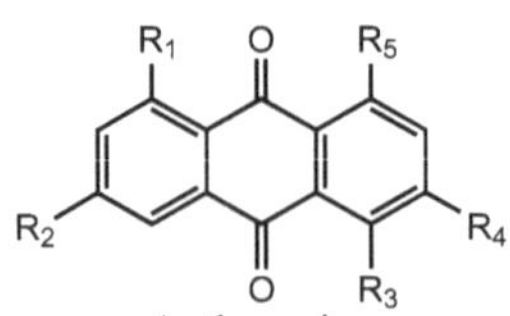

Anthraquinones

Emodin: $R^1 = R^3 = R^4 = R^5 = OH$: $R^2 = CH_3$

Physcion: $R^1 = R^5 = OH$; $R^2 = CH_3$; $R^3 = H$; $R^4 = OCH_3$

Rhein: $R^1 = R^5 = OH$; $R^2 = R^3 = H$; $R^4 = COOH$

2.2.3.2 Cannabinoids

Cannabinoids are phytochemicals that act on cannabinoid receptors and play important role in CNS (central nervous system) and immune system by targeting CR1 and 2 receptors. Cannabidiol and Δ^9-tetrahydro cannabinol —the major constituents of genus *Cannabis* have proven efficacy in reducing T1D incidences in NOD mice through T cell apoptosis induction and myeloid-derived suppressor cells (MDSCs) (Hegde et al., 2010; Lombard et al., 2007). Recent research on cannabinoids have revealed the protective action in mice by reducing the number of $CD3^+$ cells in pancreas and cytokines like interleukin (IL)-12, interferon (IFN)-γ, and Tumor necrosis factor (TNF)-α (Li et al., 2001).

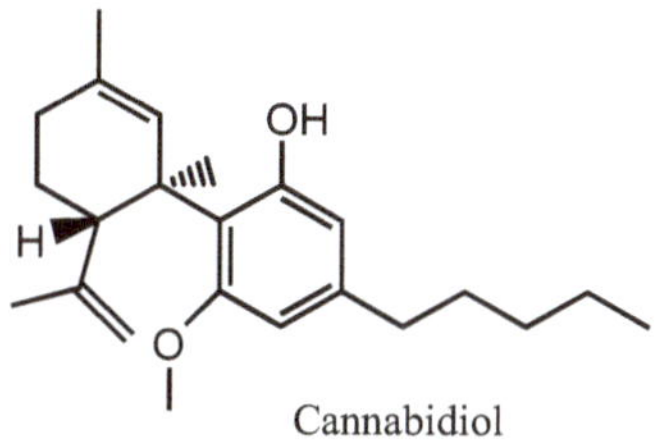

Cannabidiol

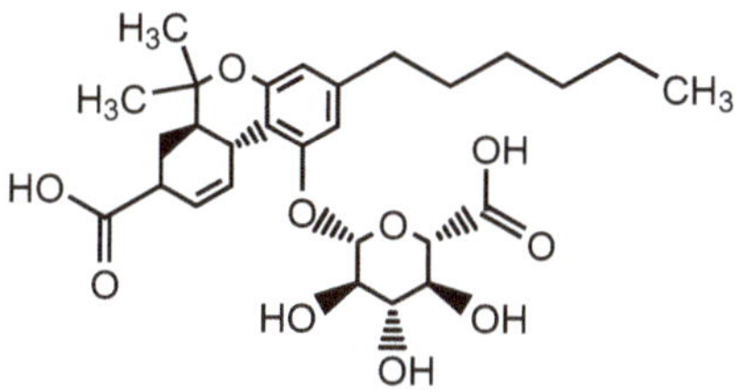

9-Tetrahydroxy cannabinol

2.2.3.3 Stilbene

Resveratrol (*trans*-3,4,5-trihydroxystilbene), a polyphenolic compound found in many fruits and vegetables mostly in grapes, has potent immunosuppressive activity used in autoimmune disease treatment (Das and Das, 2007). The major antiinflammatory effect of resveratrol is due to suppression of NF-κB-mediated production of inflammatory cytokines and inducing apoptosis. This compound inhibits chemokine receptor 6 production on immune cells leading to reduced islets infiltration and suppression of T1D in NOD mice (Lee and Yang, 2011). Reports suggest the inhibitory effect of resveratrol on macrophages expression of CCR6 and IL-17-producing cells migration to pancreas from peripheral lymph nodes is most important step in treating T1D.

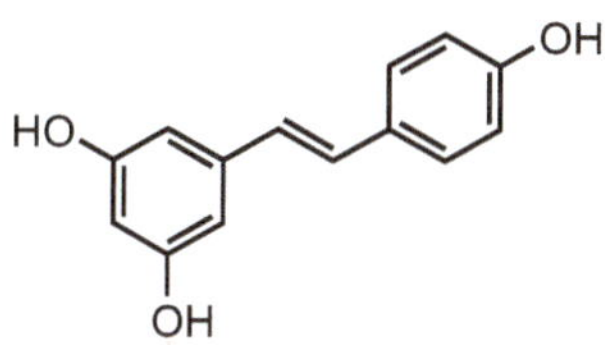

Resveratrol

2.2.3.4 *Tripterygium wilfordii* (Family: Celastraceae)

Triptolide, active compound of plant *Tripterygium wilfordii* belonging to family Celastraceae has concealed the development of T1D by suppressing T cell proliferation and also inducing T cell apoptosis (Wang et al., 2006; Yang et al., 1998). Reports suggest that crude extract of this plant has reducing effect on T1D in NOD mice through downregulation of Th1 cytokines and T cell activation (Shi et al., 2006). The number of CD4, CD8, CD4CD69, CD8CD69, and

IFN-γ-producing T cells in spleen was drastically reduced by triptolide treatment in NOD mice (Huang et al., 2013).

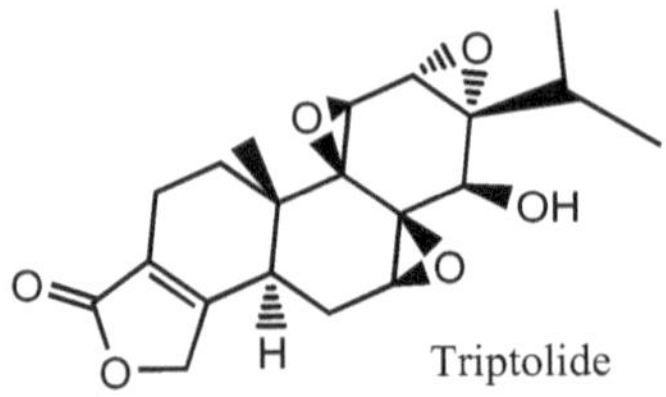

Triptolide

2.2.3.5 *Astragalus bisulcatus* (Family: Fabaceae)

Polysaccharide from *Astragalus bisulcatus* belonging to family Fabaceae has shown reduced T1D in NOD mice by impairing the ratio of CD4–CD 8 T cells, reducing proinflammatory cytokine (interleukin IL-1, IL-2, IL-6, IL-12, TNF-α, INF-γ, etc.) and increasing IL-4, IL-5, IL-10, TGF-β, Bcl-2, etc. taken as a whole these results indicate reduced T1D incidents through Th2/Th1cytokine ratio skewing (Chen et al., 2008).

2.2.3.6 *Olea europaea* (Family: Oleaceae)

Extracts of *Olea europaea* have shown inhibition of inflammation through nitric oxide (NO) production and proinflammatory cytokines (IFN-γ, TNF-α, and IL-17). Researchers have also found out that extract of this plant also has potential to induce insulin secretion and aid in peripheral glucose uptake (Cvjeticanin et al., 2010). Extracts of this plant have shown reduction in clinical signs of T1D and histopathological changes in spleen were also reduced (Cvjeticanin et al., 2010).

2.2.4 Compounds/Plants That Regulate Both Immune Cells and β-Cells

Some phytochemicals have dual qualities of regulating growth and development of β-cells along with regulation of immune system. One such class of compounds is polyenes, which regulate both immune response and also β-cell regeneration.

2.2.4.1 *Bidens pilosa* (Family: Asteraceae)

Polyyenes from the plant *B. pilosa* like 2-β-D-glucopyranosyloxy-1-hydroxytrideca-5,7,9,11-tetrayne (cytopiloyne), 2-ß-D-glucopyranosyloxy-1-hydroxy-5(E)-tridecene-7,9,11-triyne, and 3-ß-D-glucopyranosyl-1-hydroxy-6 (E)-tetradecene-8,10,12-triyne have shown modulated T cell differentiation and cytokine production and suppressed T1D (Chang et al., 2007). Moreover, *B. pilosa* extract possesses the property of protecting β-cells and enhancing insulin production in diabetic mice. Reports suggest that butanol extract of this plant inhibited Th1 cytokines and enhanced Th2 cytokine production in NOD mice. Polyacetylenic compounds and butanol fraction from *B. pilosa* can modulate the differentiation of helper T cells and prevent autoimmune diabetes in NOD mice (Chang et al., 2004). Thus, this plant has dual advantage of immune system regulation and endocrine system (Hsu et al., 2009; Chien et al., 2009).

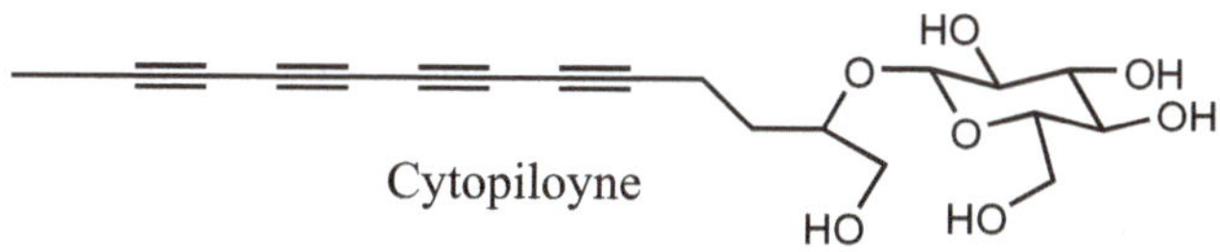

Cytopiloyne

2.2.4.2 Capsicum annuum (Family: Solanaceae)

Capsaicin, a component of capsicum plant increased population of CD11b+ F4/80+ macrophages and offered protection from T1D in NOD mice. Reports suggest that chilli pepper extract at 2% exerts an insulinotropic action on β-cells. No effect on liver glycogen, serum lipid levels were observed upon treatment with chilli extract (Islam and Choi, 2008).

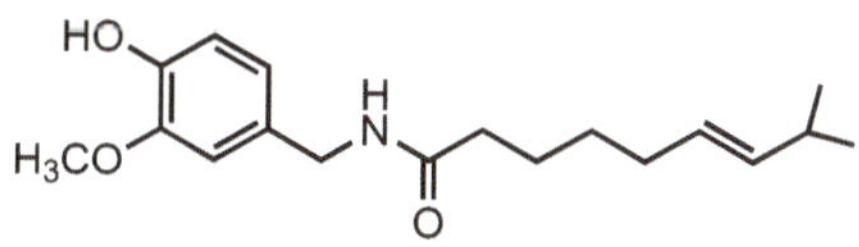

Capsicin

Apart from all the aforementioned phytoconstituents, many other plant extracts and chemical constituents as listed in Table 8.1 are also active in treating T1D. But the major drawback with these compounds is their unknown mechanism of action, which is not clear till date and much research is yet to be carried out with in vitro and in vivo experimentation.

3. OTHER OPTIONS FOR TREATMENT AND PROPHYLAXIS OF T1D

3.1 Protein/Peptide-Based Therapy

Hsp60: Cohens group identified for the first time that T cells react against mycobacterial Hsp60 along with its human and murine homolog's during pre-diabetic stage (Elias et al., 1990). Injecting with Hsp60 peptide spanning from 437 to 460 residues in incomplete Freund's adjuvant (IFA) or mineral oil prevented the progression of hyperglycemia in NOD mice by inducing antigenic shift from Th1 to Th2 response (Elias et al., 1991).

3.2 Dendritic Cells Targeted Peptide Therapy

Engagement of cognate pMHC (peptide major histocompatibility) by T cells on DCs results in tolerogenic responses along with deletion of cognate T cells and also

TABLE 8.1 Other Plant Sources of Antidiabetic Compounds (Ezuruiken and Prieto, 2014; Mohana Lakshmi et al., 2012)

Plant (Family)	Active Ingredient
Abroma augusta (put family in each case)	Volatile oil
Acacia arabica	Arabin
Annona squamosa	Acetogenins, squamoside
Allium sativum	Allin, allicin
Allium cepa	Protein, carbohydrate, vitamin A, B, C, allyl propyldisulfide
Aloe barbadensis	Barbaloin, isobarbaloin, resin
Artemisia pallens	Davanone
Azadirachta indica	Nimbin, nimbidin
Boerhavia diffusa	Punarnavine, punarnavoside
Camellia sinensis	Polyphenols
Capparis decidua	Capsaicin
Cinnamomum zeylanicum	Volatile oil, tannins
Emblica officinalis	Vitamin C, tannin
Eucalyptus	Cineol, essential oil
Ficus benghalensis	Tannin
Gymnema sylvestre	Gymnemic acid, Quercitol
Helicteres isora	Saponin, tannin, lignin
Ipomoea aquatica	Carotene
Lupinus albus	Alkaloid, fatty oil, asparagines
Mangifera indica	Mangiferin
Myrtus communis	Mirtiioleum
Momordica charantia	Momordicine alkaloid, ascorbic acid
Nelumbo nucifera	Nuciferin, Nornuciferin
Panax ginseng	Glycans, Panaxans I,J,K
Punica granatum	Vitamin C, protein, tannin, gallic acid, Pelletierine
Picrorhiza kurroa	Picrorhizin, kutkin
Swertia chirayata	Zanthone mangiferin, gentianine, Swerchirin
Trigonella foenum-graceum	Protein, fat, volatile oil, fixed oil, carbohydrate

TABLE 8.1 Other Plant Sources of Antidiabetic Compounds (Ezuruiken and Prieto, 2014; Mohana Lakshmi et al., 2012)—Cont'd

Plant (Family)	Active Ingredient
Urtica dioica	Fatty oil
Vinca rosea	Vincristine, vinblastine
Withania somnifera	Withanolides,withaferine, withanine, somnine
Xanthium strumarium	Phenolics, caffeic acid
Zingiber officinale	Sesquiterpene
Zizyphus sativa	Tannin

expansion of T regulatory cells (Mahnke et al., 2003). This strategy was exploited to target antigens to immature DCs and promote tolerance to self-antigens. This targeting is majorly done by conjugating antigen (peptide/protein) to an antibody (anti-DEC-205) against C-type lectin receptor, which is an endocytic receptor (Jiang et al., 1995). A parallel approach includes conjugation of complementarity-determining region 3 (CDR 3) of immunoglobulin molecule with antigen, which was not so effective during prediabetic stage (Greeg et al., 2004).

3.3 Altered Peptide Ligands (APLs)

These ligands can be classified depending on type of T cell response that they can induce. Altered peptide ligand (APL) of imogen 38 has shown to block antigen-induced activation of cognate human T cell clone but the mechanism remains unclear (Geluk et al., 1998). Recently, Neurocrine Biosciences, Inc. developed an APL for dominant epitope of insulin B chain (9–23 residues) referred as NBI-6024, which induced specific Th2 cells which delayed T1D in mice (Alleva et al., 2002) and it had no effects when administered to T1D patients (Walter et al., 2009).

3.4 DNA Vaccines

Plasmid-encoding autoantigens are also used to induce tolerogenic effect against specific autoantigens, as their efficiency is very low local electroporation, pretreatment with bupivacaine, cardiotoxin, or using gene gun are recommended (Prud'homme et al., 2007). Insulin DNA vaccination and GAD65 DNA vaccination are the one which are in clinical trials.

3.5 Monoclonal Antibodies

Monoclonal antibodies targeted to CD3 and CD20 have received lot of attention as this antibody transiently activates CD3 receptor, which initiates cytokine release and finally blocks T cell proliferation and differentiation. Humanized Ortho-Kung T cell antibody, rituximab, is among the one in clinical trials (Rewers and Gottlieb, 2009).

4. HERBAL FORMULATIONS IN MARKET

Many of the multinational companies are marketing combination of herbal active ingredients as formulations for diabetes treatment. All these products are based on the Ayurvedic knowledge of activity of the active ingredients. Some of the marketed products are listed in Table 8.2 (Modak et al., 2007).

TABLE 8.2 List of Marketed Herbal Formulations

Product	Company	Some of Active ingredients
Diabecon	Himalaya	*Glycyrrhiza glabra, Asparagus racemosus, Tribulus terrestris, Gmelina arborea*
Diasulin	Catalysis	*Cassia auriculata* *Gymnema sylvestre* *Syzygium cumini, Tinospora cordifolia,* *Trigonella foenum-graecum*
Pancreatic tonic 180 cp	Ayurvedic herbal supplement	*Pterocarpus marsupium* *Azadirachta indica, Ficus racemosa* *Cinnamomum tamala*
Ayurveda alternative herbal formula	Chakrapani Ayurveda	*G. sylvestre, pushkarmool, Azadirachta indica* *T. cordifolia*
Bitter gourd powder	Garry and Sun natural remedies	*Momordica charantia*
Diacare	Admark Herbals Limited	*Terminalia chebula, A. indica*
Gurmar powder	Garry and Sun natural remedies	*G. sylvestre*
Diabetes daily care	Nature's health Supply	α-Lipoic acid, *G. sylvestre* extract, *Momordica* *G. glabra*
Diabecure	Nature Beaute Sante	*Juglans regia* Taraxacum *Erytherea centaurium*
Epinsulin	Swastik Formulations	*Pterocarpus marsupium*

5. ADVANTAGES OF NATURAL IMMUNOMODULATORS

The major advantages of natural immunomodulators are as follows:

- active in very minute doses
- reduced side effects
- plenty of availability in nature
- requires less cost of production

6. CLINICAL TRIALS

Validation of biological active ingredients of plants involving human volunteers is necessary step toward possible amalgamation of traditional and modern health-care system. Table 8.3 represents few of the ongoing clinical trial lists of T1D drugs. Efforts should be made in such a way that complete profiling of the active ingredients should be done with biological and toxicological data for further preceding the molecule to clinical trials.

7. CONCLUDING REMARKS

In this chapter, we have discussed about many natural phytochemicals that are used in treatment and prophylaxis of T1D. As discussed earlier, there are many

TABLE 8.3 Ongoing Clinical Trials in T1D

Trial	Sponsor	Phase
Islet regeneration	NIH, Baylor College of Medicine/ Rubina Heptulla, MD	Ongoing
DIAGNODE-1	Johnny Ludvigsson	Phase 1
LADA	Norwegian University of Science and Technology	Ongoing
DIABGAD	Johnny Ludvigsson	Phase 2
Roles of T helper 1 cytokines in type 1 diabetes	Istanbul Medeniyet University	Ongoing
CSII in type 1 diabetes: Diet, Quality of Life & Cardiometabolic Risks—A Longitudinal study	Liverpool John Moores University	Ongoing
Profiling of Original Cellular and Humoral Biomarkers of Type 1 Diabetes (Lymphoscreen)	Nantes University Hospital	Ongoing
Transdermal Basal Insulin Patch Study in Type 1 Diabetes	Altea Therapeutics	Phase 2

advantages of using natural product-based therapies for T1D, but the disadvantages like toxicity, purity of the active ingredient, and methods of isolation are to be emphasized mainly for a complete and successful treatment. In addition to this, a combined approach like using constituents that regulate both immune system and β-cell growth is appreciated than single agents. There is an urgent need for developing and standardizing the herbal extracts using modern tools and technologies for a better product development and achieving consistency in the method developed. Further research is needed in identifying the mechanism of action of many of the known herbal ingredients, which already have antidiabetic effect. Potential drug–herb interaction studies are to be given first priority in this development process and traditional knowledge like *Ayurveda* and other ancient medicines should be treasured and used in combination with present day advanced technology for better results. Further efforts are needed in direction of traditional and natural medicines for proper structural elucidation, characterization, and assessing their potentiality in terms of toxicity and activity.

In conclusion, as large numbers of drugs have been developed for various diseases derived from natural products and also majority of the population relies on traditional and folk medicine, more emphasis on plant and natural products should be given for the development of treatment and prophylaxis of T1D. Such natural product-based approach to immune prophylaxis and therapy of diabetes can offer commercially significant entities and lead molecules for not only T1D but many other autoimmune diseases.

LIST OF ABBREVIATIONS

APLs Altered peptide ligands
Bcl-2 B cell lymphoma 2
CD Cluster differentiation
CDR 3 Complementarity-determining region
CNS Central nervous system
CXCR/CCR5 Chemokine receptor type 4
CHGA Chromogranin A
DC Dendritic cells
DEC-205 Dectin receptor
EGCG Epigallocatechin-3 gallate
GAD Glutamic acid decarboxylase
HLA Human leukocyte antigen
HSP Heat shock protein
IA-2 Insulinoma antigen 2
IAPP Islet amyloid polypeptide
IDF International Diabetes Federation
IFA Incomplete Freund's adjuvant
IFN Interferon
IGRP Islet-specific glucose 6 phosphatase catalytic subunit-related protein
IL Interleukin
iNKT invariant natural killer cells

MDSCs Myeloid-derived suppressor cells
NF-KB Nuclear factor kappa-light-chain-enhancer of activated B cells
NK Natural killer cells
NOD Nonobese diabetic mouse
PDX1 Pancreatic duodenal homeobox factor 1
pMHC Peptide major histocompatibility
STZ Streptozotocin-induced diabetes in mice
TNF Tumor necrosis factor
T1D Type 1diabetes
WHO World Health Organization
Zn 8 Zinc transporter 8

REFERENCES

Achenbach, P., Bonifacio, E., Ziegler, A.G., 2005. Predicting type 1 diabetes. Current Diabetes Reports 5, 98–103.

Alleva, D.G., Gaur, A., Jin, L., Wegmann, D., Gottlieb, P.A., Pahuja, A., Johnson, E.B., Motheral, T., Putnam, A., Crowe, P.D., Ling, N., Boehme, S.A., Conlon, P.J., 2002. Immunological characterization and therapeutic activity of an altered-peptide ligand, NBI-6024, based on the immunodominant type 1 diabetes autoantigen insulin B-chain (9–23) peptide. Diabetes 51, 2126–2134.

Atkinson, M.A., Eisenbarth, G.S., Michels, A.W., 2015. Type 1 diabetes. Lancet 383, 4–10.

Babu, P.V.A., Liu, D., Gilbert, E.R., 2013. Recent advances in understanding the anti-diabetic actions of dietary flavonoids. Journal of Nutritional Biochemistry 24, 1777–1789.

Baekkeskov, S., Aanstoot, H.J., Christgau, S., Reetz, A., Solimena, M., Cascalho, M., Folli, F., Richter-Olesen, H., De Camilli, P., 1990. Identification of the 64K autoantigen in insulin-dependent diabetes as the GABA-synthesizing enzyme glutamic acid decarboxylase. Nature 347, 151–156.

Bakay, M., Pandey, R., Hakonarson, H., 2013. Genes involved in type 1 diabetes: an update. Genes (Basel) 4, 499–521.

Bardsley, J.K., Want, L.L., 2004. Overview of diabetes. Critical Care Nursing Quaterly 27, 106–112.

Bingley, P.J., Williams, A.J., Gale, E.A., 1999. Optimized autoantibody-based risk assessment in family members. Implications for future intervention trials. Diabetes Care 22, 1796–1801.

Birk, O.S., Elias, D., Weiss, A.S., Rosen, A., Van-Der Zee, R., Walker, M.D., Cohen, I.R., 1996. NOD mouse diabetes: the ubiquitous mouse hsp60 is a beta-cell target antigen of autoimmune T cells. Journal of Autoimmunity 9, 159–166.

Chang, C.T.L., Chang, S.L., Lee, Y.M., Chiang, Y.M., Chuang, D.Y., Kuo, H.K., Yang, W.C., 2007. Cytopiloyne, a polyacetylenic plucoside, prevents type 1 diabetes in non-obese diabetic mice. Journal of Immunology 178, 6984–6993.

Chang, S.L., Chang, C.L., Chiang, Y.M., Hsieh, R.H., Tzeng, C.R., Wu, T.K., Sytwu, H.K., Shyur, L.F., Yang, W.C. 2004. Polyacetylenic compounds and butanol fraction from Bidens pilosa can modulate the differentiation of helper T cells and prevent autoimmune diabetes in non-obese diabetic mice. Planta Medica 70, 1045–1051.

Chen, W., Li, Y.M., Yu, M.H., 2008. Astragalus polysaccharides: an effective treatment for diabetes prevention in NOD mice. Experimental and Clinical Endocrinology & Diabetes 11, 468–474.

Chien, S.C., Young, P.H., Hsu, Y.J., Chen, C.H., Tien, Y.J., Shiu, S.Y., Li, T.H., Yang, C.W., Marimuthu, P., Tsai, L.F., Yang, W.C., 2009. Anti-diabetic properties of three common *Bidens pilosa* variants in Taiwan. Phyto Chemistry 70, 246–1254.

Chueh, W.-H., Lin, J.-Y., 2011. Berberine, an isoquinoline alkaloid in herbal plants, protects pancreatic islets and serum lipids in nonobese diabetic mice. Journal of Agriculture Food and Chemistry 59, 8021–8027.

Crino, A., Schiaffini, R., Ciampalini, P., Suraci, M.C., Manfrini, S., Visalli, N., Matteoli, M.C., Patera, P., Buzzetti, R., Guglielmi, C., Spera, S., Costanza, F., Fioriti, E., Pitocco, D., Pozzilli, P., 2005. A two year observational study of nicotinamide and intensive insulin therapy in patients with recent onset type 1 diabetes mellitus. Journal of Pediatric Endocrinology and Metabolism 18, 749–754.

Cvjeticanin, T., Miljkovic, D., Stojanovic, I., Dekanski, D., Stosic-Grujicic, S., 2010. Dried leaf extract of *Olea europaea* ameliorates islet-directed autoimmunity in mice. British Journal of Nutrition 103, 1413–1424.

Das, S., Das, D.K., 2007. Anti-Inflammatory Responses of resveratrol. Inflammation and Allergy Drug Targets 6, 168–173.

De Aizpurua, H.J., French, M.B., Chosich, N., Harrison, L.C., 1994. Natural history of humoral immunity to glutamic acid decarboxylase in non-obese diabetic (NOD) mice. Journal of Autoimmunity 7, 643–653.

Delong, T., Baker, R.L., Reisdorph, N., Reisdorph, R., Powell, R.L., Armstrong, M., Barbour, G., Bradley, B., Haskins, K., 2011. Islet amyloid polypeptide is a target antigen for diabetogenic $CD4^+$ T cells. Diabetes 60, 2325–2330.

Elias, D., Markovits, D., Reshef, T., Van der Zee, R., Cohen, 1990. Induction and therapy of autoimmune diabetes in the non-obese diabetic (NOD/Lt) mouse by a 65-kDa heat shock protein. Proceedings of the National Academy of Sciences 87, 1576–1580.

Elias, D., Reshef, T., Birk, O.S., Van Der Zee, R., Walker, M.D., Cohen, 1991. Vaccination against autoimmune mouse diabetes with a T-cell epitope of the human 65-kDa heat shock protein. Proceedings of the National Academy of Sciences 88, 3088–3091.

Eurodiab ACE study group, 2000. Variation and trends in incidence of childhood diabetes in Europe. Lancet 355, 873–876.

Ezuruiken, U.F., Prieto, J.M., 2014. The use of plants in the traditional management of diabetes in Nigeria: pharmacological and toxicological considerations. Journal of Ethnopharmacology 155, 857–924.

Fu, Z., Zhen, W., Yuskavage, J., Liu, D., 2011. Epigallocatechin gallate delays the onset of type 1 diabetes in spontaneous non-obese diabetic mice. British Journal of Nutrition 105, 1218–1225.

Geluk, A., Van Meijgaarden, K.E., Roep, B.O., Ottenhoff, T.H., 1998. Altered peptide ligands of islet autoantigen Imogen 38 inhibit antigen specific T cell reactivity in human type-1 diabetes. Journal of Autoimmunity 11, 353–361.

Gregg, R.K., Jain, R., Schoenleber, S.J., Divekar, R., Bell, J.J., Lee, H.H., Yu, P., Zaghouani, H., 2004. A sudden decline in active membrane-bound TGF-beta impairs both T regulatory cell function and protection against autoimmune diabetes. Journal of Immunology 173, 7308–7316.

Harjutsalo, V., Sjöberg, L., Tuomilehto, J., 2008. Time trends in the incidence of type 1 diabetes in Finnish children: a cohort study. Lancet 371, 1777–1782.

Hegde, V.L., Nagarkatti, M., Nagarkatti, P.S., 2010. Cannabinoid receptor activation leads to massive mobilization of myeloid-derived suppressor cells with potent immunosuppressive properties. European Journal of Immunology 40, 3358–3371.

Hsu, Y.J., Lee, T.H., Chang, C.L., Huang, Y.T., Yang, W.C., 2009. Antihyperglycemic effects and mechanism of *Bidens pilosa* water extract. Journal of Ethnopharmacology 122, 379–383.

Huang, S.H., Lin, G.J., Chu, C.H., Yu, J.C., Chen, T.W., Chen, Y.W., Chien, M.W., Chu, C.C., Sytwu, H.K., 2013. Triptolide ameliorates autoimmune diabetes and prolongs islet graft survival in nonobese diabetic mice. Pancreas 42, 442–451.

International Diabetes Federation, 2013. Available online: http://www.idf.org/diabetesatlas/ .

Islam, M.S., Choi, H., 2008. Dietary red chilli (*Capsicum frutescens* L.) is insulinotropic rather than hypoglycemic in type 2 diabetes model of rats. Phytotherapy Research 22, 1025–1029.

James, R., Gavin, Alberti, K.G.M.M., Davidson, M.B., Defronzo, R.A., Allan Drash, Gabbe, S.G., Saul Genuth, Harris, M.I., Richard Kahn, M.P.H., Harry Keen, Knowler, W.C., Harold Lebovitz, Maclaren, N.K., Palmer, J., Philip Raskin, P., Rizza, R.A., Michael, P., 2002. Report of the expert committee on the diagnosis and classification of diabetes mellitus. Diabetes Care 25, S5–S20.

Jiang, W., Swiggard, W.J., Heufler, C., Peng, M., Mirza, A., Steinman, R.M., Nussenzweig, M.C., 1995. The receptor DEC-205 expressed by dendritic cells and thymic epithelial cells is involved in antigen processing. Nature 375, 151–155.

Karvonen, M., Viik-Kajander, M., Moltchanova, E., Libman, I., Laporte, R., Tuomilehto, J., 2000. Incidence of childhood type 1 diabetes worldwide. Diabetes Mondiale (Dia-Mond) Project group. Diabetes Care 23, 1516–1526.

Kawakami, M., Hirayama, A., Tsuchiya, K., Ohgawara, H., Nakamura, M., Umezawa, K., 2010. Promotion of beta-cell differentiation by the alkaloid conophylline in porcine pancreatic endocrine cells. Biomedicine and Pharmacotherapy 64, 226–231.

Krishnamurthy, B., Dudek, N.L., Mckenzie, M.D., Purcell, A.W., Brooks, A.G., Gellert, S., Colman, P.G., Harrison, L.C., Lew, A.M., Thomas, H.E., Kay, T.W., 2006. Responses against islet antigens in NOD mice are prevented by tolerance to proinsulin but not IGRP. Journal of Clinical Investigation 116, 3258–3265.

Lee, S.M., Yang, H., 2011. Prevention and treatment of diabetes with resveratrol in a non-obese mouse model of type 1 diabetes. Diabetologia 54, 1136–1146.

Lernmark, Å., Elding Larsson, H., 2013. Immune therapy in type 1 diabetes mellitus. Nature Reviews Endocrinology 9, 92–103.

Li, X., Kaminski, N.E., Fischer, L.J., 2001. Examination of the immunosuppressive effect of delta 9-tetrahydrocannabinol in streptozotocin-induced autoimmune diabetes. International Journal of Immunopharmacology 1, 699–712.

Lombard, C., Nagarkatti, M., Nagarkatti, P., 2007. CB2 cannabinoid receptor agonist, JWH-015, triggers apoptosis in immune cells: potential role for CB2-selective ligands as immunosuppressive agents. Clinical Immunology 122, 259–270.

Mahnke, K., Qian, Y., Knop, J., Enk, A.H., 2003. Induction of $CD4^+/CD25^+$ regulatory T cells by targeting of antigens to immature dendritic cells. Blood 101, 4862–4869.

Meng, F., Abedini, A., Plesner, A., Verchere, C.B., Raleigh, D.P., 2010. The flavanol (-)-epigallocatechin 3-gallate inhibits amyloid formation by islet amyloid polypeptide, disaggregates amyloid fibrils, and protects cultured cells against IAPP-induced toxicity. Biochemistry 49, 8127–8133.

Michael, P., Morran, 2008. Immunology and genetics of type 1 diabetes. Mount Sinai Journal of Medicine 75, 314–327.

Mineo, D., Ricordi, C., Xu, X., Pileggi, A., Garcia-Morales, R., Khan, A., Baidal, D.A., Han, D., Monroy, K., Miller, J., Pugliese, A., Froud, T., Inverardi, L., Kenyon, N.S., Alejandro, R., 2008. Combined islet and hematopoietic stem cell allotransplantation: a clinical pilot trial to induce chimerism and graft tolerance. American Journal of Transplantation 8, 1262–1274.

Modak, M., Dixit, P., Londhe, J., Ghaskadbi, S., Devasagayam, T.P.A., 2007. Indian herbs and herbal drugs used for the treatment of diabetes: recent advances in Indian herbal drug research. Journal of Clinical Biochemistry and Nutrition 40, 163–173.

Mohana Lakshmi, S., Sandhya Rani, K.S., Usha Kiran Reddy, T., 2012. A review on diabetes milletus and the herbal plants used for its treatment. Asian Journal of Pharmaceutical and Clinical Research 5, 15–21.

Moltchanova, E.V., Schreier, N., Lammi, N., Karvonen, M., 2009. Seasonal variation of diagnosis of type 1 diabetes mellitus in children worldwide. Diabetic Medicine 26, 673–678.

Narendran, P., Mannering, S.I., Harrison, L.C., 2003. Proinsulin – a pathogenic autoantigen in type 1 diabetes. Autoimmunity Reveiw 2, 204–210.

Ogata, T., Li, L., Yamada, S., Yamamoto, Y., Tanaka, Y., Takei, I., Umezawa, K., Kojima, I., 2004. Promotion of beta-cell differentiation by conophylline in fetal and neonatal rat pancreas. Diabetes 53, 2596–2602.

Onkamo, P., Vaananen, S., Karvonen, M., Tuomilehto, J., 1999. Worldwide increase in incidence of type I diabetes— the analysis of the data on published incidence trends. Diabetologia 42, 1395–1403.

Patterson, C.C., Dahlquist, G.G., Gyürüs, E., Green, A., Soltész, G., Eurodiab Study Group, 2009. Incidence trends for childhood type 1 diabetes in Europe during 1989–2003 and predicted new cases 2005–20: a multicentre prospective registration study. Lancet 373, 2027–2033.

Petersen, J.S., Karlsen, A.E., Markholst, H., Worsaae, A., Dyrberg, T., Michelsen, B., 1994. Neonatal tolerization with glutamic acid decarboxylase but not with bovine serum albumin delays the onset of diabetes in NOD mice. Diabetes 43, 1478–1484.

Prud'homme, G.J., Draghia-Akli, R., Wang, Q., 2007. Plasmid-based gene therapy of diabetes mellitus. Gene Therapy 14, 553–564.

Rewers, M., Gottlieb, P., 2009. Immunotherapy for the prevention and treatment of type 1 diabetes human trials and a look into the future. Diabetes Care 32, 1769–1782.

Saito, R., Yamada, S., Yamamoto, Y., Kodera, T., Hara, A., Tanaka, Y., Kimura, F., Takei, I., Umezawa, K., Kojima, I., 2012. Conophylline suppresses pancreatic stellate cells and improves islet fibrosis in Goto-Kakizaki rats. Endocrinology 153, 621–630.

Sandler, S., Hellerstrom, C., Eizirk, D.L., 1993. Effects of nicotinamide supplementation on human pancreatic islet function in tissue culture. Journal of Clinical Endocrinology. Metabolism 77, 1574–1576.

Sasidharan, S., Sumathi, V., Jegathambigai, N.R., Latha, L.Y., 2011. Antihyperglycaemic effects of ethanol extracts of *Carica papaya* and *Pandanus amaryfollius* leaf in streptozotocin-induced diabetic mice. Natural Product Research 25, 1982–1987.

Seth, S.D., Sharma, B., 2004. Medicinal plants of India. Indian Journal of Medical Research 120, 9–11.

Shen, M.Y., Liu, Y.J., Don, M.J., Chen, L.H.Y., Mettling, Z.W., Pierre Corbeau, C., Chiang, C.-K., Jang, Y.-S., Li, T.-H., Young, P., Chang, C.L.T., Lin, Y.-L., Yang, W.-C., 2011. Combined phytochemistry and chemotaxis assays for identification and mechanistic analysis of anti-inflammatory phytochemicals in *Fallopia japonica*. PLoS One 6, e27480.

M.-Y.I. Shen, Y.-P. Lin, B.-C. Yang, Y.-S. Jang, C.-K. Chiang, C. Mettling, Z.-W. Chen, J.-R. Sheu, C.L. Chang, Y.-L.I.H. Lin & W.-C. Yang (2012) Catenarin prevents type 1 diabetes in non-obese diabetic mice via inhibition of leukocyte migration involving the MEK6/p38 and MEK7/JNK pathways,Evidence-Based Complementary and Alternative Medicine 2012 Article ID 982396:13 pages.

Shi, N.W., Wu, Y.J., Wang, Y.F., 2006. Study on the mechanism of *Tripterygium wilfordii* in preventing type 1 nonobese diabetic mice. Chinese Journal of Integrated Traditional and Western Medicine 26, 79–83.

Soto, C., Mena, R., Luna, J., Cerbon, M., Larrieta, E., Vital, P., URIA, E., Sanchez, M., Recoba, R., Barron, H., Favari, L., Lara, A., 2004. Silymarin induces recovery of pancreatic function after alloxan damage in rats. Life Science 75, 2167–2180.

Soto, C., Raya, L., Pérez, J., González, I., Pérez, S., 2014. Silymarin induces expression of pancreatic Nkx6.1 transcription factor and β-cells neogenesis in a pancreatectomy model. Molecules 19, 4654–4668.

Tisch, R., Yang, X.D., Singer, S.M., Liblau, R.S., Fugger, L., MC Devitt, H.O., 1993. Immune response to glutamic acid decarboxylase correlates with insulitis in non-obese diabetic mice. Nature 366, 72–75.

Tisch, R., Wang, B., Weaver, D.J., Liu, B., Bui, T., Arthos, J., Serreze, D.V., 2001. Antigen-specific mediated suppression of beta cell autoimmunity by plasmid DNA vaccination. Journal of Immunology 16, 2122–2132.

Vague, P., Picq, R., Bernal, M., Lassmann-Vague, V., Vialettes, B., 1989. Effect of nicotinamide treatment on the residual insulin secretion in type 1 (insulin-dependent) diabetic patients. Diabetologia 32, 316–321.

Visalli, N., Cavallo, M.G., Signore, A., Baroni, M.G., Buzzetti, R., Fioriti, E., Mesturino, C., Fiori, R., Lucentini, L., Matteoli, M.C., Crin, A., Corbi, S., Spera, S., Teodonio, C., Paci, F., Amoretti, R., Pisano, L., Suraci Multar, G., Sulli, N., Cervoni, M., De Mattia, G., Faldetta, M.R., Boscherini, B., Pozzilli, P., 1999. A multi-centre randomized trial of two different doses of nicotinamide in patients with recent-onset type 1 diabetes (the IMDIAB VI). Diabetes/Metabolism Research and Reviews 15, 181–185.

Waldron-Lynch, F., Herold, K.C., 2011. Immunomodulatory therapy to preserve pancreatic beta-cell function in type 1 diabetes. Nature Reviews Drug Discovery 10, 439–452.

Walter, M., Philotheou, A., Bonnici, F., Ziegler, A.G., Jimenez, R., NBI-6024 Study Group, 2009. No effect of the altered peptide ligand NBI-6024 on beta-cell residual function and insulin needs in new-onset type 1diabetes. Diabetes Care 32, 2036–2040.

Wang, X., Matta, R., Shen, G., Nelin, L.D., Pei, D., Liu, Y., 2006. Mechanism of triptolide-induced apoptosis: effect on caspase activation and Bid cleavage and essentiality of the hydroxyl group of triptolide. Journal of Molecular Medicine 84, 405–415.

Wenzlau, J.M., Juhl, K., Yu, L., Moua, O., Sarkar, S.A., Gottlieb, P., Rewers, M., Eisenbarth, G.S., Jensen, J., Davidson, H.W., Hutton, J.C., 2007. The cation efflux transporter ZnT8 (Slc30A8) is a major autoantigen in human type 1 diabetes. Proceedings of the National Academy of Sciences 104, 17040–17045.

Wenzlau, J.M., Walter, M., Gardner, T.J., Frisch, L.M., Yu, L., Eisenbarth, G.S., Ziegler, A.G., Davidson, H.W., Hutton, J.C., 2010. Kinetics of the post-onset decline in zinc transporter 8 autoantibodies in type 1 diabetic human subjects. Journal of Clinical Endocrinology. Metabolism 95, 4712–4719.

Yang, Y., Liu, Z., Tolosa, E., Yang, J., Li, L., 1998. Triptolide induces apoptotic death of T lymphocyte. Immunopharmacology 40, 139–149.

Yang, J., Danke, N.A., Berger, D., Reichstetter, S., Reijonen, H., Greenbaum, C., Pihoker, C., James, E.A., Kwok, W.W., 2006. Islet-specific glucose-6-phosphatase catalytic subunit-related protein-reactive CD4+ T cells in human subjects. Journal of Immunology 176, 2781–2789.

Yuan, H.-D., Tae Kim, J., Kim, S.H., Chung, S.H., 2012. Ginseng and diabetes: the evidences from in vitro, animal and human studies. Journal of Ginseng Research 36, 27–39.

Zhang, Y., Cai, J., Ruan, H., Pi, H., Wu, J., 2007. Antihyperglycemic activity of kinsenoside, a high yielding constituent from *Anoectochilus roxburghii* in streptozotocin diabetic rats. Journal of Ethnopharmacology 114, 141–145.

Zhang, J.G., Liu, Q., Liu, Z.L., Li, L., Yi, L.T., 2015. Antihyperglycemic activity of *Anoectochilus roxburghii* polysaccharose in diabetic mice induced by high-fat diet and streptozotocin. Journal of Ethnopharmacology 164, 180–185.

Chapter 9

α-Glucosidase Enzyme Inhibitors From Natural Products

G. Abbas, A.S. Al-Harrasi, H. Hussain
University of Nizwa, Nizwa, Sultanate of Oman

1. INTRODUCTION

1.1 Natural Products

The use of medicinal plants to treat various diseases dates back since ancient times. Medicinal plants are the major source of drugs and majority of them are derived directly or indirectly from plants. For instance, glucophage (metformin), one of the popular hypoglycemic drugs, is derived from the medicinal plant known as *Galega officinalis* (Grover et al., 2002; Hunter, 2001). The utilization of natural products for drug discovery is increasing worldwide and there is a growing interest in the promotion of traditional health-care systems. About 80% of the world's population uses plant-based traditional health remedies as these are affordable and are safer to use. During the last few decades, researchers have aimed at identifying and validating plant-derived products for the treatment of various diseases (Maciel et al., 2002).

1.2 Diabetes Mellitus and Its Management

The world is facing an imminent epidemic of diabetes mellitus. Diabetes mellitus, particularly non-insulin-dependent diabetes mellitus (NIDDM), is increasing at an alarming rate (Grey, 2008; Farag and Gaballa, 2011; Boath et al., 2012). The World Health Organization (WHO) estimated that approximately 346 million people of the world suffer from diabetes, which will be doubled by the year 2030, posing a major challenge to health-care systems (Kodra et al., 2008). Improvement in glycemic control significantly reduces the risk of type-2 diabetes mellitus (Misbah et al., 2013). The antidiabetic drugs being currently used are associated with some serious side effects, and hence there is a need to discover new potent therapeutic agents with high potency and low side effects (Park et al., 2008).

Discovery and Development of Antidiabetic Agents From Natural Products.
http://dx.doi.org/10.1016/B978-0-12-809450-1.00009-0

1.3 Role of α-Glucosidase Enzyme in Diabetes Management

The enzyme α-glucosidase plays an important role to increase blood glucose level in the body. α-Glucosidase enzyme is present in the small intestine, which is involved in the digestion and absorption of carbohydrates. It causes the hydrolysis of polysaccharides and oligosaccharides into monomers and thus increases glucose concentration in the body. The inhibitors of α-glucosidase enzyme actually slow down the digestion and absorption of carbohydrates. As a result, the postprandial blood glucose is maintained at a lower level leading to a decreased insulin demand. In this way, the sudden increase of glucose after meal can be controlled independently of insulin (Van de Laar et al., 2005; Conforti et al., 2005). α-Glucosidase is thus regarded as an important therapeutic target to treat carbohydrate-mediated diseases (Chiasson, 2006; Elya et al., 2012). By understanding the role of α-glucosidase, new pharmacologic agents from natural sources have been identified so far.

1.4 General Assay Protocol for α-Glucosidase Enzyme Inhibition

A general procedure to perform α-glucosidase enzyme inhibition assay in vitro is presented as follows:

α-Glucosidase enzyme (E.C. 3.2.1.20) solution is prepared just before its use by dissolving 0.1 mg of α-glucosidase enzyme in 10 mL of 50 mM phosphate buffer (pH 6.8). In 96-well plate, 20 μL of inhibitor sample of concentration 0.5 mg/mL is premixed with 120 μL of phosphate buffer (pH 6.8) and then 20 μL of 5 mM *p*-nitrophenyl *α*-D-glucopyranoside (PNPG) is added and thoroughly mixed together gently. The reaction mixture is then preincubated at 37°C for 5 min. After preincubation, 20 μL *α*-glucosidase enzyme solution is added in reaction wells and again incubated at 37°C for 15 min. The blank sample wells contain 40 μL of phosphate buffer instead of 20 μL *α*-glucosidase and 20 μL of inhibitor. The control sample wells contain 20 μL of phosphate buffer in place of 20 μL of inhibitor sample. After 15 min of incubation, 100 μL Na_2CO_3 (200 mM) is added to each well to terminate the reaction. *α*-Glucosidase enzyme inhibition activity of inhibitor is determined spectrophotometrically at 405 nm on spectrophotometer. Acarbose is generally used as a standard inhibitor (Mosihuzzman et al., 2013; Hong et al., 2013). Following formula is used to calculate percent inhibition:

$$\text{Inhibition } \% = 1 - (\text{OD blank sample/OD control}) \times 100$$

Different concentrations of the inhibitor are used to determine the IC_{50} (concentration at which 50% inhibition is observed).

During last two decades, various efforts have been made to discover more potent and less toxic α-glucosidase inhibitors for the management of diabetes mellitus. For this purpose, large numbers of plants have been evaluated to identify more cost-effective and easily affordable drugs. In this chapter, we have summarized the outcome of research conducted on medicinal plants collected from different parts of the world. The most promising α-glucosidase inhibitors identified in between 2000 and 2015 are presented in this chapter.

2. BRIEF METHODOLOGY, CHARACTERIZATION, AND IDENTIFICATION OF α-GLUCOSIDASE INHIBITORS FROM MEDICINAL PLANTS

2.1 *Morus Atropurpurea*

The plant *Morus atropurpurea* belongs to the family Moraceae, which is commonly known as Mulberry or Fig family. In a study, Hong et al. (2013) collected 10kg of leaves of *M. atropurpurea* randomly from Guangdong Province, China. Approximately, 9kg of air-dried *M. atropurpurea* leaves were extracted with boiling water and different fractions were prepared. High-performance liquid chromatography (HPLC) was used to determine the content of individual flavonoids. Four flavonoids including rutin (**1**), isoquercetin (**2**), kaempferol-3-*O*-rutinoside (**3**), and astragalin (**4**) were isolated from the *M. atropurpurea* leaves as shown in Fig. 9.1. The compounds were identified by comparing the MS and NMR data of four flavonoids with the literature (Chu and Yu, 2001). These four flavonoids from *M. atropurpurea* were then tested against α-glucosidase enzyme. On α-glucosidase enzyme inhibition assay, astragalin (**1**) exhibited significant inhibition with IC_{50} value of 13.19±1.10μM, isoquercetin (**2**) IC_{50} value of 116.7±1.17μM, kaempferol-3-*O*-rutinoside (**3**) IC_{50} value of 365.4±1.05μM, and rutin (**4**) showed potent inhibition with IC_{50} value of 15.82±1.11μM, as shown in Table 9.1 (Hong et al., 2013).

Flavonoid	R^1	R^2
Astragalin (**1**)	Glu	H
Isoquercetin (**2**)	Glu	OH
Kaempferol-3-*O*-rutinoside (**3**)	Glu (6→1) Rha	H
Rutin (**4**)	Glu (6→1) Rha	OH

(Glu = β-D-glucopyranosyl; Rha = α-L-rhamnopyranosyl)

FIGURE 9.1 Four flavonoids isolated from *Morus atropurpurea*.

TABLE 9.1 α-Glucosidase Enzyme Inhibition Activity of Various Compounds Isolated From Different Medicinal Plants

Compound Name	Plant Name	α-Glucosidase Inhibition (IC_{50})	References
Astragalin (**1**)	*Morus atropurpurea*	15.82 ± 1.11 μM	Hong et al. (2013)
Isoquercetin (**2**)	*M. atropurpurea*	116.7 ± 1.17 μM	Hong et al. (2013)
Kaempferol-3-*O*-rutinoside (**3**)	*M. atropurpurea*	365.4 ± 1.05 μM	Hong et al. (2013)
Rutin (**4**)	*M. atropurpurea*	13.19 ± 1.10 μM	Hong et al. (2013)
Quercetin (**5**)	*Gynura medica*	1.67.00 mg/mL	Tan et al. (2013)
Kaempferol-3-*O*-β-D-glucopyranoside (**6**)	*G. medica*	1.46.00 mg/mL	Tan et al. (2013)
3,5-Dicaffeoylquinic acid methyl ester (**7**)	*G. medica*	0.53.00 mg/mL	Tan et al. (2013)
Arborinone (**8**)	*Iris loczyi*	0.112 ± 0.008 mM	Mosihuzzman et al. (2013)
5,7-Dihydroxy-2′,6-dimethoxyisoflavone (**9**)	*I. loczyi*	0.321 ± 0.008 mM	Mosihuzzman et al. (2013)
Irisoid A (**10**)	*I. loczyi*	0.360 ± 0.006 mM	Mosihuzzman et al. (2013)
8-Methoxyeriodictyol (**11**)	*Iris unguicularis*	0.328 ± 0.004 mM	Mosihuzzman et al. (2013)
Kaempferol (**12**)	*I. unguicularis*	0.047 ± 0.003 mM	Mosihuzzman et al. (2013)
4′,5,7-Trihydroxy-6-methoxyflavanone (**13**)	*I. unguicularis*	0.097 ± 0.003 mM	Mosihuzzman et al. (2013)
Luteolin 7-*O*-glucoside (**14**)	*Salvia chloroleuca*	18.3.00 0.00 μM	Asghari et al. (2015)
Luteolin 7-*O*-glucuronide (**15**)	*S. chloroleuca*	14.7.00 μM	Asghari et al. (2015)
Diosmetin 7-*O*-glucuronide (**16**)	*S. chloroleuca*	17.1.00 μM	Asghari et al. (2015)
Apigenin 7-*O*-glucoside (**17**)	*Syzygium cumini*	8 ± 0.17 mg/mL	Alagesan et al. (2012)

TABLE 9.1 α-Glucosidase Enzyme Inhibition Activity of Various Compounds Isolated From Different Medicinal Plants—cont'd

Compound Name	Plant Name	α-Glucosidase Inhibition (IC_{50})	References
Dihydro-3,3′,4′,5,7-pentahydroxyflavone glycoside (**18**)	*Psidium guajava*	10 ± 0.04 mg/mL	Alagesan et al. (2012)
7-*O*-geranyl-6-methoxypseudobaptigenin (**19**)	*Millettia conraui*	35.5 ± 0.5 μM	Tiabou Tchinda et al. (2007)
β-Amyrine (**20**)	*M. conraui*	10.3 ± 0.5 μM	Tiabou Tchinda et al. (2007)
Vitexin (**21**)	*Vigna angularis*	0.4.00 mg/mL	Yao et al. (2011)
Isovitexin (**22**)	*V. angularis*	4.8 0.00 mg/mL	Yao et al. (2011)
Isorhamnetin 3-*O*-β-D-rutinoside (**23**)	*Microcos paniculata*	275.4 μM	Chen et al. (2013)
β-Sitosterol (**24**)	*Terminalia sericea*	54.49 μM	Nkobole et al. (2011)
β-Sitosterol-3-acetate (**25**)	*T. sericea*	129.36 μM	Nkobole et al. (2011)
Lupeol (**26**)	*T. sericea*	66.48 μM	Nkobole et al. (2011)
Stigma-4-ene-3-ol (**27**)	*T. sericea*	164.87 μM	Nkobole et al. (2011)
Salacinol (**28**)	*Salacia reticulata*	3.2 μg/mL	Yoshikawa et ál. (2002)
14-Hydroxyabieta-8,11,13-triene-11-carbaldehyde-18-oic-12-carboxy-13-(1-hydroxy-1-methylethyl)-lactone (**29**)	*Phlomis tuberosa*	0.379 ± 0.016 mM	Yang et al. (2015)
14-Hydroxyabieta-8,11,13-triene-17-oic-12-carboxy-13-(1-hydroxy-1-methylethyl)-lactone (**30**)	*P. tuberosa*	0.624 ± 0.026 mM	Yang et al. (2015)
14,16-Dihydroxyabieta-8,11,13-triene-15,17-dioic acid (**31**)	*P. tuberosa*	0.721 ± 0.037 mM	Yang et al. (2015)

Continued

TABLE 9.1 α-Glucosidase Enzyme Inhibition Activity of Various Compounds Isolated From Different Medicinal Plants—cont'd

Compound Name	Plant Name	α-Glucosidase Inhibition (IC_{50})	References
Phlomisol (15,16-eposy-8,13(16),14-labdatrien-19-ol) (**32**)	*P. tuberosa*	0.067 ± 0.003 mM	Yang et al. (2015)
Phlomisoic acid (**33**)	*P. tuberosa*	0.482 ± 0.019 mM	Yang et al. (2015)
15,16-Eposy-8,13(16), 14-labdatrien (**34**)	*P. tuberosa*	0.210 ± 0.010 mM	Yang et al. (2015)
Luteolin (**35**)	*P. tuberosa*	0.428 ± 0.018 mM	Yang et al. (2015)
Quercetin-3-rhamnoside (**36**)	*P. tuberosa*	0.562 ± 0.021 mM	Yang et al. (2015)
Flavoglaucin (**37**)	*P. tuberosa*	0.229 ± 0.017 mM	Yang et al. (2015)
Tetrahydroauroglaucin (**38**)	*P. tuberosa*	0.283 ± 0.013 mM	Yang et al. (2015)
Dihydroauroglaucin (**39**)	*P. tuberosa*	0.255 ± 0.013 mM	Yang et al. (2015)
2-(2′,3-Epoxy-1′-heptenyl)-6-hydroxy-5-(3″-methyl-2″-butenyl)benzaldehyde (**40**)	*P. tuberosa*	0.371 ± 0.015 mM	Yang et al. (2015)
3,4-Dicaffeoylquinic acid (**41**)	*Tussilago farfara*	0.90 mM	Gao et al. (2008)
3,5-Dicaffeoylquinic acid (**42**)	*T. farfara*	0.89 mM	Gao et al. (2008)
4,5-Dicaffeoylquinic acid (**43**)	*T. farfara*	0.91 mM	Gao et al. (2008)

2.2 *Gynura Medica*

The plant *Gynura medica* is a newly identified species that belongs to the family Compositae. In a study, Tan et al. (2013) collected the plant *G. medica* from the Huoshan districts, Anhui province, China. About 2 kg crude extract was prepared with 80% ethanol (*v*/*v*) under reflux at 80°C. Other fractions such as chloroform, ethyl acetate (EA), and *n*-butanol were prepared from ethanol extract. Seven phenolic compounds including five flavonols and two phenolic acids

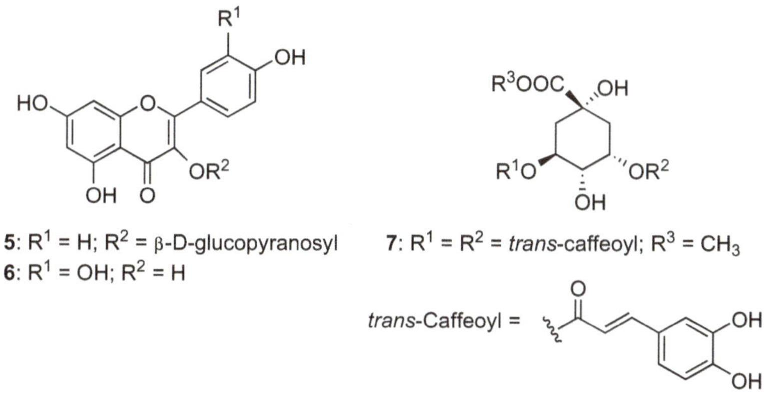

FIGURE 9.2 α-glucosidase enzyme inhibitors isolated from *Gynura medica.*

were isolated from the *G. medica* leaf. Structure elucidation and characterization were done by the extensive nuclear magnetic resonance (NMR) and mass spectral analysis. From the leaf of *G. medica*, quercetin (**5**), kaempferol-3-*O*-β-D-glucopyranoside (**6**), and 3,5-dicaffeoylquinic acid methyl ester (**7**) were isolated, which were then investigated against α-glucosidase enzyme (Fig. 9.2). On α-glucosidase assay, quercetin (**5**), kaempferol-3-*O*-β-D-glucopyranoside (**6**), and 3,5-dicaffeoylquinic acid methyl ester (**7**) exhibited inhibitory activity with the IC_{50} values of 1.67, 1.46, and 0.53 mg/mL, respectively (Tan et al., 2013).

2.3 *Iris loczyi* and *Iris unguicularis*

The plants, *Iris loczyi* and *Iris unguicularis* (Family: Iridaceae), were investigated by Mosihuzzman et al. (2013) to evaluate their potential role against α-glucosidase enzyme and antiglycation activity. In their study, the whole plant of *I. loczyi* was collected from Ziarat Valley, Quetta, Pakistan. *Iris unguicularis,* also known as Algerian iris, is a rhizomatous flowering plant and commonly found in Turkey, Western Syria, and Tunisia. In present study, rhizomes of *I. unguicularis* were collected from Antakya, Turkey.

Around 30 kg of air-dried whole plant material of *I. loczyi* was chopped and soaked in aqueous 90% ethanol (120 L) and the resulting extract was concentrated to a gum (1.8 kg). From ethanolic extract, six major fractions were obtained. Altogether 11 known compounds were isolated from these fractions by using column chromatography on silica gel repeatedly. Most active compounds against α-glucosidase enzyme were found to be arborinone (**8**), 5,7-Dihydroxy-2′,6-dimethoxyisoflavone (**9**), and irisoid A (**10**) (Fig. 9.3) with IC_{50} values of 0.112 ± 0.008 mM, 0.321 ± 0.008, and 0.360 ± 0.006, respectively. Similarly, dried rhizomes of *I. unguicularis* (1.0 kg) were soaked in ethanol (10 L) and concentrated gum (250.3 g) was obtained. This ethanolic extract was then dissolved in distilled water (3 L) and further fractionation was done. Several compounds

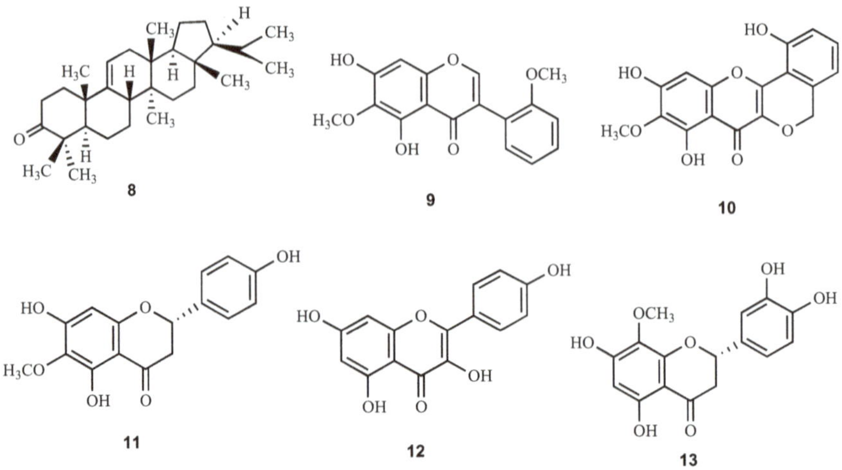

FIGURE 9.3 α-glucosidase enzyme inhibitors isolated from *Iris loczyi* and *Iris unguicularis*.

were isolated by using preparative recycling HPLC, Octadecyl silica (ODS) column, and repeated column chromatography (silica gel). All known compounds from both plants were characterized by comparison of physical and spectral data with that reported in the literature (Choudhary et al., 2001a,b,c; Abbaskhan et al., 2010). Most promising α-glucosidase enzyme inhibitors isolated from *I. unguicularis* were 8-methoxyeriodictyol (**11**), kaempferol (**12**), 4′,5,7-trihydroxy-6-methoxyflavanone (**13**) (Fig. 9.3) with IC_{50} values of 0.328±0.004, 0.047±0.003, and 0.097±0.003 mM, respectively as shown in Table 9.1.

2.4 *Salvia chloroleuca*

The plant *Salvia chloroleuca* belongs to the family Lamiaceae with huge number of species widely distributed in the Central and South America, Central Asia, Mediterranean region, and Eastern Asia. In this study, Asghari et al. (2015) collected the aerial parts of *S. chloroleuca* from Shahrestanak, Tehran, Iran. Dried leaf material (100 g) was grinded and extracted by with *n*-hexane, ethylacetate, and methanol (2 L each). About 20 g of methanol extract was obtained after evaporation to dryness under reduced pressure. Five combined fractions were prepared and most active fractions were separated by using HPLC. Altogether four compounds were isolated. Structures of these four compounds were identified by using 1-D and 2-D NMR spectroscopy and ESI-MS, further confirmation was done by comparison with published data in literature (Lu and Foo, 2000; Murata et al., 2010; Gohari et al., 2011; Ayatollahi et al., 2010).

Three compounds including luteolin 7-*O*-glucoside (**14**), luteolin 7-*O*-glucuronide (**15**), and diosmetin 7-*O*-glucuronide (**16**) (Fig. 9.4) exhibited significant inhibition against α-glucosidase enzyme with IC_{50} values of 18.3, 14.7, and 17.1 μM, respectively (Asghari et al., 2015).

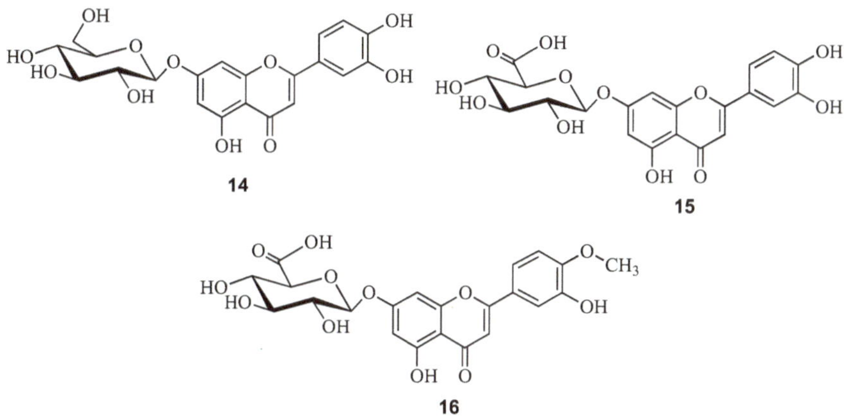

FIGURE 9.4 Three α-glucosidase enzyme inhibitors from *Salvia chloroleuca*.

2.5 *Syzygium cumini* and *Psidium guajava*

The plant *Syzygium cumini* belongs to the family Myrtaceae, which is native to the Subcontinent and bordering regions of Southeast Asia. *Psidium guajava* is also a member of the Myrtaceae family, which is widely cultivated in tropical and subtropical regions around the world. In this study, both plants *P. guajava* and *S. cumini* were collected by Alagesan et al. (2012) from botanical garden, Tamil Nadu Agricultural University, Coimbatore, India. Aqueous extracts of both plants were prepared and subjected to phenol contents estimation and α-glucosidase inhibition. Different fractions were prepared while active fractions were subjected to silica gel column chromatography. The most active components were separated and purity was confirmed by reversed phase-HPLC (RP-HPLC). Characterization of isolated components was done by using spectroscopic and liquid chromatography–mass spectrometry (LC-MS) techniques and by comparing with published literature (Gao et al., 2004; Jung et al., 2006).

One α-glucosidase inhibitor known as apigenin 7-*O*-glucoside (**17**) with IC_{50} value of 10 ± 0.04 mg/mL was identified from *S. cumini* seed extract, while other inhibitor dihydro-3,3′,4′,5,7-pentahydroxyflavone glycoside (**18**) was obtained from *P. guajava* leaf extract with IC_{50} value of 8 ± 0.17 mg/mL (Alagesan et al., 2012) (Fig. 9.5).

2.6 *Millettia conraui*

The plant *Millettia conraui* is a species of legume in the Fabaceae family. It is found in Cameroon and Nigeria. In current study, Tiabou Tchinda et al. (2007) collected the stem barks of *M. conraui* at Kumbo (Northwest province Cameroon). Methanol extract was prepared from the air-dried and powdered stem barks of *M. conraui* (2 kg). On the basis of thin-layer chromatography (TLC) profile, different fractions of 250 mL were collected and regrouped.

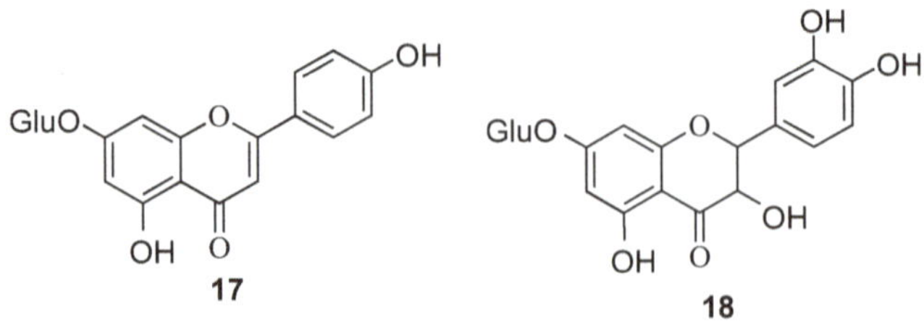

FIGURE 9.5 Compounds isolated from *Syzygium cumini* and *Psidium guajava*.

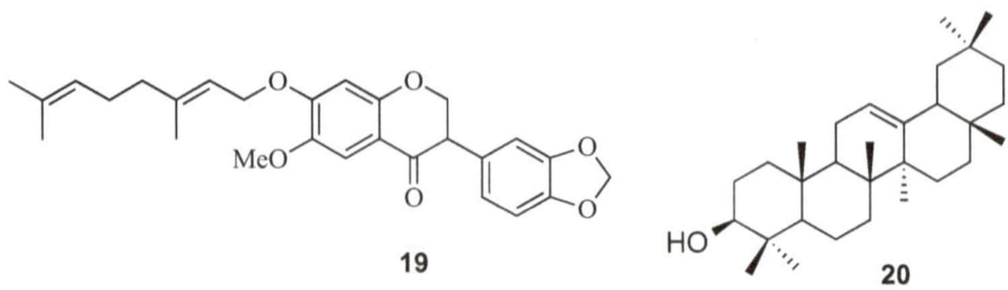

FIGURE 9.6 Inhibitors of α-glucosidase enzyme isolated from *Millettia conraui*.

Several compounds were isolated, which were characterized by using 1-D and 2-D NMR spectroscopy and ESI-MS technique. Two compounds obtained from *M. conraui* exhibited potent activity. In this study, a new geranylated isoflavone, 7-*O*-geranyl-6-methoxypseudobaptigenin (**19**, Fig. 9.6) along with the known secondary metabolites were isolated from the methanol extract of the stem barks of *M. conraui*. The new compound **19** obtained as white crystals exhibited significant activity against α-glucosidase enzyme with IC_{50} value of $35.5 \pm 0.5\,\mu M$. This inhibitory activity is probably due to hydrophobic interactions between geranyl moieties of the compound with the enzyme binding sites. While β-amyrine (**20**, Fig. 9.6) exhibited potent α-glucosidase enzyme activity with IC_{50} value of $10.3 \pm 0.5\,\mu M$ (Tiabou Tchinda et al., 2007).

2.7 *Vigna angularis* (Azuki Beans)

The plant *Vigna angularis* commonly known as Azuki beans belongs to the family Fabaceae. It is widely grown throughout East Asia and the Himalayas. In this study, Yao et al. (2011) obtained Azuki beans from the Chinese National Genebank (Beijing, China). Altogether 3 kg dried Azuki beans were crushed and 70% ethanol extract was prepared. The concentrated ethanol extract was partitioned with different solvents to obtain four extracts. All the compounds were separated by using a combination of Sephadex LH-20 column chromatography and column chromatography with methanol (MeOH) as an eluent. These isolated compounds were then characterized by using 1-D and 2-D NMR spectroscopy and ESI-MS technique. All the compounds were investigated against α-glucosidase enzyme. Two most active inhibitors were identified as vitexin

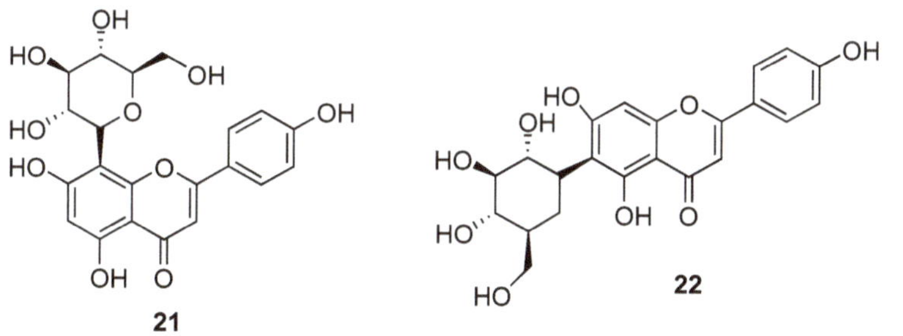

FIGURE 9.7 Compounds isolated from *Vigna angularis* (Azuki beans).

(**21**) and isovitexin (**22**) with IC_{50} values of 0.4 mg/mL and 4.8 mg/mL, respectively (Yao et al., 2011). Structures of these compounds were also confirmed as vitexin and isovitexin (Fig. 9.7) by the comparison of their spectral data with that of published in the literature (Hosoya et al., 2005; Tanaka et al., 2005).

2.8 *Microcos paniculata*

The plant *Microcos paniculata* belongs to the family Tiliaceae, which is native to China and Southeast Asia. Dried leaves (*Microctis folium*) of *M. paniculata* are important part of Chinese herbal tea and a traditional Chinese medicine.

In this study, dried leaves were collected by Chen et al. (2013) from different places in the Guangdong, Province, China. Methanol extract (70%) was obtained from dried *M. paniculata* plant material under reflux conditions. It was chromatographed on Sephadex LH-20 while prep-HPLC was used for purification. Three compounds were isolated from *M. paniculata*, however; a compound isorhamnetin 3-*O*-β-D-rutinoside (**23**, Fig. 9.8) showed significant inhibition against α-glucosidase enzyme with IC_{50} value of 275.4 μM. In this study, acarbose was used as a positive control with IC_{50} value of 1007 μM. Structure elucidation of isolated compounds was done by using ^{1}H-NMR, ^{13}C-NMR, and mass spectrometry (MS) and by comparing their data with those published in the literature (Feng et al., 2008).

2.9 *Terminalia sericea*

The plant *Terminalia sericea* belongs to the family Combretaceae. It is a species of deciduous tree that is native to Southern Africa. In a study, Nkobole et al. (2011) chopped the air-dried stem bark of *T. sericea* (1.8 kg) into powder and extracted with 100% acetone. Acetone extract was filtered first and then evaporated under reduced pressure. Altogether, 80 g of concentrated acetone extract was subjected to silica gel column chromatography. All fractions and subfractions were rechromatographed on silica gel column. Main fractions were subjected to Sephadex CC and compounds were isolated as eluted with 98% ethanol. Four compounds including β-sitosterol (**24**), β-sitosterol acetate (**25**),

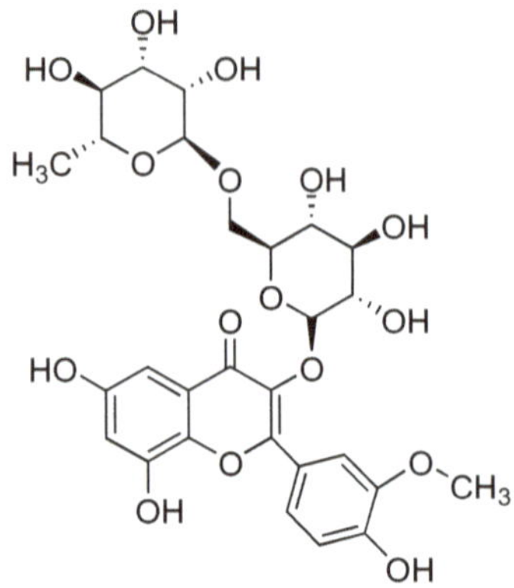

23

FIGURE 9.8 Compound isolated from *Microctis folium*.

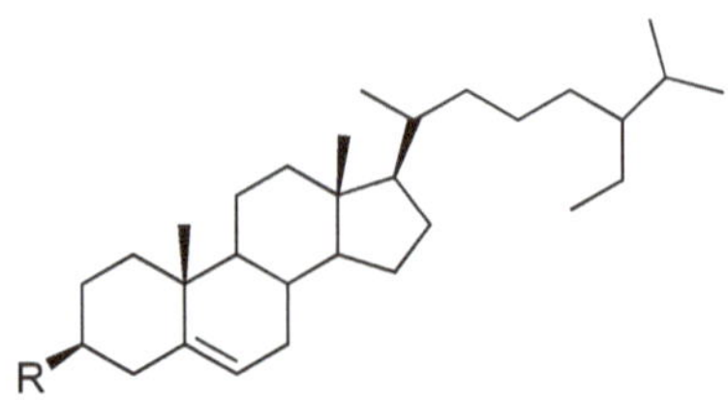

24: β-sitostierol R= OH
25: β-sitosterol-3-acetate R= OAc

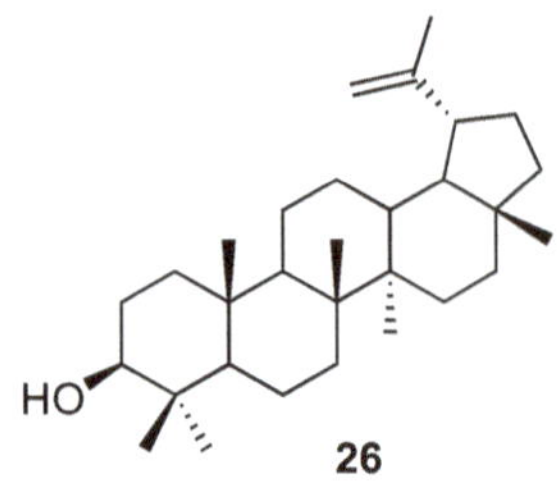

26

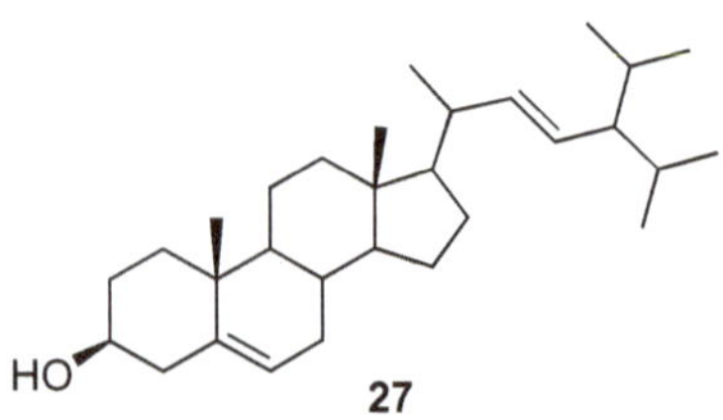

27

FIGURE 9.9 α-glucosidase enzyme inhibitors isolated from *Terminalia sericea*.

Lupeol (**26**), and stigma-4-ene-3-ol (**27**) were isolated and evaluated against α-glucosidase enzyme as shown in Fig. 9.9. These compounds exhibited significant inhibition against α-glucosidase enzyme with IC_{50} values of 54.49, 129.36, 66.48, and 164.87 μM, respectively. Structural elucidation of isolated compounds was done by their physical (melting point (mp), specific rotation: [α] D), and spectroscopic (^1H and ^{13}C NMR) data (Nkobole et al., 2011; Li et al., 2007).

2.10 *Salacia reticulata*

The plant *Salacia reticulata* belongs to the family Celastraceae, which is indigenous to dry zone forests of the Sri Lanka. *Salacia reticulata* is widely used

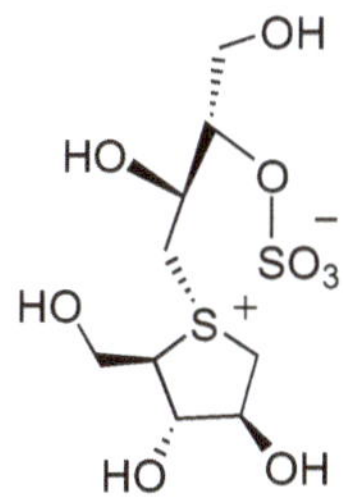

28

FIGURE 9.10 Salacinol isolated from *Salacia reticulata*.

as antidiabetic Ayurvedic traditional medicine. The plant *S. reticulata* was collected by Yoshikawa et al. (2002) in Sri Lanka. Methanol extract of the dried roots and stems (1.7 kg) powder was prepared under reflux. Methanol extract was concentrated under reduced pressure and further fractions were prepared with different solvents. In this study, salacinol (**28**) has been isolated from *S. reticulata* through bioassay-guided separation, which exhibited the most potent inhibition against α-glucosidase enzyme with IC_{50} value of 3.2 μg/mL (Fig. 9.10). On the basis of chemical and physicochemical evidence, the absolute stereostructure of salacinol was determined especially by using X-ray crystallographic analysis (Yoshikawa et al., 2001, 2002).

2.11 *Phlomis tuberosa*

The plant *Phlomis tuberosa* belongs to the family Lamiaceae (mint family). It is native to Southern Europe and Asia. In this study, roots of plant *P. tuberosa* were collected by Yang et al. (2015) from Shangdu Town, China. Ethanol extract of air-dried, chopped roots (1 kg) of *P. tuberosa* were obtained under reflux. To obtain 2-D separation, the sepbox system was coupled to HPLC/SPE/HPLC setup. Various fractions and subfractions were prepared and further purification was done by semipreparative HPLC. Altogether 20 compounds were isolated from *P. tuberosa*, which were investigated against α-glucosidase enzyme, as shown in Fig. 9.11. The most promising inhibitors included [14-hydroxyabieta-8,11,13-triene-11-carbaldehyde-18-oic-12-carboxy-13-(1-hydroxy-1-methylethyl)-lactone (**29**), 14-hydroxyabieta-8,11,13-triene-17-oic-12-carboxy-13-(1-hydroxy-1-methylethyl)-lactone (**30**), 14,16-dihydroxyabieta-8,11,13-triene-15,17-dioic acid (**31**), and phlomisol (15,16-eposy-8,13(16),14-labdatrien-19-ol) (**32**), phlomisoic acid (**33**), 15,16-eposy-8,13(16),14-labdatrien (**34**), luteolin (**35**), quercetin-3-rhamnoside (**36**), flavoglaucin (**37**), tetrahydroauroglaucin (**38**), dihydroauroglaucin (**39**), and 2-(2′,3-epoxy-1′-heptenyl)-6-hydroxy-5-(3″-methyl-2″-butenyl)benzaldehyde (**40**) with IC_{50} values of 0.379 ± 0.016 mM, 0.624 ± 0.026, 0.721 ± 0.037, 0.067 ± 0.003, 0.482 ± 0.019, 0.210 ± 0.010, 0.428 ± 0.018, 0.562 ± 0.021, 0.229 ± 0.017, 0.283 ± 0.013, 0.255 ± 0.013, and 0.371 ± 0.015 mM, respectively as shown in Table 9.1. Structures of isolated

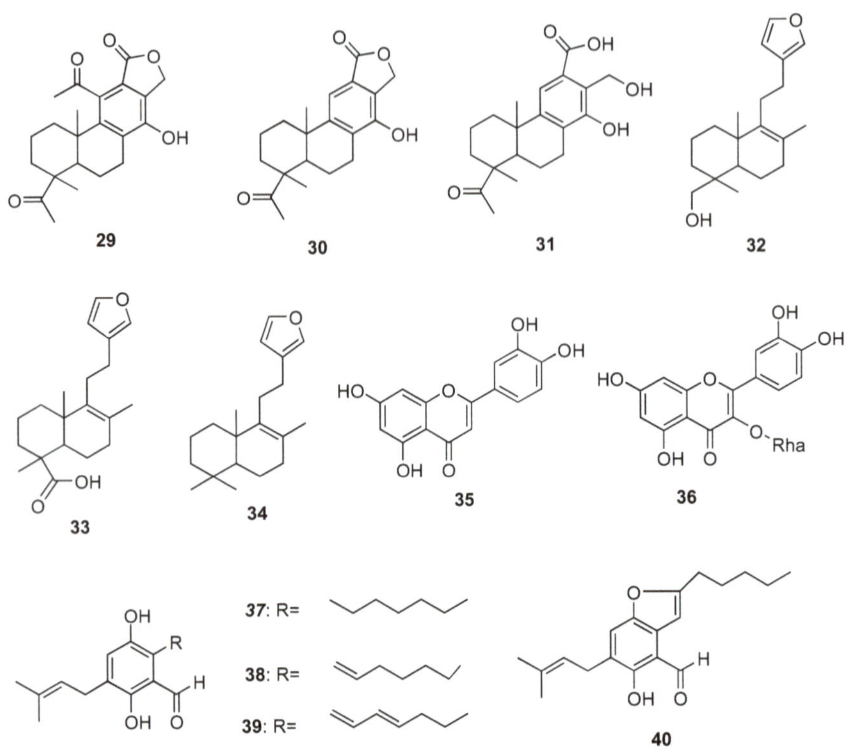

FIGURE 9.11 α-glucosidase enzyme inhibitors isolated from *Phlomis tuberosa*.

compounds were identified by ^{1}H- and ^{13}C-NMR analyses and confirmed by comparing with reported data in literature (Kanchanapoom et al., 2002; Zhang et al., 2007; Xu et al., 2009; Li et al., 2008a,b; Modaressi et al., 2009; Takeda et al., 2000).

2.12 *Tussilago farfara*

The plant *Tussilago farfara*, a member of Compositae family, also known as Kuan-Dong-Hua in China, is used as common folk medicine. In present study, Gao et al. (2008) obtained the flower buds of *T. farfara* from the local herbal market in Chengdu, China. The dried flower buds (40 g) of *T. farfara* were extracted with 70% methanol and 5.94 g methanol extract was obtained. Different fractions of methanol extract were prepared by using different solvents such as hexane, chloroform methanol, and water. These fractions were then subjected to enzyme inhibition assays. Active fractions were subjected to reverse-phase preparative HPLC for purification.

Altogether eight compounds were isolated from the dried flower buds of *T. farfara*. Methanol fraction yielded 3,4-dicaffeoylquinic acid (**41**), 3,5-dicaffeoylquinic acid (**42**), and 4,5-dicaffeoylquinic acid (**43**), which were reported

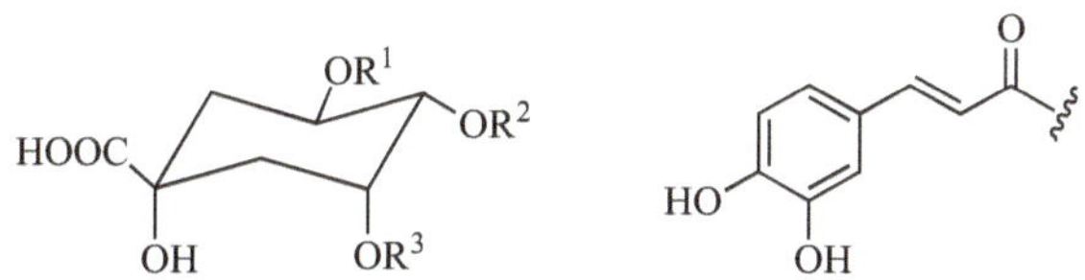

trans-Caffeoyl

41: 3,4-dicaffeoylquinic acid R^1 = H; R^2 = R^3 = *trans*-caffeoyl

42: 3,5-dicaffeoylquinic acid R^1= R^3 = *trans*-cafferoyl; R^2 = H

43: 4,5-dicaffeoylquinic acid R^1= *trans*-caffeoyl; R^2 = R^3 = H

FIGURE 9.12 α-glucosidase enzyme inhibitors isolated from *Tussilago farfara.*

first time from this plant as shown in Fig. 9.12. These three compounds exhibited significant inhibition against α-glucosidase enzyme with IC_{50} values of 0.90, 0.91, and 0.89 mM, respectively (Gao et al., 2008). The structures of these compounds were elucidated on the basis of MS and NMR data analyses and by comparing with that of published literature (Basnet et al., 1996; Ichikawa et al., 1991; Iwai et al., 2004; Zhu et al., 2004).

3. CONCLUDING REMARKS

Diabetes mellitus is the world's most challenging health problem, which is spreading very rapidly. Although the management of type 2 diabetes mellitus has been improved to a certain level during the last decade, drug adverse effects and drug resistance are still matters of great concern that need more effective approaches. α-Glucosidase enzyme plays an important role to achieve better glycemic control independent of insulin and thus is an effective therapeutic target to manage diabetes mellitus particularly NIDDM.

Over the years, several medicinal plants belonging to different parts of the world have been evaluated against α-glucosidase enzyme. Bioassay-guided isolation, structure elucidation, and characterization of secondary metabolites obtained from the medicinal plants resulted in the identification of some classes of natural products such as flavonoids, terpenes, alkaloids, quinones, and polyphenols with enhanced antidiabetic activity. Several new promising inhibitors of α-glucosidase enzyme have been identified both in vitro and in vivo. Furthermore, detailed studies can be conducted on these α-glucosidase enzyme inhibitors, which might have the potential role as drug candidates for the management of diabetes mellitus in future.

LIST OF ABBREVIATIONS

1-D NMR One-dimensional NMR
1H-NMR Proton NMR
2-D NMR Two-dimensional NMR

^{13}C-NMR Carbon isotope (^{13}C) NMR
HPLC High-performance liquid chromatography
IC_{50} Inhibitory concentration 50%
LC-MS Liquid chromatography–mass spectrometry
MS Mass spectrometry
NIDDM Non-insulin-dependent diabetes mellitus
NMR Nuclear magnetic resonance
ODS Octadecyl silica
PNPG *p*-Nitrophenyl *α*-D-glucopyranoside
RP-HPLC Reversed phase HPLC
SPE Solid-phase extraction
TLC Thin-layer chromatography
WHO World Health Organization

REFERENCES

Abbaskhan, A., Siddiqui, H., Anjum, S., Orhan, I., Gurbuz, I., Ayanoglud, F., 2010. New and known constituents from *Iris unguicularis* and their antioxidant activity. Heterocycles 82 (1), 813–824.

Alagesan, K., Thennarasu, P., Kumar, V., Sankarnarayanan, S., Balsamy, T., 2012. Identification of α-glucosidase inhibitors from *Psidium guajava* leaves and *Syzygium cumini* Linn. seeds. International Journal of Pharma Sciences and Research 3 (2), 316–322.

Asghari, B., Salehi, P., Sonboli, A., Ebrahimi, S.N., 2015. Flavonoids from *Salvia chloroleuca* with α-amylsae and α-glucosidase inhibitory effect. Iranian Journal of Pharmaceutical Research 14 (2), 609–615.

Ayatollahi, S.A., Shojaii, A., Kobarfard, F., Mohammadzadeh, M., Choudhary, M.I., 2010. Two flavones from *Salvia leriaefolia*. Iranian Journal of Pharmaceutical Research 8 (3), 179–184.

Basnet, P., Katsumichi, K., Hase, K., Kadota, S., Namba, T., 1996. Four Di-*O*-caffeoyl quinic acid derivatives from propolis. Potent hepatoprotective activity in experimental liver injury models. Biological and Pharmaceutical Bulletin 19 (11), 1479–1484.

Boath, A.S., Stewart, D., Mcdougall, G.J., 2012. Berry components inhibit α-glucosidase in vitro: synergies between acarbose and polyphenols from black currant and rowanberry. Food Chemistry 135 (3), 929–936.

Chen, Y.G., Li, P., Li, P., Yan, R., Zhang, X.Q., Wang, Y., Zhang, X.T., Ye, W.C., Zhang, Q.W., 2013. α-Glucosidase inhibitory effect and simultaneous quantification of three major flavonoid glycosides in *Microctis folium*. Molecules 18 (4), 4221–4232.

Chiasson, J.L., 2006. Acarbose for the prevention of diabetes, hypertension, and cardiovascular disease in subjects with impaired glucose tolerance: the Study to Prevent Non-Insulin-Dependent Diabetes Mellitus (STOP-NIDDM) Trial. Endocrine Practice 12 (Suppl. 1), 25–30.

Choudhary, M.I., Baig, I., Nur-E-Alam, M., Shahzad-Ul-Hussan, S., Öndognii, P., Bunderya, M., Oyun, Z., 2001a. New α-glucosidase inhibitors from the Mongolian medicinal plant *Ferula mongolica*. Helvetica Chimica Acta 84 (8), 2409–2416.

Choudhary, M.I., Nur-e-Alam, M., Akhtar, F., Ahmad, S., Baig, I., Oendoegnii, P., Gombosurengyin, P., Rahman, A.U., 2001b. Five new peltogynoids from underground parts of *Iris bungei*: a Mongolian medicinal plant. Chemical and Pharmaceutical Bulletin 49 (10), 1295–1298.

Choudhary, M.I., Nur-e-Alam, M., Baig, I., Akhtar, F., Khan, A.M., Ndögnii, P.Ö., Badarchiin, T., Purevsuren, G., Nahar, N., Atta-ur-Rahman, 2001c. Four new flavones and a new isoflavone from Iris bungei. Journal of Natural Products 64 (7), 857–860.

Chu, Y., Yu, M., 2001. Study on the ecological distribution of morus germplasm resources in Yunnan province. Acta Sericologica Sinica 28 (1), 1–7.

Conforti, F., Statti, G., Loizzo, M.R., Sacchetti, G., Poli, F., Menichini, F., 2005. In Vitro antioxidant effect and inhibition of alpha-amylase of two varieties of *Amaranthus caudatus* seeds. Biological and Pharmaceutical Bulletin 28 (6), 1098–1102.

Elya, B., Basah, K., Mun'im, A., Yuliastuti, W., Bangun, A., Septiana, E.K., 2012. Screening of α-glucosidase inhibitory activity from some plants of Apocynaceae, Clusiaceae, Euphorbiaceae, and Rubiaceae. Journal of Biomedicine and Biotechnology 2012:281078. http://dx.doi.org/10.1155/2012/281078.

Farag, Y.M., Gaballa, M.R., 2011. Diabesity: an overview of a rising epidemic. Nephrology Dialysis Transplantation 26 (1), 28–35.

Feng, S.X., Liu, M.F., Wei, X.Y., Lin, L.D., 2008. Triterpenoids and flavonoids from the leaves of Microcos paniculata. Journal of Tropical and Subtropical Botany 16 (1), 51–56.

Gao, H., Nishioka, T., Kawabata, J., Kasai, T., 2004. Structure-activity relationships for α-glucosidase inhibition of baicalein, 5,6,7-trihydroxyflavone: the effect of A-ring substitution. Bioscience, Biotechnology and Biochemistry 68 (2), 369–375.

Gao, H., Huang, Y.N., Gao, B., Xu, P.Y., Inagaki, C., Kawabata, J., 2008. α-Glucosidase inhibitory effect by the flower buds of *Tussilago farfara* L. Food Chemistry 106 (3), 1195–1201.

Gohari, A.R., Ebrahimi, H., Saeidnia, S., Foruzani, M., Ebrahimi, P., Ajani, Y., 2011. Flavones and flavone glycosides from *Salvia macrosiphon* boiss. Iranian Journal of Pharmaceutical Research 10 (2), 247–251.

Grey, A., 2008. Skeletal consequences of thiazolidinedione therapy. Osteoporosis International 19 (2), 129–137.

Grover, J.K., Yadav, S., Vats, V., 2002. Medicinal plants of India with anti-diabetic potential. Journal of Ethnopharmacology 81 (1), 81–100.

Hong, H.C., Li, S.L., Zhang, X.Q., Ye, W.C., Zhang, Q.W., 2013. Flavonoids with α-glucosidase inhibitory activities and their contents in the leaves of *Morus atropurpurea*. Chinese Medicine 8 (1), 1–7.

Hosoya, T., Yun, Y.S., Kunugi, A., 2005. Five novel flavonoids from *Wasabia japonica*. Tetrahedron 61 (29), 7037–7044.

Hunter, D., 2001. Life in the fast lane: high-throughput chemistry for lead generation and optimisation. Journal of Cellular Biochemistry 84 (S37), 22–27.

Ichikawa, K., Sakurai, Y., Akiyama, T., Yoshioka, S., Shiraki, T., Horikoshi, H., Kuwano, H., Kinoshita, T., Boriboon, M., 1991. Isolation and structure determination of aldose reductase inhibitors from traditional Thai medicine and syntheses of their derivatives. Sankyo Kenkyusho Nempo 43, 99–110.

Iwai, K., Kishimoto, N., Kakino, Y., Mochida, K., Fujita, T., 2004. In vitro antioxidative effects and tyrosinase inhibitory activities of seven hydroxycinnamoyl derivatives in green coffee beans. Journal of Agricultural and Food Chemistry 52 (15), 4839–4898.

Jung, M., Park, M., Lee, H.C., Kang, Y.H., Kang, E.S., Kim, S.K., 2006. Antidiabetic agents from medicinal plants. Current Medicinal Chemistry 13 (10), 1203–1218.

Kanchanapoom, T., Kasai, R., Yamasaki, K., 2002. Phenolic glycosides from *Markhamia stipulata*. Phytochemistry 59 (5), 557–563.

Kodra, J.T., Jorgensen, A.S., Andersen, B., Behrens, C., Brand, C.L., Christensen, I.T., Guldbrandt, M., Jeppesen, C.B., Knudsen, L.B., Madsen, P., Nishimura, E., 2008. Novel glucagon receptor antagonists with improved selectivity over the glucose-dependent insulinotropic polypeptide receptor. Journal of Medicinal Chemistry 51 (17), 5387–5396.

Li, T., Liu, J., Zhang, X., Ji, G., 2007. Antidiabetic activity of lipophilic (−)-epigallocatechin-3-gallate derivative under its role of α-glucosidase inhibition. Biomedicine & Pharmacotherapy 61 (1), 91–96.

Li, Y.L., Li, J., Wang, N.L., Yao, X.S., 2008a. Flavonoids and a new polyacetylene from *Bidens parviflora* Willd. Molecules 13 (8), 1931–1941.

Li, D.L., Li, X.M., Li, T.G., Dang, H.Y., Proksch, P., Wang, B.G., 2008b. Benzaldehyde derivatives from *Eurotium rubrum*, an endophytic fungus derived from the mangrove plant *Hibiscus tiliaceus*. Chemical and Pharmaceutical Bulletin 56 (9), 1282–1285.

Lu, Y., Foo, L.Y., 2000. Flavonoid and phenolic glycosides from *Salvia officinalis*. Phytochemistry 55 (3), 263–267.

Maciel, M.A.M., Pinto, A.C., Veiga Jr., V.F., Grynberg, N.F., Echevarria, A., 2002. Medicinal plants: the need for multidisciplinary scientific studies. Química Nova 25 (3), 429–438.

Misbah, H., Aziz, A.A., Aminudin, N., 2013. Antidiabetic and antioxidant properties of *Ficus deltoidea* fruit extracts and fractions. BMC Complementary and Alternative Medicine 13 (1), 118–129.

Modaressi, M., Delazar, A., Nazemiyeh, H., Fathi-Azad, F., Smith, E., Rahman, M.M., Gibbons, S., Nahar, L., Sarker, S.D., 2009. Antibacterial iridoid glucosides from *Eremostachys laciniata*. Phytotherapy Research 23 (1), 99–103.

Mosihuzzman, M., Naheed, S., Hareem, S., Talib, S., Abbas, G., Khan, S.N., Choudhary, M.I., Sener, B., Tareen, R.B., Israr, M., 2013. Studies on α-glucosidase inhibition and anti-glycation potential of *Iris loczyi* and *Iris unguicularis*. Life Sciences 92 (3), 187–192.

Murata, T., Miyase, T., Yoshizaki, F., 2010. Cyclic spermidine alkaloids and flavone glycosides from *Meehania fargesii*. Chemical and Pharmaceutical Bulletin 58 (5), 696–702.

Nkobole, N., Houghton, P.J., Hussein, A., Lall, N., 2011. Antidiabetic activity of *Terminalia sericea* constituents. Natural Product Communications 6 (11), 1585–1588.

Park, H., Hwang, K.Y., Kim, Y.H., Oh, K.H., Lee, J.Y., Kim, K., 2008. Discovery and biological evaluation of novel α-glucosidase inhibitors with in vivo antidiabetic effect. Bioorganic & Medicinal Chemistry Letters 18 (13), 3711–3715.

Takeda, Y., Matsumura, H., Masuda, T., Honda, G., Otsuka, H., Takaishi, Y., Sezik, E., Yesilada, E., 2000. Phlorigidosides A–C, iridoid glucosides from *Phlomis rigida*. Phytochemistry 53 (8), 931–935.

Tan, C., Wang, Q., Luo, C., Chen, S., Li, Q., Li, P., 2013. Yeast α-glucosidase inhibitory phenolic compounds isolated from *Gynura medica* leaf. International Journal of Molecular Sciences 14 (2), 2551–2558.

Tanaka, J.C.A., Silva, C.C.D., Dias Filho, B.P., Nakamura, C.V., Carvalho, J.E.D., Foglio, M.A., 2005. Chemical constituents of *Luehea divaricata* Mart. (Tiliaceae). Quimica Nova 28 (5), 834–837.

Tiabou Tchinda, A., Nahar Khan, S., Fuendjiep, V., Ngandeu, F., Ngono Ngane, A., Choudhary, M.I., 2007. Alpha-glucosidase inhibitors from *Millettia conraui*. Chemical and Pharmaceutical Bulletin 55 (9), 1402–1403.

Van De Laar, F.A., Lucassen, P.L., Akkermans, R.P., Van De Lisdonk, E.H., Rutten, G.E., Van Weel, C., 2005. α-Glucosidase inhibitors for Patients with type 2 diabetes results from a Cochrane systematic review and meta-analysis. Diabetes Care 28 (1), 154–163.

Xu, M., Shen, L., Wang, K., 2009. A new biflavonoid from *Daphniphyllum angustifolium* Hutch. Fitoterapia 80 (8), 461–464.

Yang, Y., Gu, L., Xiao, Y., Liu, Q., Hu, H., Wang, Z., Chen, K., 2015. Rapid identification of α-glucosidase inhibitors from *Phlomis tuberosa* by sepbox chromatography and thin-layer chromatography bioautography. PLoS One 10 (2), e0116922.

Yao, Y., Cheng, X., Wang, L., Wang, S., Ren, G., 2011. A determination of potential α-glucosidase inhibitors from azuki beans (*Vigna angularis*). International Journal of Molecular Sciences 12 (10), 6445–6451.

Yoshikawa, M., Nishida, N., Shimoda, H., Takada, M., Kawahara, Y., Matsuda, H., 2001. Polyphenol constituents from Salacia species: quantitative analysis of mangiferin with alpha-glucosidase and aldose reductase inhibitory activities. Yakugaku Zasshi: Journal of the Pharmaceutical Society of Japan 121 (5), 371–378.

Yoshikawa, M., Morikawa, T., Matsuda, H., Tanabe, G., Muraoka, O., 2002. Absolute stereostructure of potent α-glucosidase inhibitor, salacinol, with unique thiosugar sulfonium sulfate inner salt structure from *Salacia reticulata*. Bioorganic & Medicinal Chemistry 10 (5), 1547–1554.

Zhang, T., Zhou, J., Wang, Q., 2007. Flavonoids from aerial part of Bupleurum chinense DC. Biochemical Systematics and Ecology 35 (11), 801–804.

Zhu, X., Zhang, H., Lo, R., 2004. Phenolic compounds from the leaf extract of artichoke (*Cynara scolymus* L.) and their antimicrobial activities. Journal of Agricultural and Food Chemistry 52 (24), 7272–7278.

Chapter 10

Treasures of Indigenous Indian Herbal Antidiabetics: An Overview

I. Mohanram, J.S. Meshram
Rashtrasant Tukadoji Maharaj Nagpur University, Nagpur, India

1. INTRODUCTION

Plants are gift of nature to all living beings as they put up with remedies for various diseases in addition to other beneficial concerns. The plant kingdom warrants our keen attention in present time as most of the valuable ones are becoming endangered reported by Kala (2000). In India, indigenous medicine has been used in the treatment of illness as reported in Vedic literature, Charaka Samhita, and Sushruta Samhita since ancient times. Loukas et al. (2010) emphasized that Ayurveda is a medicine practised primarily in India, which promotes self-healing, good health, and durability. It includes diet and herbal remedies, while accentuate the body, mind, and spirit in disease prevention and treatment. Traditional herbal medicine provides a rich source for new drug discovery. A number of medicinal herbs have been assessed for their therapeutic potentials and majorly have shown defensive action against various chronic diseases reported by Vedavathy (2003). The presences of bioactive chemicals in herbal medicines provide the nutrients and therapeutic ingredients to prevent, mitigate, or treat many diseases or conditions including diabetes. Management of diabetes with fewer side effects is still a challenge to modern medicine. Treatment of diabetes may include the drugs like insulin, tolbutamide, glibenclamide, and metformin. According to Stein et al. (2013) and Safavi et al. (2013), these drugs generally show hypersensitivity, liver problems, lactic acidosis, diarrhea, and other side effects but herbal medicines were less toxic and free from side effects than synthetic one. In diabetes, the amount of insulin produced by pancreas is get reduced quantitatively, which results in the increase of sugar in blood and excess sugar eliminate outside through urine. Hence, there has been a requirement for exponential growth in the field of herbal medicine and gaining popularity both in developing and developed countries because of their natural origin and less side effects.

Discovery and Development of Antidiabetic Agents From Natural Products.
http://dx.doi.org/10.1016/B978-0-12-809450-1.00010-7

Several herbs have shown antidiabetic activity when assessed using presently available experimental techniques. Mohammed et al. (2013) describes a wide array of plant-derived active principles representing numerous chemical compounds has demonstrated activity consistent with their possible use in the treatment of insulin-dependent diabetes mellitus. The plants contain various chemical constituents that are efficient in the treatment of various diseases. Therefore, the herbal medicines were considered as the potential source to cure many diseases including diabetes as reported by Cooke and Plotnick (2008).

Herein, traditional antidiabetic medicinal herbs that are available in different regions of India are compiled. These folklore medicines are useful to the health professionals, scientists, and research scholars working in the field of pharmacology and therapeutics to develop evidence-based alternative medicines to cure different kinds of diabetes.

2. ANTIDIABETIC INDIAN MEDICINAL PLANTS

A good number of Indian plant extracts are found useful in treating diabetic systems. Lakshmi et al. (2014) studied the antidiabetic potential of *Musa paradisiaca* in streptozotocin (STZ)-induced diabetic rats; the ethanolic extracts of leaves and fruit peels of *M. paradisiaca* showed promising antidiabetic activity with lowering of blood sugar level up to 20.5% and 25.5%, respectively at 500 mg/kg body weight. Similarly, Thomson et al. (2007) studied the hypoglycemic, hypocholesterolemic, and hypotriglyceridemic effects of *Allium sativum* in STZ-induced diabetic rats. In comparison to nondiabetic rats, STZ-induced diabetic rats possess approximately 200% higher serum glucose, 50% higher serum cholesterol, and 30% higher serum triglyceride levels as well as 86% higher urinary protein levels which on treatment with extract of raw garlic (*A. sativum*) significantly lowered serum glucose to 57%, cholesterol 40%, and triglyceride 35%. In addition, urinary protein levels in garlic-treated diabetic rats were 50% lower compared to the diabetic controls. This chapter primarily enlists a list of medicinal plants reported for their potential antidiabetic potential. Family, active parts, major chemical constituents responsible for their pharmacological profile, and regions where these plant species mostly grow in India are summed up in Table 10.1. These herbal medicines have been practised traditionally to cure various life-threatening diseases among which diabetes mellitus is one of them. Fig. 10.1 presents chemical structures of most promising phytochemical constituents isolated from these plant species.

The chemistry of these phytoconstituents suggests that the functional groups such as amide, methyl, and hydroxyl adjacent to different other heteroatoms enhance the medicinal potency. Similarly, conjugated double bonds also play active role to make these constituents biologically active.

TABLE 10.1 List of Antidiabetic Medicinal Herbs Found in Various Parts of India

S. No	Botanical Name	Family	Active Parts	Chemical Constituents	Regions	References
1	*Abelmoschus moschatus*	Malvaceae	Root, leaves, seed	Myricetin, Farnesol	Andhra Pradesh, Himalayas, Maharashtra, Gujarat, Madhya Pradesh, Karnataka, Tamil Nadu	Liu et al. (2010)
2	*Abroma augusta*	Malvaceae	Root, bark, leaves	Friedelin, Choline, Betaine	Sikkim, Darjeeling, Punjab, Uttar Pradesh	Mir et al. (2013)
3	*Acacia catechu*	Mimosaceae	Stem, bark, leaves	α, β, γ catechin, Alkaloids, Quercetin, Kaempferol	Punjab, Uttarakhand, Bihar, Odisha, Uttar Pradesh, Jammu & Kashmir, Himachal Pradesh, Kerala, Madhya Pradesh, Gujarat, Maharashtra	Ikarashi et al. (2011)
4	*Acacia nilotica*	Mimosaceae	Gum, leaves, seed, root, stem	Polyphenols, Tannins, Gallic acid, Octacosanol, Betulin, Flavonoids, α-amyrin, β-sitosterol	Maharashtra, Rajasthan, Kerala, Karnataka, Punjab	Ali et al. (2012)
5	*Acacia polyacantha*	Mimosaceae	Seed	Tryptamines, β-Carbolines	Kerala, Andhra Pradesh, Delhi, Goa, Gujarat, Karnataka, Maharashtra, Tamil Nadu, West Bengal	Okpanachi et al. (2010)
6	*Achyranthes aspera*	Amaranthaceae	Leaves, stem, seed, root	Achyranthine, Betaine, Pentatriacontane, 6-pentatriacontanone, Hexatriacontane, Tritriacontane	Madhya Pradesh, Odisha, Bihar, Karnataka, Assam, Tamil Nadu, Maharashtra	Srivastav et al. (2011)
7	*Aconitum palmatum*	Ranunculaceae	Root	Alkaloids, Flavonoids	Sikkim, Darjeeling, Gujarat, Rajasthan	Chhetri et al. (2005)

Continued

TABLE 10.1 List of Antidiabetic Medicinal Herbs Found in Various Parts of India—cont'd

S. No	Botanical Name	Family	Active Parts	Chemical Constituents	Regions	References
8	*Aegle marmelos*	Rutaceae	Leaves, root, fruit, seed	Skimmianine, Aegeline, Lupeol, Cineole, Citral, Citronellal, Cuminaldehyde, Eugenol, Marmesinin	Assam, Maharashtra, Puducherry, Andhra Pradesh, Bihar, Uttaranchal, Himachal Pradesh, Uttar Pradesh, Bihar, Odisha, Madhya Pradesh, West Bengal, Tripura	Maity et al. (2009)
9	*Agrimonia eupatoria*	Rosaceae	Whole plant, leaves	Volatile oil, Apigenin, Luteolin, Quercetin, Kaempferol, Triterpene, Glycosides, Tannins	Himalayas, Odisha, Kashmir to North Bengal	Al-snafi (2015)
10	*Albizia odoratissima*	Mimosaceae	Bark, seed, leaves	Carbohydrates, Glycosides, Flavonoids, Steroids, Tannins, Fixed oils	Andhra Pradesh, Kerala, Tamil Nadu, Karnataka	Rajan et al. (2011a)
11	*Allium cepa*	Amaryllidaceae	Seed, bulb	Allicin, Arginine, Ascorbic acid, Choline, Diphenylamine, Fiber, Fructose, Pectin, Rutin, S-methyl-cysteine-sulfoxide, Xylose	Uttar Pradesh, Maharashtra, Odisha, Karnataka, Gujarat, Andhra Pradesh, Bihar, Tamil Nadu, New Delhi, Punjab, Rajasthan	Kelkar et al. (2001)
12	*Allium sativum*	Amaryllidaceae	Bulb	Allicin, Peptides, Steroids, Terpenoids, Flavonoids, Phenols	Tamil Nadu, Andhra Pradesh, Uttar Pradesh, Gujarat, Madhya Pradesh, Arunachal Pradesh, Kerala, West Bengal, Maharashtra, Himachal Pradesh	Thomson et al. (2007)
13	*Aloe vera*	Xanthorrhoeaceae	Leaves	Amino Acids, Anthraquinones, Enzymes, Lignin, Hormones, Minerals, Salicylic acid, Saponins, Sterols, Sugars, Vitamins	Rajasthan, Andhra Pradesh, Gujarat, Tamil Nadu, Maharashtra,	Sharma et al. (2013)

14	*Alpinia calcarata roscoe*	Zingiberaceae	Rhizome	Protocatechinic acid, 1,8-Cineole, Quercetin, β-pinene, 4-O-methyl-syringic acid, Methyl cinnamate, Vanillic, Diterpenes	Tamil Nadu, Assam, Nagaland, Manipur, Mizoram, Tripura, Kerala	Rahman and Islam (2015)
15	*Alpinia galangal*	Zingiberaceae	Rhizome	Galangin, Alpinin, Kaemperol, Quercetin	West Bengal, Bihar, Goa, Assam, Karnataka, Gujarat	Shetty and Monisha (2015)
16	*Anacardium occidentale*	Anacardiaceae	Bark, leaves	Alkaloids, Oxalate, Kaemperol, Polyphenol, Saponin, Tannin	Andhra Pradesh, Goa, Karnataka, Kerala, Madhya Pradesh, Maharashtra, Odisha, Puducherry, West Bengal, Tamil Nadu	Obaineh and Shadrach (2013)
17	*Andrographis paniculata*	Acanthaceae	Root	Flavonoids, β-Sitosterol	Andhra Pradesh, Puducherry	Jayakumar et al. (2013)
18	*Annona squamosa*	Annonaceae	Bark, leaves, seed	Alkaloids, Coumarins, Flavanoids, Saponins, Terpenoids	Assam, Bihar, Madhya Pradesh, Maharashtra, Odisha, Rajasthan, Uttar Pradesh, Andhra Pradesh, Tamil Nadu	Biba et al. (2013)
19	*Anogeissus latifolia*	Combretaceae	Bark, root	Tannin, Glycosides Alanine, Myricetin, Phenylalanine, Quercetin, Trimethyl ellagic acid	Himalayas, Madhya Pradesh	Ramchandran et al. (2012)
20	*Areca catechu*	Arecaceae	Flowers, leaves, fruit	Alkaloids, Quercetin, β-sitosterol, Tannins, Polyphenols	Assam, Karnataka, Kerala, Maharashtra, West Bengal, Tripura	Mondal et al. (2012)
21	*Artemisia pallens*	Asteraceae	Aerial part	Davana oil, Methyl cinnamate, Ethyl cinnamate, Sesquiterpene lactones	Andhra Pradesh, Himalayas, Karnataka, Maharashtra, Tamil Nadu	Haider et al. (2014)

Continued

TABLE 10.1 List of Antidiabetic Medicinal Herbs Found in Various Parts of India—cont'd

S. No	Botanical Name	Family	Active Parts	Chemical Constituents	Regions	References
22	*Artocarpus heterophyllus*	Moraceae	Seed, fruit, bark, root, leaves	Flavonoids, Amino acids, β-sitosterol	Goa, Kerala, Odisha, Tamil Nadu, Karnataka, Maharashtra	Chackrewarthy et al. (2010)
23	*Asparagus racemosus*	Asparagaceae	Stem, leaves, root	Essential oils, Asparagine, Arginine, Tyrosine, Flavonoids, Resin, Tannin	Himalayas, Uttar Pradesh, Rajasthan, Gujarat, Sikkim, Darjeeling, Tamil Nadu	Hannan et al. (2012)
24	*Averrhoa carambola*	Oxalidaceae	Leaves, bark	Saponins, Tannins, Alkaloids, Flavonoids	Kerala up to West Bengal.	Thomas et al. (2008)
25	*Azadirachta indica*	Meliaceae	Bark, root, leaves	Nimbinene, Azadirachtin, Azadirachtol, Nimbandiol, Nimbolide, Quercetin, β-sitosterol, n-hexacosanol, Nimocin	Uttar Pradesh, Haryana, Punjab, Himachal Pradesh, Orissa, Andhra Pradesh, Kerala, Karnataka, Tamil Nadu, Puducherry	Patil et al. (2013)
26	*Bacopa monnieri*	Plantaginaceae	Aerial part	Alkaloid brahmine, Nicotinine, Herpestine, Bacosides, Triterpenoid saponins, Betulinic acid, D-mannitol, Stigmastanol, β-sitosterol, Glycoside	Andhra Pradesh, Kerala, Puducherry	Al-Snafi (2013)
27	*Benincasa hispida*	Cucurbitaceae	Fruit	Glycosides, Sacchrides, Proteins, Carotenes, β-sitosterin, Uronic acid, Volatile oils, Flavonoids	Andhra Pradesh, Uttar Pradesh, Delhi, Kerala, Tamil Nadu	Mahatma et al. (2014)
28	*Berberis aristata*	Berberidaceae	Root, stem	Alkaloids, Glycosides, Carbohydrates, Saponins	Himalayas, Himachal Pradesh, Sikkim, Darjeeling, Uttaranchal	Upwar et al. (2011)
29	*Biophytum sensitivum*	Oxalidaceae	Flower, leaves	Amentoflavone, Polysaccharide, Polyphenolic compounds, Saponin, Essential oil, Pectin	Kerala	Bharati and Sahu (2012)

30	*Bixa orellana*	Bixaceae	Leaves, seed	Tocotrienols, β-carotene, Essential oil, Fatty acids, Flavonoids, Vitamin C, Norbixin, Sesquiterpene	Karnataka, Tamil Nadu, Andhra Pradesh, Odisha, Assam, West Bengal, Maharashtra	Vilar et al. (2014)
31	*Boerhaavia diffusa*	Nyctaginaceae	Leaves, root	Rotenoids, Flavonoids, Xanthones, Purine nucleoside, Lignans, Steroids	West. Bengal, Kerala, Assam Puducherry	Mishra et al. (2014)
32	*Boenninghausenia albiflora*	Rutaceae	Leaves, root	β-myrcene, β-guaiene, β-ocimene, β-caryophyllene, Bicyclogermacren, α-terpinyl acetate, Geijerene, β-copaene, Coumarins	Sikkim, Darjeeling, Uttarakhand, Himalayas, Jammu and Kashmir, Uttar Pradesh	Chhetri et al. (2005)
33	*Bombax ceiba*	Malvaceae	Seed, root, bark, leaves	Lupeol, Steroids, Tannins, Flavonoids, Saponins, Stigmasterol, Campesterol, Amyrin	Rajasthan, Madhya Pradesh, Uttar Pradesh, Kerala	Antil et al. (2013)
34	*Borassus flabellifer*	Arecaceae	Root, fruit, flower	Flavonoids, Saponins, Tannins, Butyric acid, Albuminoids, Steroidal Glycosides, Carbohydrates	Tamil Nadu, West Bengal, Andhra Pradesh, Bihar, Kerala	Sandhya et al. (2010)
35	*Brassica juncea*	Brassicaceae	Seed, flower, leaves, stem	Isothiocyanates, Essential oils, β-sitosterol, Campesterol, Brassicasterol, Glucosinolates, Ascorbate	Punjab, Darjeeling, Sikkim, Rajasthan, Haryana, Uttar Pradesh	Thirumalai et al. (2011)
36	*Butea monosperma*	Fabaceae	Leaves, bark, fruit	Kino-oil, Cyanidin, Histidine, Lupenone, Lupeol, Flavonoids, Steroids, Triterpene, Butein, Coreopsin, Proteolytic and Lypolytic enzymes, Tannin	Puducherry, Tamil Nadu, Jharkhand, West Bengal	Sharma et al. (2009)
37	*Caesalpinia bonducella*	Caesalpiniaceae	Seed, leaves, root, bark	Steroidal saponin, Fatty acids, Phytosterols, Isoflavones, Amino acids, Phenols	Kerala, Gujarat, Uttar Pradesh	Singh and Raghav (2012)

Continued

TABLE 10.1 List of Antidiabetic Medicinal Herbs Found in Various Parts of India—cont'd

S. No	Botanical Name	Family	Active Parts	Chemical Constituents	Regions	References
38	*Caesalpinia digyna*	Caesalpiniaceae	Root	Tannin, Alkaloids, Bonducin, Saponin, Friedelin, Hexacosanoic acid, β-sitosterol, Stigmasterol	Andhra Pradesh, Madhya Pradesh	Srinivasan et al. (2011)
39	*Cajanus cajan*	Fabaceae	Leaves, seed	Carbohydrate, Amino acids, Fats and oils, Protein, Phenols, Glycosides, Anthraquinone, Saponin glycoside	Madhya Pradesh, Uttar Pradesh, Maharashtra, Bihar, Andhra Pradesh, Punjab, Haryana, West Bengal, Assam, Odisha, Rajasthan, Himachal Pradesh, Gujarat, Jammu & Kashmir, Karnataka, Tamil Nadu, Kerala	Nahar et al. (2014)
40	*Calamus rotang*	Arecaceae	Fruit, leaves, root	Phenylpropanoids, Sesquiterpenes, Monoterpenes, Xanthone glycosides, Flavones, Lignans, Steroids, Inorganic constituents, Triterpenoid saponins	Andhra Pradesh, Sikkim, Darjeeling, Tamil Nadu, Kerala, Himalayas, West Bengal, Assam	Ripa et al. (2015)
41	*Callicarpa macrophylla*	Verbenaceae	Leaves, heart wood, flower, fruit	Glycosides, flavonoids, Tannins, Alkaloids, Clerodane, Phyllocladane diterpenes, Fatty acids, Lignans, Monoterpenes, Phenylpropanoids, Phytosterols, Sesquiterpenes, Triterpenes	Jammu & Kashmir, Uttaranchal, Himalayas, Tamil Nadu	Pandey et al. (2014)
42	*Cannabis sativa*	Cannabaceae	Leaves, root	Tetrahydrocannabinol, Cannabinoids, α-Pinene, Myrcene, Linalool, Limonene, β-ocimene, α-Terpinolene, Caryophyllene, α-Humulene	Assam, Darjeeling, Sikkim, Bihar, Madras, Uttar Pradesh, Punjab, West Bengal, Jammu and Kashmir, Rajasthan, Himachal Pradesh, Manipur, Tripura	Levendal and Frost (2006)

43	*Capparis decidua*	Capparaceae	Leaves, flower, fruit, seed	Alkaloids, Terpenoids, Glycosides, Fatty acids, β-Sitosterol	Punjab, Gujarat, Arunachal Pradesh, Himalayas, Andhra Pradesh, Karnataka, Maharashtra	Zia-Ul-Haq et al. (2011)
44	*Carica papaya*	Caricaceae	Leaves, fruit	β-carotene, Crytoxanthin, Polyphenols, Papain, Carposide, Chymopapain, Benzyl isothiocynate, Papaya oil, Myricetin, Kaemferol, Carpinine,Vitamin C & E	Assam, Himachal Pradesh, Rajasthan, Gujarat, Bihar, Odisha, Chhattisgarh, Telangana, Kerala, Tamil Nadu, Karnataka, Maharashtra, Uttar Pradesh, Puducherry, Tripura, Mizoram, Nagaland, Arunachal Pradesh	Maniyar and Bhixavatimath (2012)
45	*Carissa carandas*	Apocynaceae	Fruit	Steroids, Tannin, Saponin, Emodins, Anthocyanin, Coumarin, Alkaloids, Proteins, Amino acids, Carbohydrates, Flavonoids, Diterpenes, Phytosterol	Himalayas, Rajasthan, Gujarat, Bihar, West Bengal, Uttar Pradesh	Godghate et al. (2012)
46	*Cassia auriculata*	Caesalpiniaceae	Flower, fruit, leaves	Flavonoids, Tannins, Steroids, Terpenoids, Alkaloids, Carbohydrates, Coumarins, Phenols, Saponins, Glycosides	Punjab, Haryana, Uttar Pradesh, Kerala, west Bengal, Puducherry	Gupta et al. (2011)
47	*Cassia fistula*	Caesalpiniaceae	Bark, flower	Anthraquinones, Fistulic acid, Rhein glycoside, Sennosides A & B, Phlobaphenes, Emodin, Lupeol, Chrysophanic acid, Fistuacacidin, β-sitosterol, Hexacosanol	Kerala, Puducherry	Jarald et al. (2013)
48	*Catharanthus roseus*	Apocynaceae	Leaves	Carbohydrate, Flavinoid, Saponin, Alkaloids	Tamil Nadu, Karnataka, Gujarat, Odisha, West Benagal, Darjeeling, Puducherry, Sikkim, Andhra Pradesh	Yadav et al. (2013)

Continued

TABLE 10.1 List of Antidiabetic Medicinal Herbs Found in Various Parts of India—cont'd

S. No	Botanical Name	Family	Active Parts	Chemical Constituents	Regions	References
49	*Cedrus deodara*	Pinaceae	Heart wood, stem	Alkaloids, Saponins, Tannins, Glycosides, α-terpineol, Flavonoids, Fixed oils & fats	Jammu & Kashmir, Himachal Pradesh, Sikkim, Darjeeling, Uttarakhand, Himalayas	Singh et al. (2013)
50	*Ceiba pentandra*	Malvaceae	Root, seed	Tannins, Flavonoids, Glycosides, Reducing sugar, Triterpenes	Maharashtra, West Bengal, Tamilnadu, Karnataka, Kerala	Bairwa et al. (2010)
51	*Cinnamomum tamala*	Lauraceae	Stem, leaves	Polyphenols, Monoterpenoids, Sesquiterpenoids, Phellandrene, Eugenol, Linalool	Assam, Darjeeling, Sikkim, Uttarakhand, Himalayas	Bisht and Sisodia (2011)
52	*Citrullus colocynthis*	Cucurbitaceae	Root, fruit, seed, leaves	Glycosides, Flavonoids, Alkaloids, Fatty acids, Essential oils	Punjab, Gujarat, Haryana, Rajasthan, Madhya Pradesh, Maharashtra	Hussain et al. (2014)
53	*Coccinia indica*	Cucurbitaceae	Root, leaves, fruit	Resins, Alkaloids, Saponins, Sterols, Tannins, Flavonoids, Proteins, Amino acids, Glycosides, Carbohydrates	Kerala, Tamil Nadu, Maharashtra, Uttar Pradesh, Bengal, Bihar, Odisha	Hussain et al. (2011)
54	*Cocculus hirsutus*	Menispermaceae	Leaves	Carbohydrates, Steroids, Alkaloids, Glycosides, Tannins, Saponins, Flavonoids	West Bengal, Bihar, Punjab, Tamil Nadu, Himalayas, Gujarat	Meena et al. (2014)
55	*Coscinium fenestratum*	Menispermaceae	Stem	Alkaloids, Ceryl-alcohol, Saponin, Hentriacontane, Sitosterol glucoside, Palmitic acid, Oleic acid	Himalayas, Jammu & Kashmir, Kerala, Karnataka, Tamil Nadu, Uttar Pradesh, Odisha, Punjab	Shirwaikar et al. (2005)
56	*Costus speciosus*	Costaceae	Rhizome	Saponins, Steroids, Alkaloids, α-tocopherol, Tricontanoic acids, Curcumin, Tricontanol, Triterpenes	Himalayas, Himachal Pradesh, Assam, Andhra Pradesh, Maharashtra, Karnataka, Tamil Nadu, Kerala, Sikkim, Darjeeling	Eliza et al. (2009)

57	*Cressa cretica*	Convolvulaceae	Whole plant	Alkaloids, Glycosides, Flavonoids, β-sitosterol, Quercetin	Assam, Tamil Nadu, Gujarat, Rajasthan, Andhra Pradesh, Puducherry	Verma et al. (2014)
58	*Cucumis sativus*	Cucurbitaceae	Fruit	Glycosides, Steroids, Flavonoids, Carbohydrates, Terpenoids, Tannins	Rajasthan, Madhya Pradesh, Uttar Pradesh, Karnataka, Himachal Pradesh, Maharashtra, Puducherry	Karthiyayini et al. (2009)
59	*Cucumis trigonus*	Cucurbitaceae	Fruit	Steroids, Triterpenoids, Alkaloids, Saponins, Flavonoids	Maharashtra, Goa, Tamil Nadu, Andhra Pradesh, Himalayas	Salahuddinand Jalalpure (2010)
60	*Curcuma longa*	Zingiberaceae	Rhizome	Curcumin, Sesquiterpenoids, Diaryl heptanoids, Tumerone	Andhra Pradesh, Tamil Nadu, Kerala, Bihar, Odisha, Maharashtra, Puducherry	Nishiyama et al. (2005)
61	*Dillenia indica*	Dilleniaceae	Leaves	Triterpenoids, Flavonoids, Lupeol, Betulinaldehyde, Stigmasterol	Assam, West Bengal, Maharashtra, Meghalaya, Arunachal Pradesh, Andhra Pradesh	Parvin et al. (2009)
62	*Dioscorea alata*	Dioscoreaceae	Tuber	Carbohydrates, γ-tocopherol, Ipomeanol	Kerala, Odisha, Assam, Punjab, Gujarat, Himachal Pradesh, Maharashtra	Maithili et al. (2011)
63	*Diospyros* peregrina	Ebenaceae	Bark	Triterpenes, Steroids, Betulinic acid, Ursolic acid	West Bengal, Odisha, Bihar, Jharkhand, Assam, Tripura, Gujarat, Sikkim	Dewanjee et al. (2008)
64	*Elephantopus scaber*	Asteraceae	Root, leaves	Sesquiterpene Lactones, Steroid, Triterpenoids, Isoscabertopin	Kerala, Karnataka, Tamil Nadu, Madhya Pradesh	Daisy et al. (2009)
65	*Eugenia jambolana*	Myrtaceae	Stem, seed	Betulinic acid, β-sitosterol, Eugenin, Quercetin, Kaempferol, Myricetin, Gallic acid, Ellagic acid, Tannins	Maharashtra, Karnataka, Andhra Pradesh, Tamil Nadu, Kerala	Ayyanar and Subash-babu (2012)

Continued

TABLE 10.1 List of Antidiabetic Medicinal Herbs Found in Various Parts of India—cont'd

S. No	Botanical Name	Family	Active Parts	Chemical Constituents	Regions	References
66	*Ficus religiosa*	Moraceae	Bark	Tannins, Saponins, Flavonoids, Steroids, Terpenoids, Glycosides	Uttar Pradesh, West Bengal, Andhra Pradesh, Kerala	Pandit et al. (2010)
67	*Ficus benghalensis*	Moraceae	Bark	Quercetin, Lupeol, β-amyrin, β-sitosterol-alpha-D-glucose, Meso-inositol	Andhra Pradesh, Maharashtra, Puducherry, Karnataka	Gayathri and Kannabiran (2008)
68	*Ficus carica*	Moraceae	Stem bark	Campesterol, Hentriacontanol, Stigmasterol, Euphorbol, Ingenol, Taraxerone	Uttar Pradesh, Karnataka, Tamil Nadu, Maharashtra, Gujarat, Kerala	Ahmad et al. (2013)
69	*Ficus racemosa*	Moraceae	Bark, root, fruit	Lupenol, β-sistosterol, Stigmasterol, Gluanol acetate, α-amyrin acetate	Haryana, Bihar, Kerala, Karnataka, Madhya Pradesh Puducherry, Sikkim, Darjeeling	Gul-E-Rana et al. (2013)
70	*Flacourtia jangomas*	Salicaceae	Leaves, stem	Tannins, Phenols, Carbohydrates, Fats, Proteins, Amino acids,	Meghalaya, Chhattisgarh, Uttar Pradesh	Singh and Singh (2010)
71	*Garcinia gummi-gutta*	Clusiaceae	Seed, fruit	Xanthones, Benzophenones, Organic acids, Amino acids	Assam, Kerala, Karnataka	Semwal et al. (2015)
72	*Girardinia diversifolia*	Urticaceae	Root	Alkaloids, Phenols, Carbohydrates, Glycosides, Saponins, Proteins, Flavonoids, Tannins	Himalayas, Uttarakhand, Himachal Pradesh, Jammu & Kashmir.	Baig et al. (2014)
73	*Gmelina arborea*	Lamiaceae	Bark, leaves	Alkaloids, Carbohydrate, Glycoside, Proteins, Amino acid, Tannins, Steroids, Triterpenoids, Saponins, Flavonoids	Assam, West Bengal, Tamil Nadu, Puducherry, Himalayas, Odisha, Bihar	Kumaresan et al. (2014)

74	*Helicteres isora*	Sterculiaceae	Root, fruit	Phytosterols, Fixed oils and fats, Tannins, Amino acids, Flavones, Carbohydrates, Rosmarinic acid	Odisha, Jharkhand, Madhya Pradesh, Puducherry	Suthar et al. (2009)
75	*Hemidesmus indicus*	Apocynaceae	Root	Hexatriacontane, Lupeol, Rutin, Octacosanoate, α-amyrin, β-amyrin, Sitosterol, Tannins, Flavonoids, Hyperoside, Coumarin	Karnataka, Himalayas, Arunachal Pradesh, Odisha, Kerala, Maharashtra, Tamil Nadu, Bihar, Puducherry	Rajan et al. (2011b)
76	*Hibiscus rosa-sinensis*	Malvaceae	Leaves, flower	β-sitosterol, Stigmasterol, Taraxeryl acetate, Cyanidin diglucoside, Flavonoids, Thiamine, Riboflavin, Niacin, Ascorbic acid	Tamil Nadu, Maharashtra, Uttar Pradesh, Kerala, West Bengal, Bihar	Venkatesh et al. (2008)
77	*Holoptelea integrifolia*	Ulmaceae	Bark, leaves	Terpenoid, Steroids, Tannins, Saponins, Proteins, Carbohydrates, 2-Aminonaphthoquinone	Rajasthan, Andhra Pradesh, Karnataka, Maharashtra, Gujarat	Khalid et al. (2013)
78	*Hybanthus enneaspermus*	Violaceae	Aerial part	Alkaloids, Flavanoids	Puducherry, Tamil Nadu, Andhra Pradesh	Patel et al. (2011)
79	*Hydnocarpus laurifolia*	Flacourtiaceae	Seed	Triglycerides of fatty acids, Sterols, Flavonoids, Flavonolignans	West Bengal, Maharashtra to Kerala, Assam, Tripura	Rao et al. (2014)
80	*Hygrophila auriculata*	Acanthaceae	Leaves, root, seed, flower	Alkaloids, Phytosterol, Essential oils, Terpenoids	Puducherry, Tamil Nadu, Karnataka, Andhra Pradesh, Kerala	Rastogi et al. (2015)
81	*Ichnocarpus frutescens*	Apocynaceae	Root	Flavonoids, Triterpene glycosides	Karnataka, Uttar Pradesh, Tripura, Tamil Nadu, Andhra Pradesh, Puducherry	Barik et al. (2008)

Continued

TABLE 10.1 List of Antidiabetic Medicinal Herbs Found in Various Parts of India—cont'd

S. No	Botanical Name	Family	Active Parts	Chemical Constituents	Regions	References
82	*Ipomoea batatas*	Convolvulaceae	Root, leaves	Flavonoids, Anthocyanins, Carbohydrates, Caffeoylquinic acid	Uttar Pradesh, West Bengal, Bihar, Assam, Odisha, Andhra Pradesh, Maharashtra, Sikkim, Darjeeling	Panda and Sonkamble (2012)
83	*Ipomoea carnea*	Convolvulaceae	Leaves	Phnols, Flavonoids, Tannins, Alkaloids	Andhra Pradesh, Maharashtra, Jharkhand, Assam, Jammu	Khan et al. (2014)
84	*Jatropha curcas*	Euphorbiaceae	Leaves, stem, aerial part	Saponins, Steroids, Tannins, Glycosides, Alkaloids, Flavonoids Coumarin, Catechin, Quercetin	Andhra Pradesh, Maharashtra, Assam, Karnataka, Rajasthan, Puducherry	Najda et al. (2013)
85	*Lagenaria siceraria*	Cucurbitaceae	Seed, fruit, aerial part	Flavonoids, Saponins, Triterpenes, Volatile essential oils	Gujarat, Kerala, Maharashtra,	Prajapati et al. (2010)
86	*Lantana camara*	Verbenaceae	Fruit, leaves	Flavonoids, Carbohydrates, Proteins, Quinine, Glycosides, Oligosaccharides, Saponins, Steroids, Triterpene, Tannin	Maharashtra, Kerala, Uttarakhand, Himachal Pradesh, Tamil Nadu, Jammu & Kashmir	Kazmi et al. (2012)
87	*Lawsonia inermis*	Lythraceae	Leaves	2-hydroxy-1, 4-napthoquinone, Flavonoids, Coumarins, Steroids, Tannins, Napthoquinone	Rajasthan, Gujarat, Madhya Pradesh, Tamil Nadu, Karnataka, Kerala, Maharashtra, Uttar Pradesh, Bengal, Bihar, Odisha, Puducherry	Chauhan (2011)
88	*Lepidium sativum*	Brassicaceae	Seed	Carbohydrates, Lipids, Flavonoids, Alkaloids, Proteins, Saponins	Uttar Pradesh, Rajasthan, Gujarat, Maharashtra, Punjab, Karnataka, Tamil Nadu, Madhya Pradesh	Ahmad et al. (2015)

89	*Limonia acidissima*	Rutaceae	Leaves	Alkaloids, Flavonoids, Phnols, Terpenoids, Tannins, Saponins, Glycosides, Gum, Mucilage, Fixed oils	Maharashtra, Andhra Pradesh, Himalayas, Punjab, Assam, Tamil Nadu, Gujarat, Delhi, Rajasthan	Vijayvargia and Vijayvargia (2014)
90	*Linum usitatissimum*	Linaceae	Seed, leaves, root	Flavonoids, Terpenoids, β-sitosterol, Stigmasterol	Maharashtra, Madhya Pradesh, Chhattisgarh, Uttar Pradesh, Bihar, Rajasthan, Karnataka, West Bengal	Borhade (2013)
91	*Litchi chinensis*	Sapindaceae	Seed	Flavonoids, Sterols, Triterpenens, Phenolic compounds, Vitamin C, Organic acids	Bihar, Tripura, Assam, Kerala, West Bengal, Uttar Pradesh	Ibrahim and Mohamed (2015)
92	*Luffa acutangula*	Cucurbitaceae	Fruit	Triterpene, Saponins, Alkaloids, Anthraquinone glycosides	Manipur, Himalayas, Maharashtra, Karnataka, Jammu & Kashmir, Delhi, Haryana, Puducherry	Pimple et al. (2011)
93	*Madhuca indica*	Sapotaceae	Leaves	Carbohydrates, Proteins, Sterols, Flavonoids, Tannins, Saponins, Triterpenoids	Madhya Pradesh, Uttar Pradesh, Kerala, Bihar, Telangana, Puducherry, Chhattisgarh, Jharkhand	Chaudhary et al. (2015)
94	*Mangifera indica*	Anacardiaceae	Peel, leaves	Saponin, Steroids, Tannin, Flavonoid, Reducing sugars, Cardiac glycosides, Anthraquinone	Uttar Pradesh, Gujarat, Andhra Pradesh, Bihar, Karnataka, Maharashtra, Puducherry	Nwankwo and Osaro-Mathew (2014)
95	*Manilkara zapota*	Sapotaceae	Flower, leaves	Alkaloids, Flavonoids, Tannins, Saponins, Proteins, Carbohydrates, Glycosides, Carotenoids, Terpenoids	Maharashtra, Gujarat, Andhra Pradesh, Tamil Nadu, West Bengal, Karnataka, Madhya Pradesh	Fayek et al. (2012)
96	*Mentha piperita*	Lamiaceae	Leaves	Tannins, Flavanoids, Phenolic acids, Alkaloids, Saponins	Uttar Pradesh, Maharashtra	Jose et al. (2013)

Continued

TABLE 10.1 List of Antidiabetic Medicinal Herbs Found in Various Parts of India—cont'd

S. No	Botanical Name	Family	Active Parts	Chemical Constituents	Regions	References
97	*Mimosa pudica*	Fabaceae	Leaves	Terpenoids, Flavonoids, Glycosides, Alkaloids, Quinines, Tannins, Saponins, Coumarins	Assam, Kerala, Uttar Pradesh, Maharashtra,	Joseph et al. (2013)
98	*Momordica charantia*	Cucurbitaceae	Fruit	Glycosides, Saponins, Alkaloids, Reducing sugars, Resins, Phenolic compounds, Fixed oils	Assam, Maharashtra, Uttar Pradesh, Bihar, Darjeeling, Puducherry Sikkim, Tamil Nadu	Joesph and Jini (2013)
99	*Moringa oleifera*	Moringaceae	Leaves	Alkaloids, Flavonoid, Tannins, Saponins, Glycosides, Volatile oils	Assam, Andhra Pradesh, West Bengal, Tamil Nadu, Karnataka, Puducherry	Gupta et al. (2012)
100	*Mucuna pruriens*	Fabaceae	Seed	Alkaloids, Saponins, Sterols, Proteins, Carbohydrates, Lipids, Fiber, Minerals	Gujarat, Punjab, Kerala, Karnataka, Telangana	Kumar and Saha (2013)
101	*Murraya koenigii*	Rutaceae	Leaves	Linalool, Elemol, Geranyl acetate, Myrcene, Allo-Ocimene, α-Terpinene, Neryl acetate	Maharashtra, Himalayas, Kerala, Himachal pradesh, West Bengal, Madhya Pradesh, Puducherry	Arulselvan et al. (2006)
102	*Musa paradisiaca*	Musaceae	Stem, fruit	Alkaloids, Glycosides, Saponins, Tannins, Anthraquinones, Flavonoids	Kerala, Tamil Nadu, Karnataka, Andhra Pradesh, Maharashtra, Gujarat, Assam, Puducherry	Onyenekwe et al. (2013)
103	*Nardostachys jatamansi*	Caprifoliaceae	Rhizome, root	Sesquiterpene, Coumarins, Lignans, Neolignans, Alkaloids	Uttarakhand, Himachal Pradesh, Arunachal Pradesh, Himalayas, Darjeeling, Sikkim	Bhatt and Kothiyal (2015)
104	*Nelumbo nucifera*	Nelumbonaceae	Rhizome	Quercetin-3-O-glucopyranoside, Kaempferol-3-O-glucopyranoside, Myricetin-3-O-glucopyranoside	Andhra Pradesh, Maharashtra, Haryana, Manipur, Puducherry, Kashmir, Karnataka	Mukherjee et al. (1997)

105	*Ocimum sanctum*	Lamiaceae	Aerial part	Essential oils, Methyl eugenol, Caryophyllene, Urosolic acid, Apigenin, Luteolin, Stigmasterol, Triacontanol ferulate, Isovitexin, Aesculetin, Luteolin, Circineol, Procatechuic acid, Phenylpropane glucosides	Puducherry, Karnataka, Maharashtra, Rajasthan, West Bengal, Himalayas, Himachal Pradesh	Patil et al. (2011)
106	*Oroxylum indicum*	Bignoniaceae	Stem, bark, leaves	β-Sitosterol, Ellagic acid, Biochanin A, Oroxylin A, Ursolic acid	Sikkim, Darjeeling, Uttarakhand, Assam, Rajasthan, Arunachal Pradesh, Kerala, Himalayas, Karnataka, Andhra Pradesh	Singh and Kakkar (2013)
107	*Ougeinia dalbergioides*	Fabaceae	Bark	Genistein, Ougenin, Kaempferol, Lupeol, Ferreirin, Neophellamuretin,Betulin	Himalayas, Punjab, Maharashtra, Karnataka, Tamil Nadu, Visakhapatnam	Velmurugan et al. (2011)
108	*Paederia foetida*	Rubiaceae	Leaves	Iridoid Glycosides, Sitosterol, Stigmasterol, Alkaloids, Carbohydrates, Protein, Amino acid, Volatile oil	Andhra Pradesh, Arunachal Pradesh, Assam, Bihar, Odisha, Manipur, Meghalaya, Mizoram, Nagaland, Sikkim, Darjeeling Telangana, West Bengal	Kumar et al. (2014)
109	*Panax pseudoginseng*	Araliaceae	Rhizome, root	Triterpene saponins, Essential oil, Phenols, Ginsenosides	Himalayas, Sikkim, Darjeeling, Assam, Arunachal Pradesh, Manipur, Meghalaya, Nagaland	Jia and Zhao (2009)
110	*Pandanus odoratissimus*	Pandanaceae	Root, aerial part	Alkaloids, Sterols, Carbohydrates, Proteins, Steroids, Tannins, Terpenes, Gums, Saponins, Mucilage, Flavonoids, Glycosides	Orissa, Punjab, Madhya Pradesh, Maharashtra, Kerala, West Bengal	Sama et al. (2012)

Continued

TABLE 10.1 List of Antidiabetic Medicinal Herbs Found in Various Parts of India—cont'd

S. No	Botanical Name	Family	Active Parts	Chemical Constituents	Regions	References
111	*Paspalum scrobiculatum*	Poaceae	Leaves, flower	Steroids, Lipids, Amino acids, Carbohydrates	Odisha, Kerala, Karnataka, Tamil Nadu, Maharashtra, Rajasthan, Andhra Pradesh, Puducherry	Jain et al. (2010)
112	*Phyllanthus amarus*	Phyllanthaceae	Leaves	Alkaloids, Flavonoids, Tannins, Lignins, Triterpenoids	Jharkhand, Bihar, Chhattisgarh, Karnataka, Kerala	Raphael et al. (2002)
113	*Phyllanthus emblica*	Phyllanthaceae	Fruit, leaves	Coumarins, Alkaloids, Phenols, Steroids, Flavonoids	Odisha, Andhra Pradesh, Tamil Nadu, Puducherry, Maharashtra, Assam, Meghalaya	Nain et al. (2012)
114	*Picrorhiza kurroa*	Plantaginaceae	Rhizome, leaves	Iridoids, Acetophenones, Cucurbitacins, Pikuroside, Veronicoside, Phenol glycosides, Monoterpenes	Himalayas, Kashmir, Sikkim, Darjeeling, New Delhi, Punjab, Himalayas, Manipur, Meghalaya,	Husain et al. (2009)
115	*Pistia stratiotes*	Araceae	Leaves, seed	Alkaloids, Glycosides, Flavonoids, Steroids	Madhya Pradesh, Chhattisgarh, Puducherry,	Tripathi et al. (2010)
116	*Pithecellobium dulce*	Fabaceae	Seed	Sapogenin, Sterol glucoside, Flavone, Praline, Leucine, Valine, Quercitrin, Asparagines	Punjab, Rajasthan, Andhra Pradesh, Karnataka, Maharashtra, Madhya Pradesh, Tamil Nadu, Uttar Pradesh	Nagmoti et al. (2015)
117	*Plumbago indica*	Plumbaginaceae	Leaves	Naphthoquinones, Coumarins, Meroterpenes, Triterpenoids, Amino acids, Anthraquinones, Steroids, Sugars, Glucosides	Puducherry, Andhra Pradesh, Kerala, West Bengal, Uttar Pradesh, Manipur	Sunil et al. (2012)
118	*Pongamia pinnata*	Fabaceae	Leaves	β-sitosteryl acetate, Galactoside, Stigma sterol, Flavonoids, Tannins	Himalayas, Puducherry, Rajasthan, Maharashtra	Sikarwar and Patil (2010)

119	*Portulaca oleracea*	Portulacaceae	Stem, leaves, seed	Flavonoids, Alkaloids, Fatty acids, Terpenoids, Polysaccharides, Vitamins, Sterols, Proteins, Minerals	Gujarat, Karnataka, Puducherry, Himalayas, Rajasthan	Zhou et al. (2015)
120	*Potentilla fulgens*	Rosaceae	Root	Tannins, Flavonoids, Triterpenoids	Sikkim, Darjeeling, Meghalaya, Himalayas, Meghalaya,	Syiem et al. (2002)
121	*Psidium guajava*	Myrtaceae	Fruit	Tannins, Lectins, Triterpenes, Flavonoids, Essential oils, Saponins, Carotenoids, Vitamins C, Fiber & Fatty acids	Uttar Pradesh, Bihar, West Bengal, Madhya Pradesh, Puducherry, Punjab, Haryana	Huang et al. (2011)
122	*Pterocarpus marsupium*	Fabaceae	Heart wood	Pterostilbene, 7-hydroxyflavanone, Isoliquiritigenin, Liquiritigenin, Marsupsin, Pterosupin, *p*-hydroxybenzaldehyde, (2R)-3-(*p*-hydroxyphenyl)-lactic acid	Maharashtra, Puducherry, Andhra Pradesh, Karnataka, Madhya Pradesh, Tamil Nadu, Gujarat	Mishra et al. (2013)
123	*Punica granatum*	Lythraceae	Fruit	Ellagic acid, Gallotannins, Anthocyanins, Flavonoids	Puducherry	Das and Barman (2012)
124	*Rotula aquatica*	Boraginaceae	Root, flower	Alkaloids, Flavonoids, Tannins, Saponins, Steroids, Fixed oils & fats	Uttarakhand, Assam, Karnataka, Andhra Pradesh, Kerala, Maharashtra	Tulliballi et al. (2013)
125	*Saraca asoca*	Fabaceae	Flower, bark	Carbohydrates, Proteins, Amino acids, Flavonoids, Tannin, Saponin, Steroids, Glycoside	Sikkim, Darjeeling, Karnataka, Maharashtra, Rajasthan, Delhi, Tamil Nadu, Goa, Gujarat, Uttar Pradesh, Haryana	Jain et al. (2013)
126	*Semecarpus anacardium*	Anacardiaceae	Fruit	Alkaloids, Biflavonoids, Anacardic acids, Bhilawanols, Minerals, Vitamins, Amino acids	Himalayas, Madhya pradesh, Gujarat, Maharashtra, Himachal Pradesh	Hedayathullah et al. (2012)

Continued

TABLE 10.1 List of Antidiabetic Medicinal Herbs Found in Various Parts of India—cont'd

S. No	Botanical Name	Family	Active Parts	Chemical Constituents	Regions	References
127	*Solanum melongena*	Solanaceae	Fruit	Alkaloids, Saponins, Coumarin, α-amylase, α-glucosidase, Hydroxycinnamic acid, Anthracenocides	Karnataka, Himalayas, Mizoram, Maharashtra, Uttar Pradesh	Nwanna et al. (2013)
128	*Sphaeranthus indicus*	Asteraceae	Whole plant, flower, seed, root	Sesquiterpene lactone, Flavanoid C-glycoside, Essential oil	Puducherry, Tamil Nadu, Himalayas, Andhra Pradesh, Chhattisgarh	Galani et al. (2010)
129	*Strychnos potatorum*	Loganiaceae	Seed	Terpenoids, Tannins, Capsacin, Alkaloids, Saponins, β-Sitosterol, Stigmasterol	West Bengal, Maharashtra, Karnataka, Mizoram	Biswas et al. (2012)
130	*Swertia chirayita*	Gentianaceae	Whole plant	Xanthone, Triterpenoids, Sterol, Tannins, Carbohydrates, Glycosides, Alkaloids, Flavonoids	Sikkim, Darjeeling, Uttarakhand, West Bengal, Himalayas, UP, Gujarat, Rajasthan, Maharashtra, Madhya Pradesh, Rajasthan	Roy et al. (2015)
131	*Tamarindus indica*	Fabaceae	Leaves, seed	Tannins, Glycosides, Mallic acid, Tartaric acid, Mucilage, Pectin, Arabinose, Xylose, Galactose, Uronic acid, Triterpenes, Lupanone, Lupeol	Bihar, Odisha, Karnataka, Maharashtra, Tamil Nadu, Andhra Pradesh, Madhya Pradesh, Kerala, Uttar Pradesh, Himalayas	Maiti et al. (2004)
132	*Terminalia arjuna*	Combretaceae	Bark	Tannins, Alkaloids, Carbohydrates, Terpenoids, Steroids, Flavonoids	West Bengal, Madhya Pradesh, Bihar, Karnataka, Tamil Nadu, Gaujarat, Himalayas, Maharashtra	Ragavan et al. (2006)
133	*Terminalia catappa*	Combretaceae	Fruit, leaves	Flavonoids, Tannins, β-carotene, Glycosides, Alkaloids, Steroids, Saponins, Carbohydrate	Puducherry, Kerala, Karnataka, Tamil Nadu	Offor et al. (2015)

134	*Terminalia chebula*	Combretaceae	Fruit, leaves	Triterpenoid, Tannins, Luminol, Pyrogallol-Luminol	Puducherry, Madhya Pradesh, Gujarat, Tamil Nadu, Odisha, Karnataka	Rao and Nammi (2006)
135	*Tinospora cordifolia*	Menispermaceae	Stem, root	Alkaloids, Glycosides, Diterpenoid Lactones, Steroids	Puducherry, Maharashtra, Himalayas, Kerala, Punjab, Karnataka	Nadig et al. (2012)
136	*Tragia involucrata*	Euphorbiaceae	Root	Steroids, Coumarins, Lignin, Triterpenoids, Saponins, Tannins, Emodins, Reducing sugars Anthraquinons	Maharashtra, Kerala, Tamil Nadu, Karnataka	Farook and Atlee (2011)
137	*Tribulus terrestris*	Zygophyllaceae	Whole plant	Di-*p*-coumaroylquinic acid, Saponins, Steroids, Flavonoids, Glycosides, Alkaloids, Tannins	Puducherry, Karnataka, Kashmir, Andhra Pradesh, Kerala, Tamil Nadu	Chhatre et al. (2014)
138	*Trigonella foenum-graecum*	Fabaceae	Seed	Carbohydrates, Flavonoids, Polysaccharides, Saponins, Fixed oils, Alkaloids	Sikkim, Gujarat, Darjeeling, Rajasthan, Uttarakhand, Maharashtra, Karnataka, Madhya Pradesh, Uttar Pradesh, Punjab	Toppo et al. (2009)
139	*Withania somnifera*	Solanaceae	Root, leaves	Steroidal lactones, Alkaloids, Flavonoids, Tannins, Triterpenoids	Rajasthan, Punjab, Haryana, Uttar Pradesh, Gujarat, Maharashtra, Andhra Pradesh, Karnataka, Madhya Pradesh	Udayakumar et al. (2009)
140	*Woodfordia fruticosa*	Lythraceae	Flower	Tannins, Flavonoids, Anthraquinone, Glycosides	Kerala, Karnataka, Chhattisgarh, Punjab, Arunachal Pradesh, Tamil Nadu	Verma et al. (2012)
141	*Zanthoxylum armatum*	Rutaceae	Bark, fruit	Alkaloids, Saponins, Tannins, Flavonoids, Sterols, Lignins, Coumarins, Terpenoids, Glycosides, Amino acids	Maharashtra, Karnataka, Goa, Himalayas, Meghalaya, Manipur	Karki et al. (2014)
142	*Ziziphus nummularia*	Rhamnaceae	Leaves	Alkaloids, Saponins, Tannins, Flavonoids	Madhya Pradesh, Punjab, Rajasthan, Tamil Nadu, Andhra Pradesh, West Bengal, Bihar, Uttar Pradesh, Maharashtra	Rajasekaran et al. (2012)

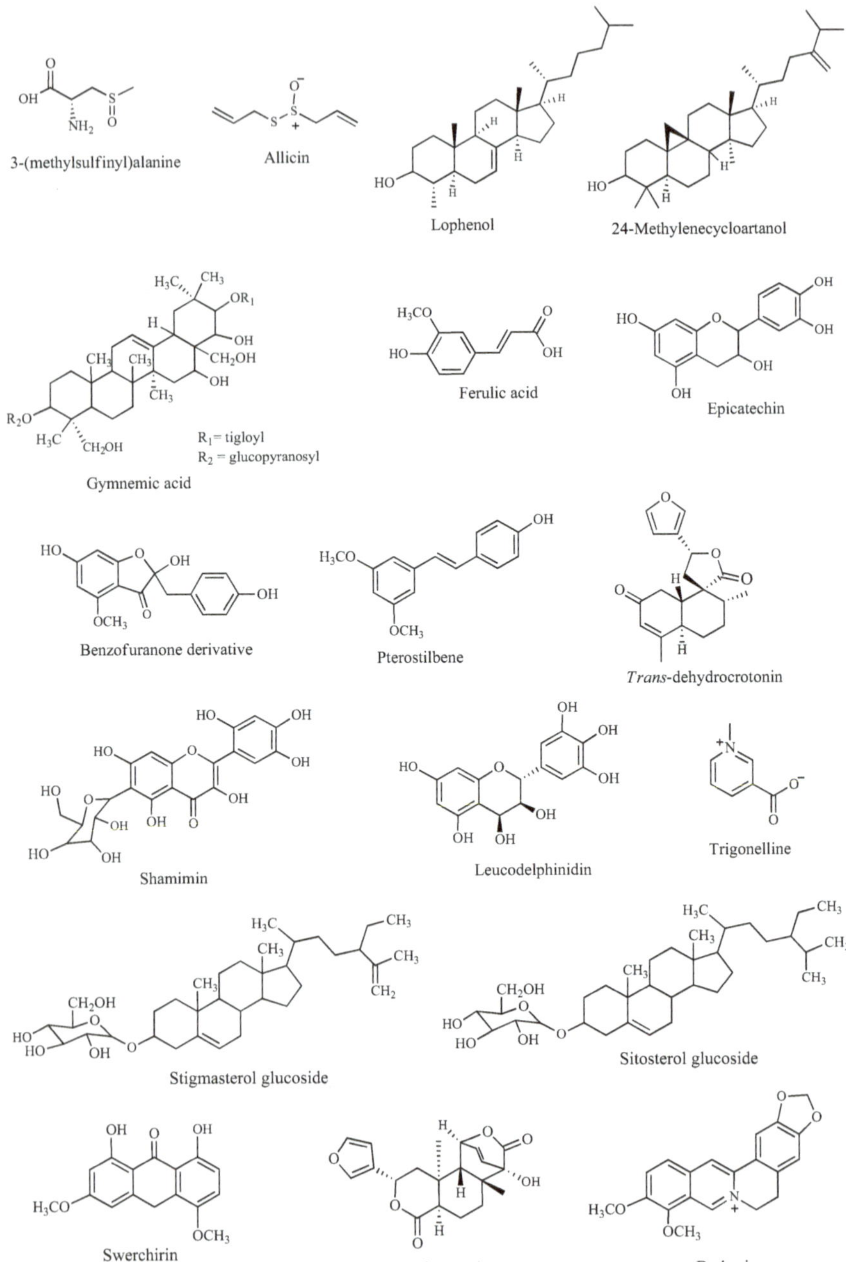

FIGURE 10.1 Some major phytoconstituents of Indian antidiabetic plant species.

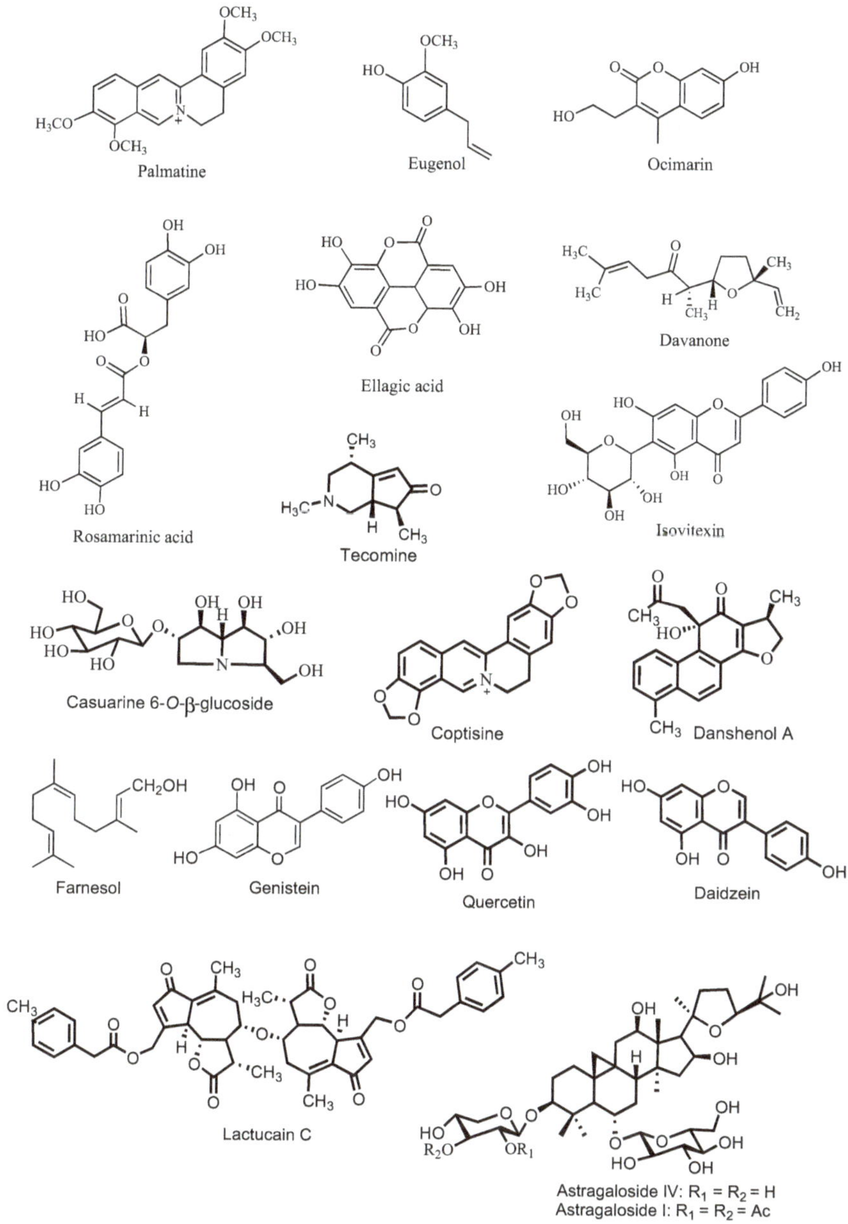

FIGURE 10.1 Cont'd.

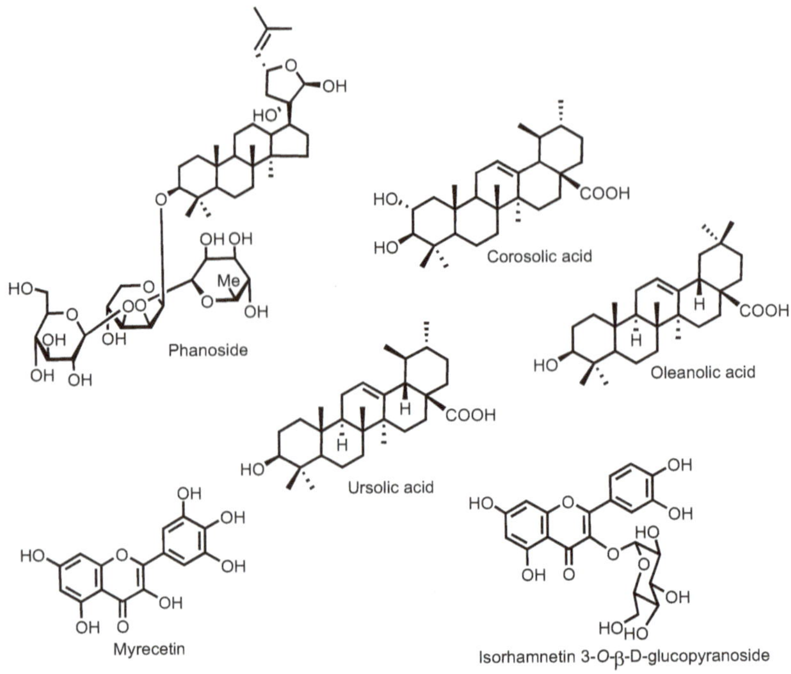

FIGURE 10.1 Cont'd.

3. CONCLUDING REMARKS

The mounting prevalence of diabetes mellitus across the globe is an issue of major socioeconomic concern. Scientific interest in plant-derived medicine is steadily rising, yet there is often a wide disparity in the caliber of information available. A detailed compilation of scientific information regarding traditional medicines is elucidated in Vedic literature, Charaka Samhita, and Sushruta Samhita. Ethnomedicinal plants of modern times and their phytochemical constituents play a vital role in the treatment of diabetes mellitus. These medicinal herbs are practised traditionally and most of them are cultivated in different regions of India. The potency of herbal medicines is significant because they have negligible side effects than chemically derived hypoglycemic agents. The phytoconstituents isolated from plants parts possess hypoglycemic effect, which can cure diabetes mellitus naturally and their activity is equivalent to synthetically derived drugs. The research for alternate remedies from the plant kingdom for diabetes mellitus is progressing all over the world as the disease exerts many challenges not only to the physician but also to the researchers. This chapter provides a huge list of plant species grown abundantly in India to the researchers deeply involved in searching new and potent antidiabetic agents from natural sources with a view to explore and exploit these natural resources to the cause.

REFERENCES

Ahmad, M.Z., Ali, M., Mir, S.R., 2013. Anti-diabetic activity of *Ficus carica* L. stem barks and isolation of two new flavonol esters from the plant by using spectroscopical techniques. Asian Journal of Biomedical and Pharmaceutical Sciences 3 (18), 22–28.

Ahmad, R., Mujeeb, M., Anwar, F., Husain, A., Ahmad, A., Sharma, S., 2015. Pharmacognostical and phytochemical analysis of *Lepidium sativum* L. seeds. International Current Pharmaceutical Journal 4 (10), 442–446.

Ali, A., Akhtar, N., Khan, B.A., Khan, M.S., Rasul, A., Zaman, S.U., Khalid, N., Waseem, K., Mahmood, T., Ali, L., 2012. *Acacia nilotica*: a plant of multipurpose medicinal uses. Journal of Medicinal Plants Research 6 (9), 1492–1496.

Al-Snafi, A.E., 2013. The pharmacology of *Bacopa monniera*: a review. International Journal of Pharmaceutical Sciences and Research 4 (12), 154–159.

Al-Snafi, A.E., 2015. The pharmacological and therapeutic importance of *Agrimonia eupatoria* – a review. Asian Journal of Pharmaceutical Science and Technology 5 (2), 112–117.

Antil, V., Sinha, B.N., Pandey, A., Diwan, A., Saini, P., 2013. *Bombax malabaricum* DC: a salutary boon. International Journal Pharmaceutical Innovation 3 (2), 17–28.

Arulselvan, P., Senthilkumar, G.P., Kumar, S.D., Subramanian, S., 2006. Anti-diabetic effect of *Murraya koenigii* leaves on streptozotocin induced diabetic rats. Pharmazie 61 (10), 874–877.

Ayyanar, M., Subash-Babu, P., 2012. *Syzygium cumini* (L.) skeels: a review of its phytochemical constituents and traditional uses. Asian Pacific Journal of Tropical Biomedicine 2 (3), 240–246.

Baig, Y., Rabbani, G., Imran, A., Sayeed, A., 2014. Antidiabetic activity of *Girardinia diversifolia* on alloxan-induced diabetes mellitus in rats – a preliminary study. International Journal of Advances in Pharmacy Medicine and Bioallied Sciences 2 (1), 15–19.

Bairwa, N.K., Sethiya, N.K., Mishra, S.H., 2010. Protective effect of stem bark of *Ceiba pentandra* Linn. against paracetamol-induced hepatotoxicity in rats. Pharmacognosy Research 2 (1), 26–30.

Barik, R., Jain, S., Qwatra, D., Joshi, A., Tripathi, G.S., Goyal, R., 2008. Antidiabetic activity of aqueous root extract of *Ichnocarpus frutescens* in streptozotocin-nicotinamide induced type-II diabetes in rats. Indian Journal of Pharmacology 40 (1), 19–22.

Bharati, A.C., Sahu, A.N., 2012. Ethnobotany, phytochemistry and pharmacology of *Biophytum sensitivum* DC. Pharmacognosy Reviews 6 (11), 68–73.

Bhatt, P.M., Kothiyal, P., 2015. A review article on phytochemistry and pharmacological profiles of *Nardostachys jatamansi* DC-medicinal herb. Journal of Pharmacognosy and Phytochemistry 3 (5), 102–106.

Biba, V.S., Lakshmi, S., Dhanya, G.S., Remani, P., 2013. Phytochemical analysis of *Annona squamosa* seed extracts. International Research Journal of Pharmaceutical and Applied Sciences 3 (4), 29–31.

Bisht, S., Sisodia, S.S., 2011. Assessment of antidiabetic potential of *Cinnamomum tamala* leaves extract in streptozotocin induced diabetic rats. Indian Journal of Pharmacology 43 (5), 582–585.

Biswas, A., Chatterjee, S., Chowdhury, R., Sen, S., Sarkar, D., Chatterjee, M., Das, J., 2012. Antidiabetic effect of seeds of *Strychnos potatorum* Linn. in a streptozotocin-induced model of diabetes. Acta Poloniae Pharmaceutica 69 (5), 939–943.

Borhade, S.S., 2013. Physico-chemical & phytochemical investigation of linseed (*Linum usitatissimum* L.) oil grown in Maharashtra (India) and analytical study by HPLC. International Journal of Drug Discovery and Herbal Research 3 (1), 584–589.

Chackrewarthy, S., Thabrew, M.I., Weerasuriya, M.K.B., Jayasekera, S., 2010. Evaluation of the hypoglycemic and hypolipidemic effects of an ethylacetate fraction of *Artocarpus heterophyllus* (jak) leaves in streptozotocin-induced diabetic rats. Pharmacognosy Magazine 6 (23), 186–190.

Chaudhary, A., Bhandari, A., Pandurangan, A., Koul, S., 2015. *Madhuca indica* J. F. Gmel. (Sapotaceae): an overview. International Journal of Pharmaceutical Sciences Letters 5 (2), 539–545.

Chauhan, H.V., 2011. In-vivo antidiabetic, lipid lowering and antioxidant activities of methanolic extract of *Lawsonia inermis* leaves. Research Journal of Pharmacy and Technology 4 (5), 764–767.

Chhatre, S., Nesari, T., Somani, G., Kanchan, D., Sathaye, S., 2014. Phytopharmacological overview of *Tribulus terrestris*. Pharmacognosy Reviews 8 (15), 45–51.

Chhetri, D.R., Parajuli, P., Subba, G.C., 2005. Antidiabetic plants used by Sikkim and Darjeeling Himalayan tribes. Indian Journal of Ethnopharmacology 99, 199–202.

Cooke, D.W., Plotnick, L., 2008. Type 1 diabetes mellitus in pediatrics. Pediatrics in Review 29 (11), 374–384.

Daisy, P., Jasmine, R., Ignacimuthu, S., Murugan, E., 2009. A novel steroid from *Elephantopus scaber* L. an ethnomedicinal plant with antidiabetic activity. Phytomedicine 16 (2–3), 252–257.

Das, S., Barman, S., 2012. Antidiabetic and antihyperlipidemic effects of ethanolic extract of leaves of *Punica granatum* in alloxan-induced non-insulin-dependent diabetes mellitus albino rats. Indian Journal of Pharmacology 44 (2), 219–224.

Dewanjee, S., Bose, S.K., Sahu, R., Mandal, S.C., 2008. Antidiabetic effect of matured fruits of *Diospyros* peregrina in alloxan-induced diabetic rats. International Journal of Green Pharmacy 2, 95–99.

Eliza, J., Diasy, P., Ignacimuthu, S., Duraipandiyan, V., 2009. Antidiabetic and antilipidemic effect of eremanthin from *Costus speciosus* (Koen.) Sm., in STZ-induced diabetic rats. Chemico-Biological Interactions 182, 67–72.

Farook, S.M., Atlee, W.C., 2011. Antidiabetic and hypolipidemic potential of *Tragia involucrata* Linn. in streptozotocin-nicotinamide induced type II diabetic rats. International Journal of Pharmacy and Pharmaceutical Sciences 3 (4), 103–109.

Fayek, N.M., Monem, A.R.A., Mossa, M.Y., Meselhy, M.R., Shazly, A.H., 2012. Chemical and biological study of *Manilkara zapota* (L.) Van Royen leaves (Sapotaceae) cultivated in Egypt. Pharmacognosy Research 4 (2), 85–91.

Galani, V.J., Patel, B.G., Rana, D.G., 2010. *Sphaeranthus indicus* Linn.: a phytopharmacological review. International Journal of Ayurveda Research 1 (4), 247–253.

Gayathri, M., Kannabiran, K., 2008. Antidiabetic and ameliorative potential of *Ficus bengalensis* bark extract in streptozotocin induced diabetic rats. Indian Journal of Clinical Biochemistry 23 (4), 394–400.

Godghate, A., Sawant, R., Sutar, A., 2012. Phytochemical analysis of ethanolic extract of roots of *Carrisa carandus* Linn. Rasayan Journal of Chemistry 5 (4), 456–459.

Gul-E-Rana, Karim, S., Khurhsid, R., Saeed-Ul-Hassan, S., Tariq, I., Sultana, M., Rashid, A.J., Shah, S.H., Murtaza, G., 2013. Hypoglycemic activity of *Ficus racemosa* bark in combination with oral hypoglycemic drug in diabetic human. Acta Poloniae Pharmaceutica 70 (6), 1045–1049.

Gupta, R., Mathur, M., Bajaj, V.K., Katariya, P., Yadav, S., Kamal, R., Gupta, R.S., 2012. Evaluation of antidiabetic and antioxidant activity of *Moringa oleifera* in experimental diabetes. Journal of Diabetes 4 (2), 164–171.

Gupta, S., Sharma, S.B., Singh, U.R., Bansal, S.K., 2011. Salutary effect of *Cassia auriculata* L. leaves on hyperglycemia-induced atherosclerotic environment in streptozotocin rats. Cardiovascular Toxicology 11, 308–315.

Haider, S.Z., Mohan, M., Andola, H.C., 2014. Constituents of *Artemisia indica* wild from Uttarakhand Himalaya: a source of davanone. Pharmacognosy Research 6 (3), 257–259.

Hannan, J.M., Ali, L., Khaleque, J., Akhter, M., Flatt, P.R., Abdel-Wahab, Y.H., 2012. Antihyperglycaemic activity of *Asparagus racemosus* roots is partly mediated by inhibition of carbohydrate digestion and absorption, and enhancement of cellular insulin action. British Journal of Nutrition 107 (9), 1316–1323.

Hedayathullah, K.H.B., Vinayagam, K.S., Palanivelu, S., Panchanatham, S., 2012. Anti-diabetic effect of *Semecarpus anacardium* Linn nut milk extract in a high fat diet STZ induced type 2 diabetic rat model. Comparative Clinical Pathology 21 (6), 1395–1400.

Huang, C.S., Yin, M.C., Chiu, L.C., 2011. Antihyperglycemic and antioxidative potential of *Psidium guajava* fruit in streptozotocin-induced diabetic rats. Food and Chemical Toxicology 41, 2189–2195.

Husain, G.M., Singh, P.N., Kumar, V., 2009. Antidiabetic activity of standardized extract of *Picrorhiza kurroa* in rat model of NIDDM. Drug Discovery and Therapeutics 3 (3), 88–92.

Hussain, A.I., Rathore, H.A., Sattar, M.Z., Chatha, S.A., Sarker, S.D., Gilani, A.H., 2014. *Citrullus colocynthis* (L.) schrad (bitter apple fruit): a review of its phytochemistry, pharmacology, traditional uses and nutritional potential. Journal of Ethnopharmacology 155 (1), 54–66.

Hussain, A., Wahab, S., Rizvi, A.M., Husssain, S., 2011. Macroscopical anatomical and physicochemical studies on leaves of *Coccinia indica* Wight & Arn., growing wildly in eastern Uttar Pradesh region of India. Indian Journal of Natural Products Resources 2 (1), 74–80.

Ibrahim, S.R., Mohamed, G.A., 2015. *Litchi chinensis*: medicinal uses, phytochemistry, and pharmacology. Journal of Ethnopharmacology 174, 492–513.

Ikarashi, N., Toda, T., Okaniwa, T., Ito, K., Ochiai, W., Sugiyama, K., 2011. Anti-obesity and anti-diabetic effects of Acacia polyphenol in obese diabetic kkay mice fed high-fat diet. Evidence-Based Complementary and Alternative Medicine 1–10.

Jain, A., Jasmine, Sharma, S., Saini, V., 2013. Hypolipidemic, hypoglycemic and antioxidant potential of *Saraca asoca* ethanolic leaves extract in streptozotocin induced – experimental diabetes. International Journal of Pharmacy and Pharmaceutical Sciences 5 (1), 302–305.

Jain, S., Bhatia, G., Barik, R., Kumar, P., Jain, A., Dixit, V.K., 2010. Antidiabetic activity of *Paspalum scrobiculatum* Linn. in alloxan induced diabetic rats. Journal of Ethnopharmacology 127 (2), 325–328.

Jarald, E.E., Joshi, S.B., Jain, D.C., Edwin, S., 2013. Biochemical evaluation of the hypoglycemic effects of extract and fraction of *Cassia fistula* Linn. in alloxan-induced diabetic rats. Indian Journal of Pharmaceutical Sciences 75 (4), 427–434.

Jayakumar, T., Hsieh, C.E.Y., Lee, J.J., Sheu Jr., 2013. Experimental and clinical parmacology of *Andrographis paniculata* and its major bioactive phytoconstituent andrographolide. Evidence-Based Complementary and Alternative Medicine 1–16.

Jia, L., Zhao, Y., 2009. Current evaluation of the millennium phytomedicine-ginseng (I): etymology, pharmacognosy, phytochemistry, market and regulations. Current Medicinal Chemistry 16 (19), 2475–2484.

Jose, A., Kumar, S.S., Mukkadan, J., 2013. A study on anti-diabetic effect of peppermint in alloxan induced diabetic model of Wistar rats. Journal of Clinical and Biomedical Sciences 3 (4), 177–181.

Joseph, B., Jini, D., 2013. Antidiabetic effects of *Momordica charantia* (bitter melon) and its medicinal potency. Asian Pacific Journal of Tropical Disease 3 (2), 93–102.

Joseph, B., George, J., Mohan, J., 2013. Pharmacology and traditional uses of *Mimosa pudica*. International Journal of Pharmaceutical Sciences and Drug Research 5 (2), 41–44.

Kala, C.P., 2000. Status and conservation of rare and endangered medicinal plants in the Indian trans-Himalaya. Biological Conservation 93 (3), 371–379.

Karki, H., Upadhayay, K., Pal, H., Singh, R., 2014. Antidiabetic potential of Zanthoxylum armatum bark extract on streptozotocin-induced diabetic rats. International Journal of Green Pharmacy 8 (2), 77–83.

Karthiyayini, T., Kumar, R., Kumar, K.L.S., Sahu, R.k., Roy, A., 2009. Evaluation of antidiabetic and hypolipidemic effect of *Cucumis sativus* fruit in streptozotocin-induced-diabetic rats. Biomedical and Pharmacology Journal 2 (2), 351–355.

Kazmi, I., Rahman, M., Afzal, M., Gupta, G., Saleem, S., Afzal, O., Shaharyar, M.A., Nautiyal, U., Ahmed, S., Anwar, F., 2012. Anti-diabetic potential of ursolic acid stearoyl glucoside: a new triterpenic gycosidic ester from *Lantana camara*. Fitoterapia 83 (1), 142–146.

Kelkar, S.M., Kaklij, G.S., Bapat, V.A., 2001. Determination of antidiabetic activity in *Allium cepa* (onion) tissue cultures. Indian Journal Biochemistry and Biophysics 38 (4), 277–279.

Khalid, S., Rizwan, G.H., Yasin, H., Perveen, R., Abrar, H., Shareef, H., Fatima, K., Ahmed, M., 2013. Medicinal importance of *Holoptelea integrifolia* (Roxb). Planch – its biological and pharmacological activities. Natural Products Chemistry and Research 2 (1), 1–4.

Khan, T.Y., Raina, R., Verma, P.K., Sultana, M., Jyoti, M.K., 2014. Phytochemical constituents and antidiabetic potential of *Ipomoea carnea Jacq* leaves extracts. Journal of Experimental and Integrative Medicine 4 (2), 137–142.

Kumar, P., Saha, S., 2013. An updated review on taxonomy, phytochemistry, pharmacology and toxicology of *Macuna pruriens*. Journal of Pharmacognosy and Phytochemistry 2 (1), 306–314.

Kumar, V., Anwar, F., Ahmed, D., Verma, A., Ahmed, A., Damanhouri, Z.A., Mishra, V., Ramteke, P.W., Bhatt, P.C., Mujeeb, M., 2014. *Paederia foetida* Linn. leaf extract: an antihyperlipidemic, antihyperglycaemic and antioxidant activity. BMC Complementary and Alternative Medicine 25, 14–76.

Kumaresan, P., Jeyanthi, K.A., Kalaivani, R., 2014. Biochemical evaluation of anti diabetic activity of aqueous extract of *Gmelina arborea* in alloxan induced albino rats. International Journal of Herbal Medicine 2 (2), 90–94.

Lakshmi, V., Agarwal, S.K., Ansari, J.A., Mahdi, A.A., Srivastava, A.K., 2014. Antidiabetic potential of *Musa paradisiaca* in streptozotocin- induced diabetic rats. The Journal of Phytopharmacology 3 (2), 77–81.

Levendal, R.A., Frost, C.L., 2006. In vivo effects of *Cannabis sativa* L. extract on blood coagulation, fat and glucose metabolism in normal and streptozotocin-induced diabetic rats. African Journal of Traditional, Complementary and Alternative Medicines 3 (4), 1–12.

Liu, I.M., Tzeng, T.F., Lious, S.S., 2010. *Abelmoschus moschatus* (Malvaceae), an aromatic plant suitable for medical or food uses to improve insulin sensitivity. Phytotherapy Research 24 (2), 233–239.

Loukas, M., Lanteri, A., Ferrauiola, J., Tubbs, R.S., Maharaja, G., Shoja, M.M., Yadav, A., Rao, V.C., 2010. Anatomy in ancient India: a focus on the susruta samhita. Journal of Anatomy 217 (6), 646–650.

Mahatma, A., Kumar, S.M., Sonowal, A., 2014. Evaluation of antidiabetic potential of methanolic extract of *Benincasa hispida* in dexamethasone induced diabetic rats. IOSR Journal of Dental Medical Sciences 13 (12), 118–125.

Maithili, V., Dhanabal, S.P., Mahendran, S., Vadivelan, R., 2011. Antidiabetic activity of ethanolic extract of tubers of *Dioscorea alata* in alloxan induced diabetic rats. Indian Journal of Pharmacology 43 (4), 455–459.

Maiti, R., Jana, D., Das, U.K., Ghosh, D., 2004. Antidiabetic effect of aqueous extract of seed of *Tamarindus indica* in streptozotocin-induced diabetic rats. Journal of Ethnopharmacology 92 (1), 85–91.

Maity, P., Hansda, D., Bandyopadhyay, U., Mishra, D.K., 2009. Biological activities of crude extracts and chemical constituents of Bael, *Aegle marmelos* (L.) corr. Indian Journal of Experimental Biology 47, 849–861.

Maniyar, Y., Bhixavatimath, P., 2012. Antihyperglycemic and hypolipidemic activities of aqueous extract of *Carica papaya* Linn. leaves in alloxan-induced diabetic rats. Journal of Ayurveda and Integrative Medicine 3 (2), 70–74.

Meena, M.K., Singh, N., Patni, V., 2014. Determination of bioactive components of the leaves of *Cocculus hirsutus* (L.) Diels using GC-MS analysis. International Journal of Pharmacy and Pharmaceutical Sciences 6 (2), 327–329.

Mir, S.H., Darzi, M.M., Mir, M.S., 2013. Efficacy of *Abroma augusta* on biochemical and histomorphological features of alloxan-induced diabetic rabbits. Iranian Journal of Pathology 8 (3), 153–158.

Mishra, A., Srivastava, R., Srivastava, S.P., Gautam, S., Tamrakar, A.K., Maurya, R., Srivastava, A.K., 2013. Antidiabetic activity of heart wood of *Pterocarpus marsupium* Roxb. and analysis of phytoconstituents. Indian Journal of Experimental Biology 51 (5), 363–374.

Mishra, S., Aeri, V., Gaur, P.K., Jachak, S.M., 2014. Phytochemical, therapeutic, and ethnopharmacological overview for a traditionally important herb: *Boerhavia diffusa* Linn. BioMed Research International 1–19.

Mohammed, S.A., Yaqub, A.G., Sanda, K.A., Nicholas, A.O., Arastus, W., Muhammad, M., Abdullahi, S., 2013. Review on diabetes, synthetic drugs and glycemic effects of medicinal plants. Journal of Medicinal Plants Research 7 (36), 2628–2637.

Mondal, S., Bhattacharya, S., Biswas, M., 2012. Antidiabetic activity of *Areca catechu* leaf extracts against streptozotocin induced diabetic rats. Journal of Advanced Pharmacy Education and Research 2 (1), 10–17.

Mukherjee, P.K., Saha, K., Pal, M., Saha, B.P., 1997. Effect of *Nelumbo nucifera* rhizome extract on blood sugar level in rats. Journal of Ethnopharmacology 58 (3), 207–213.

Nadig, P.D., Revankar, R.R., Dethe, S.M., Narayanswamy, S.B., Aliyar, M.A., 2012. Effect of *Tinospora cordifolia* on experimental diabetic neuropathy. Indian Journal of Pharmacology 44 (5), 580–583.

Nagmoti, D.M., Kothavade, P.S., Bulani, V.D., Gawali, N.B., Juvekar, A.R., 2015. Antidiabetic and antihyperlipidemic activity of *Pithecellobium dulce* (Roxb.) Benth seeds extract in streptozotocin-induced diabetic rats. European Journal of Integrative Medicine 7 (3), 263–273.

Nahar, L., Nasrin, F., Zahan, R., Haque, A., Haque, E., Mosaddik, A., 2014. Comparative study of antidiabetic activity of *Cajanus cajan* and *Tamarindus indica* in alloxan-induced diabetic mice with a reference to in vitro antioxidant activity. Pharmacognosy Research 6 (2), 180–187.

Nain, P., Saini, V., Sharma, S., Nain, J., 2012. Antidiabetic and antioxidant potential of *Emblica officinalis* gaertn leaves extract in streptozotocin-induced type-2 diabetes mellitus (T2DM) rats. Journal of Ethnopharmacology 142 (1), 65–71.

Najda, A., Almehemdi, A.F., Zabar, A.F., 2013. Chemical composition and nutritional value of Jatropha *Jatropha curcas* L. leaves. Journal of Genetic and Environmental Resources Conservation 1 (3), 221–226.

Nishiyama, T., Mae, T., Kishida, H., Tsukagawa, M., Mimaki, Y., Kuroda, M., Sashida, Y., Takahashi, K., Kawada, T., Nakagawa, K., Kitahara, M., 2005. Curcuminoids and sesquiterpenoids in *turmeric* (*Curcuma longa* L.) suppress an increase in blood glucose level in type 2 diabetic KK-Ay mice. Journal of Agricultural and Food Chemistry 53 (4), 959–963.

Nwankwo, I.U., Osaro-Mathew, R.C., 2014. Assessment of the phytochemical components of *Mangifera indica* (leaf) and *Musa paradisiaca* (roots) extracts and their antibacterial activity against some common pathogenic bacteria. IOSR Journal of Pharmacy and Biological Sciences 9 (1), 8–11.

Nwanna, E.E., Ibukun, E.O., Oboh, G., 2013. Inhibitory effects of methanolic extracts of two eggplant species from South-western Nigeria on starch hydrolysing enzymes linked to type-2 diabetes. African Journal of Pharmacy and Pharmacology 7 (23), 1575–1584.

Obaineh, O.M., Shadrach, A., 2013. Phytochemical constituents and medicinal properties of different extracts of *Anacardium occidentale* and *Psidium guajava*. Asian Journal of Biomedical and Pharmaceutical Sciences 3 (16), 20–23.

Offor, C.E., Ugwu, P.C., Okechukwu, Aja, P.M., Igwenyi, I.O., 2015. Proximate and phytochemical analyses of *Terminalia catappa* leaves. European Journal of Applied Sciences 7 (1), 9–11.

Okpanachi, A.O., Adelaiye, A.B., Dikko, A.A.U., Kabiru, M., Mohammed, A., Tanko, Y., 2010. Evaluation of the effect of aqueous-methanolic stem bark extract of Acacia polyacantha on blood glucose levels of alloxan induced diabetic Wistar rats. International Journal of Animal and Veterinary Advances 2 (3), 59–62.

Onyenekwe, P.C., Okereke, O.E., Owolewa, S.O., 2013. Phytochemical screening and effect of *Musa paradisiaca* stem extrude on rat haematological parameters. Current Research Journal of Biological Sciences 5 (1), 26–29.

Panda, V., Sonkamble, M., 2012. Phytochemical constituents and pharmacological activities of *Ipomoea batatas* I. (Lam) – a review. International Journal of Research in Phytochemistry & Pharmacology 2 (1), 25–34.

Pandey, A.S., Srivastava, B., Wanjari, M.M., Pandey, N.K., Jadhav, A.D., 2014. *Callicarpa macrophylla*: a review of its phyto-chemistry, pharmacology, folklore claims and ayurvedic studies. Global Journal of Research on Medicinal Plants and Indigenous Medicine 3 (3), 91–100.

Pandit, R., Phadke, A., Jagtap, A., 2010. Antidiabetic effect of *Ficus religiosa* extract in streptozotocin-induced diabetic rats. Journal of Ethnopharmacology 128 (2), 462–466.

Parvin, M.N., Rahman, M.S., Islam, M.S., Rashid, M.A., 2009. Chemical and biological investigations of *Dillenia indica* Linn. Bangladesh Journal of Pharmacology 4, 122–125.

Patel, D.K., Kumar, R., Prasad, S.K., Sairam, K., Hemalatha, S., 2011. Antidiabetic and in vitro antioxidant potential of *Hybanthus enneaspermus* (Linn.) F. Muell in streptozotocin-induced diabetic rats. Asian Pacific Journal of Tropical Medicine 4, 316–322.

Patil, P., Patil, S., Mane, A., Verma, S., 2013. Antidiabetic activity of alcoholic extract of Neem (*Azadirachta indica*) root bark. National Journal of Physiology, Pharmacy and Pharmacology 3 (2), 142–146.

Patil, R., Patil, R., Ahirwar, B., Ahirwar, D., 2011. Isolation and characterization of anti-diabetic component (bioactivity guided fractionation) from *Ocimum sanctum* L. (*Lamiaceae*) aerial part. Asian Pacific Journal of Tropical of Medicine 278–282.

Pimple, B.P., Kadam, P.V., Patil, M.J., 2011. Antidiabetic and antihyperlipidemic activity of *Luffa acutangula* fruit extracts in streptozotocin induced niddm rats. Asian Journal of Pharmaceutical and Clinical Research 4 (2), 156–163.

Prajapati, R.P., Kalariya, M., Parmar, S.K., Sheth, N.R., 2010. Phytochemical and pharmacological review of *Lagenaria sicereria*. Journal of Ayurveda and Integrative Medicine 1 (4), 266–272.

Ragavan, B., Krishnakumari, S., 2006. Antidiabetic effect of *T. Arjuna* bark extract in alloxan induced diabetic rats. Indian Journal Clinical Biochemistry 21 (2), 123–128.

Rahman, M., Islam, M., 2015. *Alpinia calcarata* Roscoe: a potential phytopharmacological source of natural medicine. Pharmacognosy Reviews 9, 55–62.

Rajan, M., Kishor, V., Kumar, P., Kumar, S., Venkatachalam, T., Anbarasan, V., 2011a. Pharmacognostical and phytochemical studies of the leaves of *Albizia odoratissima* (L.F) Benth. International Journal of Pharmacognosy Phytochemical Research 3 (3), 47–55.

Rajan, S., Shalini, R., Bharathi, C., Aruna, V., Elgin, A., Brindha, P., 2011b. Pharmacognostical and phytochemical studies on *Hemidesmus indicus* root. International Journal of Pharmacognosy and Phytochemical Research 3 (3), 74–79.

Rajasekaran, S., Velmurugan, C., Vetriselvan, S., Basir, R., Muthuramu, T., 2012. Phytochemical and pharmacological evaluation of leaves of Zizyphus nummularia (Burm.F.) Wight & Arn. International Journal of Phytopharmacology 3 (3), 299–303.

Ramachandran, S., Naveen, K.R., Rajinikanth, B., Akbar, M., Rajasekaran, A., 2012. Antidiabetic, antihyperlipidemic and in vivo antioxidant potential of aqueous extract of *Anogeissus latifolia* bark in type 2 diabetic rats. Asian Pacific Journal of Tropical Diseases 2, 596–602.

Rao, N.K., Nammi, S., 2006. Antidiabetic and renoprotective effects of the chloroform extract of *Terminalia chebula* Retz. seeds in streptozotocin-induced diabetic rats. BMC Complementary and Alternative Medicine 6–17.

Rao, S., Mohan, G.K., Srinivas, P., 2014. Evaluation of anti-diabetic activity of *Hydnocarpus laurifolia* in streptozotocin induced diabetic rats. Asian Journal of Pharmaceutical and Clinical Research 7 (5), 62–64.

Raphael, K.R., Sabu, M.C., Kuttan, R., 2002. Hypoglycemic effect of methanol extract of *Phyllan thus* amarus Schum & Thonn on alloxan induced diabetes mellitus in rats and its relation with antioxidant potential. Indian Journal of Experimental Biology 40 (8), 905–909.

Rastogi, A., Shankar, S., Mahalingam, G., 2015. Antidiabetic activity of methanolic extract of *Hygrohila auriculata* in adult male Wistar rats. Journal Pharmaceutical Sciences and Research 7 (3), 98–102.

Ripa, F.A., Dash, P.R., Faruk, M.O., 2015. CNS depressant, analgesic and anti-inflammatory activities of methanolic seed extract of *Calamus rotang* Linn. fruits in rat. Journal of Pharmacognosy and Phytochemistry 3 (5), 121–125.

Roy, P., Abdulsalam, F.I., Pandey, D.K., Bhattacharjee, A., Eruvaram, N.R., Malik, T., 2015. Evaluation of antioxidant, antibacterial, and antidiabetic potential of two traditional medicinal plants of India: *Swertia cordata* and *Swertia chirayita*. Pharmacognosy Research 7 (1), 57–62.

Safavi, M., Foroumadi, A., Abdollahi, M., 2013. The importance of synthetic drugs for type 2 diabetes drug discovery. Expert Opinion on Drug Discovery 8 (11), 1339–1363.

Salahuddin, M., Jalalpure, S.S., 2010. Antidiabetic activity of aqueous fruit extract of *Cucumis trigonus* Roxb. in streptozotocin-induced-diabetic rats. Journal of Ethnopharmacology 127 (2), 565–567.

Sama, V., Kusuma, R., Sateesh, V., Reddy, M.B., Mullangi, R., 2012. Antidiabetic activity of *Pandanus odoratissimus* root extract. Indian Journal of Pharmaceutical Education and Research 46 (4), 340–345.

Sandhya, S., Sudhakar, K., David, B., Rao, K.N.V., 2010. Pharmacognostical standardization of *Borassus flabellifer* root. Annals of Biological Research 1 (4), 85–94.

Semwal, R.B., Semwal, D.K., Vermaak, I., Viljoen, A.A., 2015. A comprehensive scientific overview of *Garcinia cambogia*. Fitoterapia 102, 134–148.

Sharma, B., Siddiqui, S., Ram, G., Chaudhary, M., Sharma, G., 2013. Hypoglycemic and hepatoprotective effects of processed Aloe vera gel in a mice model of alloxan induced diabetes mellitus. Journal of Diabetes and Metabolism 4 (303), 1–6.

Sharma, N., Garg, V., 2009. Antidiabetic and antioxidant potential of ethanolic extract of *Butea monosperma* leaves in alloxan-induced diabetic mice. Indian Journal of Biochemistry and Biophysics 46 (1), 99–105.

Shetty, R.G., Monisha, S., 2015. Pharmacology of an endangered medicinal plant *Alpinia galangal* – a review. Research Journal of Pharmaceutical, Biological and Chemical Sciences 6 (1), 499–511.

Shirwaikar, A., Rajendran, K., Punitha, I.S.R., 2005. Antidiabetic activity of alcoholic stem extract of *Coscinium fenestratum* in streptozotocin-nicotinamide induced type 2 diabetic rats. Journal of Ethnopharmacology 97 (2), 369–374.

Sikarwar, M.S., Patil, M.B., 2010. Antidiabetic activity of *Pongamia pinnata* leaf extracts in alloxan-induced diabetic rats. International Journal of Ayurveda Research 1 (4), 199–204.

Singh, A.K., Singh, J., 2010. Evaluation of anti-diabetic potential of leaves and stem of *Flacourtia jangomas* in streptozotocin-induced diabetic rats. Indian Journal of Pharmacology 42 (5), 301–305.

Singh, J., Kakkar, P., 2013. Modulation of liver function, antioxidant responses, insulin resistance and glucose transport by *Oroxylum indicum* stem bark in STZ induced diabetic rats. Food and Chemical Toxicology 62, 722–731.

Singh, P., Khosa, R.L., Mishra, G., 2013. Evaluation of antidiabetic activity of ethanolic extract of *Cedrus deodara* (Pinaceae) stem bark in streptozotocin induced diabetes in mice. Nigerian Journal of Experimental Clinical Biosciences 1 (1), 33–38.

Singh, V., Raghav, P.K., 2012. Review on pharmacological properties of *Caesalpinia bonduc* L. International Journal of Medicinal and Aromatic Plants 2 (3), 514–530.

Srinivasan, R., Chandrasekar, M.J.N., Nanjan, M.J., 2011. Phytochemical investigations of *Caesalpinia digyna* root. E-Journal of Chemistry 8 (4), 1843–1847.

Srivastav, S., Singh, P., Mishra, G., Jha, K.K., Khosa, R.L., 2011. *Achyranthes aspera* – an important medicinal plant: a review. Journal of Natural Products and Plant Resources 1 (1), 1–14.

Stein, S.A., Lamos, E.M., Davis, S.N., 2013. A review of the efficacy and safety of oral antidiabetic drugs. Expert Opinion on Drug Safety 12 (2), 153–175.

Sunil, C., Duraipandiyan, V., Agastian, P., Ignacimuthu, S., 2012. Antidiabetic effect of plumbagin isolated from *Plumbago zeylanica* L. root and its effect on GLUT4 translocation in streptozotocin-induced diabetic rats. Food and Chemical Toxicology 50 (12), 4356–4363.

Suthar, M., Rathore, G.S., Pareek, A., 2009. Antioxidant and antidiabetic activity of *Helicteres isora* (L.) fruits. Indian Journal of Pharmaceutical Sciences 71 (6), 695–699.

Syiem, D., Syngai, G., Khup, P.Z., Khongwir, B.S., Kharbuli, B., Kayang, H., 2002. Hypoglycemic effects of *Potentilla fulgens* L in normal and alloxan-induced diabetic mice. Journal of Ethnopharmacology 83 (1–2), 55–61.

Thirumalai, T., Therasa, V.S., Elumalai, E.K., David, E., 2011. Hypoglycemic effect of *Brassica juncea* (seeds) on streptozotocin induced diabetic male albino rat. Asian Pacific Journal of Tropical Biomedicine 4, 323–325.

Thomas, S., Patil, D.A., Patil, A., Chandra, G.N., 2008. Pharmacognostic evalution & physiochemical analysis of *Averrhoa Carambola* L. fruit. Journal of Herbal Medicine and Toxicology 2 (2), 51–54.

Thomson, M., Al-Amin, Z.M., Al-Qattan, K.K., Shaban, L.H., Ali, M., 2007. Anti-diabetic and hypolipidaemic properties of garlic (*Allium sativum*) in streptozotocin-induced diabetic rats. International Journal of Diabetes and Metabolism 15, 108–115.

Toppo, F.A., Akhand, R., Pathak, A.K., 2009. Pharmacological actions and potential uses of *Trigonella foenum-graecum*: a review. Asian Journal of Pharmaceutical and Clinical Research 2 (4), 29–38.

Tripathi, P., Kumar, R., Sharma, A.K., Mishra, A., Gupta, R., 2010. *Pistia stratiotes* (Jalkumbhi). Pharmacognosy Reviews 4 (8), 153–160.

Tulliballi, S., Seru, G., Balakrishnaiah, P., 2013. Antidiabetic activity of *Rotula aquatica* lour roots in streptozotocin induced diabetic rats. American Journal of Phytomedicine and Clinical Therapeutics 1 (7), 530–535.

Udayakumar, R., Kasthurirengan, S., Mariashibu, T.S., Rajesh, M., Anbazhagan, V.R., Kim, S.C., Ganapathi, A., Choi, C.W., 2009. Hypoglycaemic and hypolipidaemic effects of *Withania somnifera* root and leaf extracts on alloxan-induced diabetic rats. International Journal of Molecular Sciences 10 (5), 2367–2382.

Upwar, N.K., Patel, R., Waseem, N., Mahobia, N.K., 2011. Hypoglycemic effect of methanolic extract of *Berberis aristata* DC stem on normal and streptozotocin induced diabetic rats. International Journal of Pharmacy and Pharmaceutical Sciences 3 (1), 222–224.

Vedavathy, S., 2003. Scope and importance of traditional medicine. Indian Journal of Traditional Knowledge 2 (3), 236–239.

Velmurugan, C., Sundaram, T., Kumar, S., Vivek, B., Sheshadrishekar, D., Ashok Kumar, B.S., 2011. Antidiabetic and hypolipidemic activity of bark of ethanolic extract of *Ougeinia oojeinensis* (Robx.). Medical Journal of Malaysia 66 (1), 22–26.

Venkatesh, S., Thilagavathi, J., Sundar, D.S., 2008. Anti-diabetic activity of flowers of *Hibiscus rosasinensis*. Fitoterapia 79 (2), 79–81.

Verma, N., Amresh, G., Sahu, P.K., Rao, ch V., Singh, A.P., 2012. Antihyperglycemic activity of *Woodfordia fruticosa* (Kurz) flowers extracts in glucose metabolism and lipid peroxidation in streptozotocin-induced diabetic rats. Indian Journal of Experimental Biology 50 (5), 351–358.

Verma, N., Jha, K.K., Sharma, R., Chaudhary, S., Singh, A.K., Kumar, A., 2014. Biological properties, phytochemistry and traditional uses of Rudravanti (*Cressa cretica*): a review. Pharma Tutor 2 (8), 154–161.

Vijayvargia, P., Vijayvergia, R., 2014. A review on *Limonia acidissima* L.: multipotential medicinal plant. International Journal of Pharmaceutical Sciences Review and Research 36, 191–195.

Vilar, D.A., Vilar, S.A., Moura, A.L., Raffin, F.N., Oliveira, M.R., Franco, F.O., Athayde-Filho, P.F., Diniz, F.M., Barbosa-Filho, J.M., 2014. Traditional uses, chemical constituents, and biological activities of *Bixa orellana* L.: a review. The Scientific World Journal 1–11.

Yadav, P.D., Bharadwaj, S.P., Yedukondalu, M., Methushala, C.H., Kumar, A.R., 2013. Phytochemical evaluation of *Nyctanthes arbortristis*, *Nerium foleander* and *Catharathnus roseus*. Indian Journal of Research in Pharmacy and Biotechnology 1 (3), 333–338.

Zhou, Y.X., Xin, h. L., Rahman, K., Wang, s. J., Peng, C., Zhang, H., 2015. *Portulaca oleracea* L.: a review of phytochemistry and pharmacological effects. BioMed Research International 1–11.

Zia-Ul-Haq, M., Cavar, S., Qayum, M., Imran, I., Feo, D., 2011. Compositional studies: antioxidant and antidiabetic activities of *Capparis decidua* (Forsk.) Edgew. International Journal of Molecular Sciences 12 (12), 8846–8861.

Index

'*Note*: Page numbers followed by "f" indicate figures and "t" indicate tables.'

D

E

F

G

MIX
Papier | Fördert gute Waldnutzung
FSC® C083411

Kolibri 360 GmbH
Ferdinand-Jühlke-Straße 7
99095 Erfurt, Deutschland
produktsicherheit@kolibri360.de